Rheology of Fresh Cement-Based Materials

This book introduces fundamentals, measurements, and applications of rheology of fresh cement-based materials. The rheology of a fresh cement-based material is one of its most important aspects, characterizing its flow and deformation, and governing the mixing, placement, and casting quality of a concrete.

This is the first book that brings the field together on an increasingly important topic, as new types of cement-based materials and new concrete technologies are developed. It describes measurement equipment, procedures, and data interpretation of the rheology of cement paste and concrete, as well as applications such as self-compacting concrete, pumping, and 3D printing. A range of other cement-based materials such as fiber-reinforced concrete, cemented paste backfills, and alkali-activated cement are also examined.

Rheology of Fresh Cement-Based Materials serves as a reference book for researchers and engineers, and a textbook for advanced undergraduate and graduate students.

Rheology of Fresh Cement-Based Materials
Fundamentals, Measurements, and Applications

Qiang Yuan, Caijun Shi, and Dengwu Jiao

CRC Press
Taylor & Francis Group
Boca Raton London New York

CRC Press is an imprint of the
Taylor & Francis Group, an **informa** business

MATLAB® is a trademark of The MathWorks, Inc. and is used with permission. The MathWorks does not warrant the accuracy of the text or exercises in this book. This book's use or discussion of MATLAB® software or related products does not constitute endorsement or sponsorship by The MathWorks of a particular pedagogical approach or particular use of the MATLAB® software.

First edition published 2023
by CRC Press
4 Park Square, Milton Park, Abingdon, Oxon, OX14 4RN

and by CRC Press
6000 Broken Sound Parkway NW, Suite 300, Boca Raton, FL 33487-2742

© 2023 Qiang Yuan, Caijun Shi and Dengwu Jiao

CRC Press is an imprint of Informa UK Limited

The right of Qiang Yuan, Caijun Shi and Dengwu Jiao to be identified as authors of this work has been asserted in accordance with sections 77 and 78 of the Copyright, Designs and Patents Act 1988.

All rights reserved. No part of this book may be reprinted or reproduced or utilised in any form or by any electronic, mechanical, or other means, now known or hereafter invented, including photocopying and recording, or in any information storage or retrieval system, without permission in writing from the publishers.

For permission to photocopy or use material electronically from this work, access www.copyright.com or contact the Copyright Clearance Center, Inc. (CCC), 222 Rosewood Drive, Danvers, MA 01923, 978-750-8400. For works that are not available on CCC please contact mpkbookspermissions@tandf.co.uk

Trademark notice: Product or corporate names may be trademarks or registered trademarks, and are used only for identification and explanation without intent to infringe.

British Library Cataloguing-in-Publication Data
A catalogue record for this book is available from the British Library

Names: Yuan, Qiang, author. | Shi, Caijun, author. | Jiao, Dengwu, author.
Title: Rheology of fresh cement-based materials : fundamentals, measurements, and applications / Qiang Yuan, Caijun Shi, and Dengwu Jiao.
Description: First edition. | Boca Raton : CRC Press, 2023. | Includes bibliographical references and index.
Identifiers: LCCN 2022033778 | ISBN 9781032208015 (hbk) | ISBN 9781032208022 (pbk) | ISBN 9781003265313 (ebk)
Subjects: LCSH: Concrete—Viscosity. | Concrete—Mixing. | Rheology.
Classification: LCC TA440 .Y833 2023 | DDC 620.1/36—dc23/eng/20221011
LC record available at https://lccn.loc.gov/2022033778

ISBN: 978-1-032-20801-5 (hbk)
ISBN: 978-1-032-20802-2 (pbk)
ISBN: 978-1-003-26531-3 (ebk)

DOI: 10.1201/9781003265313

Typeset in Sabon
by codeMantra

Contents

Preface xiii
Authors xv

1 Introduction to rheology 1

1.1 The Subject and Object of Rheology 1
1.2 Basic Principles of Rheology 7
    1.2.1 Definition of viscosity 7
    1.2.2 Newtonian flow 8
    1.2.3 Non-Newtonian flow 9
    1.2.4 Thixotropy 11
    1.2.5 Anti-thixotropy (rheopexy) 13
1.3 Cement-Based Materials 13
    1.3.1 History of cement and concrete 13
    1.3.2 Fresh properties of cement-based materials 16
1.4 The Scope of This Book 20
References 21

2 Rheology for cement paste 25

2.1 Interaction between Particles in the Paste 25
    2.1.1 Colloidal interaction 25
        2.1.1.1 Van der Waals force 26
        2.1.1.2 Electrostatic repulsion 26
        2.1.1.3 Steric hinder force 26
    2.1.2 Brownian forces 26
    2.1.3 Hydrodynamic force 27
2.2 Effect of Compositions on Rheology 27
    2.2.1 Volume fraction 27
    2.2.2 Interstitial solution 29
    2.2.3 Cement 30
    2.2.4 Mineral admixture 31
        2.2.4.1 Fly ash 31
        2.2.4.2 Ground blast furnace slag 33
        2.2.4.3 Silica fume 34

                2.2.4.4   Limestone powder 36
                2.2.4.5   Ternary binder system 37
        2.2.5   Chemical admixtures 38
                2.2.5.1   Superplasticizer 38
                2.2.5.2   Viscosity-modifying agent 39
                2.2.5.3   Air-entraining agent 40
2.3   Effect of Temperature on Rheology 41
2.4   Effect of Shearing on Rheology 41
2.5   Effect of Pressure on Rheology 42
2.6   Summary 43
References 43

3 Rheological properties of fresh concrete materials — 51

3.1   General Considerations for Granular Materials 51
3.2   Flow Regimes of Concrete 51
        3.2.1   Relationships between aggregate volume fraction and concrete rheology 52
                3.2.1.1   Viscosity vs aggregate volume fraction 52
                3.2.1.2   Yield stress vs aggregate volume fraction 53
        3.2.2   Excess paste theory 56
3.3   Influence of Aggregate Characteristics 59
        3.3.1   Aggregate volume fraction 59
        3.3.2   Gradation and particle size 60
        3.3.3   Particle morphology 62
3.4   Effect of External Factors 67
        3.4.1   Mixing process 67
        3.4.2   Shear history 69
        3.4.3   Measuring geometry 69
3.5   Summary 69
References 70

4 Empirical techniques evaluating concrete rheology — 75

4.1   Introduction 75
4.2   Slump: ASTM Abrams Cone 75
        4.2.1   Geometry 76
        4.2.2   Testing procedure and parameters 76
        4.2.3   Data interpretation 77
4.3   Slump Flow and T_{50} 80
        4.3.1   Geometry and testing procedure 81
        4.3.2   Data interpretation 82
4.4   V-Funnel Test Flow Time 82
        4.4.1   Geometry 83
        4.4.2   Testing procedure 84
        4.4.3   Data interpretation 84

4.5 Other Methods 85
    4.5.1 L-box 85
    4.5.2 LCPC box 87
    4.5.3 V-funnel coupled with a horizontal channel 88
    4.5.4 J-ring 89
4.6 Summary 91
References 91

5 Paste rheometers   95

5.1 Introduction to the Rheology of Cement Paste 95
5.2 Rheometers for Cement Paste 96
    5.2.1 Narrow gap coaxial cylinder rheometer 96
        5.2.1.1 Geometry 96
        5.2.1.2 Measurement principle 96
        5.2.1.3 Measuring errors and artifacts 98
    5.2.2 Plate–plate rheometer 99
        5.2.2.1 Geometry 99
        5.2.2.2 Measurement principle 101
        5.2.2.3 Measuring errors and artifacts 102
    5.2.3 Other rheometers 104
        5.2.3.1 Capillary viscometer 104
        5.2.3.2 Falling sphere viscometer 107
5.3 Measuring Procedures 109
    5.3.1 Flow curves test 109
    5.3.2 Static yield stress test 110
    5.3.3 Oscillatory shear test 111
        5.3.3.1 Description of SAOS and LAOS 111
        5.3.3.2 Measurement principle 112
        5.3.3.3 Application to cement paste 115
5.4 Summary 117
References 117

6 Concrete rheometers   123

6.1 Introduction 123
6.2 Tests Methods and Principles 124
    6.2.1 Coaxial cylinder rheometer 125
        6.2.1.1 Searle rheometer 126
        6.2.1.2 Couette rheometer 126
        6.2.1.3 Principle 128
        6.2.1.4 Measuring errors and artifacts 131
    6.2.2 Parallel-plate rheometer 137
        6.2.2.1 Geometry 137
        6.2.2.2 Principle 138
        6.2.2.3 Measuring errors and artifacts 139

            6.2.3   Other rheometers 140
                    6.2.3.1   CEMAGREF-IMG rheometer 140
                    6.2.3.2   Viskomat XL 140
                    6.2.3.3   The IBB rheometer 140
                    6.2.3.4   Rheometer developed in China 142
                    6.2.3.5   The modifications of the BTRHEOM rheometer 142
                    6.2.3.6   Other instruments 143
    6.3   Measuring Procedures 143
            6.3.1   Preparation of specimen 143
            6.3.2   The testing procedures of ICAR 143
            6.3.3   The testing procedures of ConTec Viscometer 5 145
            6.3.4   The testing procedures of the BTRHEOM rheometer 147
    6.4   Data Collection and Processing 148
            6.4.1   Static yield stress test 148
            6.4.2   The flow curve test 149
            6.4.3   Thixotropy test 150
    6.5   Relation of Rheological Parameters Measured by Different Rheometers 150
    6.6   Summary 152
    References 152

7 Mixture design of concrete based on rheology   155

    7.1   Introduction 155
    7.2   Principles of Mixture Design Methods Based on Rheology 156
            7.2.1   Vectorized-rheograph approach 156
            7.2.2   Paste rheology criteria 158
            7.2.3   Concrete rheology method 164
            7.2.4   Excess paste theory 166
            7.2.5   Simplex centroid design method 167
    7.3   Typical Examples of Mixture Design 169
            7.3.1   Paste rheology criteria proposed by Wu and An 169
            7.3.2   Paste rheology model proposed by Ferrara et al. 169
            7.3.3   Concrete rheology method of Abo Dhaheer et al. 173
            7.3.4   Simplex centroid design method proposed by Jiao et al. 176
    7.4   Summary 178
    References 178

8 Rheology and self-compacting concrete   183

    8.1   Introduction to SCC 183
            8.1.1   Brief history of SCC 183
            8.1.2   Raw materials of SCC 184
                    8.1.2.1   Powder 184
                    8.1.2.2   Chemical admixtures 185
                    8.1.2.3   Aggregates 185
                    8.1.2.4   Water 186

        8.1.3   Mix proportion of SCC 186
                 8.1.3.1   Laboratory experiments and empirical parameters 186
                 8.1.3.2   Statistical method 187
                 8.1.3.3   Maximum packing density 187
                 8.1.3.4   Other methods 188
        8.1.4   Application of SCC 188
   8.2 Rheology of SCC 190
        8.2.1   Factors affecting rheology of SCC 190
                 8.2.1.1   Fly ash 190
                 8.2.1.2   Rice husk ash 190
                 8.2.1.3   Silica fume 191
                 8.2.1.4   Metakaolin 192
                 8.2.1.5   Blast furnace slag 192
                 8.2.1.6   Fibers 192
                 8.2.1.7   Air-entraining agent 193
                 8.2.1.8   Superplasticizer 193
                 8.2.1.9   Recycled concrete aggregates 195
                 8.2.1.10 Binary and ternary binder system 197
                 8.2.1.11 Other constituents 197
        8.2.2   Special rheological behaviors 198
                 8.2.2.1   Thixotropy 198
                 8.2.2.2   Shear-thinning or shear-thickening behavior 199
   8.3 Formwork Pressure of SCC 200
        8.3.1   Factors affecting formwork pressure 201
        8.3.2   Formwork pressure prediction 201
                 8.3.2.1   Method proposed by Gardner (Gardner et al., 2012) 202
                 8.3.2.2   Method proposed by Khayat (Khayat and Omran, 2010) 202
   8.4 Stability of SCC 204
        8.4.1   Static stability 204
        8.4.2   Dynamic stability 206
   8.5 Summary 208
   References 209

9 Rheology of other cement-based materials   215

   9.1 Rheology of Alkali-Activated Materials (AAMs) 215
        9.1.1   Introduction 215
        9.1.2   Effect of alkaline activators on rheology of AAMs 217
                 9.1.2.1   Na/KOH 217
                 9.1.2.2   Na/K-silicates 219
        9.1.3   Effect of precursors on the rheology of AAMs 223
                 9.1.3.1   Chemical and physical properties of precursors 223
        9.1.4   Effects of chemical admixtures on the rheology of AAMs 225
                 9.1.4.1   Water-reducing admixtures 225
                 9.1.4.2   Other chemical admixtures 231

   9.1.5 Effects of mineral additions on the rheology of AAMs 232
     9.1.5.1 Reactive mineral additions 232
     9.1.5.2 Inert mineral additions 233
   9.1.6 Effect of aggregates on the rheology of AAMs 233
 9.2 Rheology of Cement Paste Backfilling (CPB) 234
   9.2.1 Introduction 234
   9.2.2 Factors affecting the rheological properties of CPB 235
     9.2.2.1 Cement 235
     9.2.2.2 Solid concentration 236
     9.2.2.3 Mixing intensity 237
     9.2.2.4 Particle size 237
     9.2.2.5 High-range water reducer (HRWR) 238
     9.2.2.6 Temperature 239
     9.2.2.7 Other constituents 239
 9.3 Rheology of Fiber-Reinforced, Cement-Based Materials 240
   9.3.1 Introduction 240
   9.3.2 Influence of fiber on the rheology of FRCs 242
     9.3.2.1 Fiber orientation 242
     9.3.2.2 Fiber length 243
     9.3.2.3 Types of fiber 244
   9.3.3 Effect of fibers on the rheology of AAMs 246
   9.3.4 Prediction of the yield stress of FRC 248
   9.3.5 Prediction of plastic viscosity 248
 9.4 Summary 249
   9.4.1 AAMs 249
   9.4.2 Cement paste backfilling 250
   9.4.3 Fiber-reinforced, cement-based materials 251
 References 251

10 Rheology and Pumping 259

 10.1 Introduction 259
 10.2 Characterization of Pumpability 260
   10.2.1 Definition of pumpability 260
   10.2.2 Determination of pumpability 261
 10.3 Lubrication Layer 261
   10.3.1 The formation of the lubrication layer 261
   10.3.2 The determination of the lubrication layer 262
     10.3.2.1 Tribometer 262
     10.3.2.2 Sliding pipe 263
     10.3.2.3 Other methods 264
 10.4 Prediction of Pumping 264
   10.4.1 Empirical model for pumping prediction 265
   10.4.2 Numerical model for pumping prediction 266
   10.4.3 Computer simulations for pressure loss predictions 267

    10.5 *Effect of Pumping on the Fresh Properties of Concrete* 268
          10.5.1 *Air content* 268
          10.5.2 *Rheology* 273
                10.5.2.1 *Yield stress and plastic viscosity* 273
                10.5.2.2 *Thixotropy* 278
                10.5.2.3 *Supplementary cementitious materials (SCMs)* 281
                10.5.2.4 *Water absorption by aggregates* 283
                10.5.2.5 *Water-reducing agent* 284
    10.6 *Summary* 284
    References 285

11 Rheology and 3D printing   291

    11.1 *Introduction to 3D-Printing Concrete* 291
          11.1.1 *Development of 3D printing technology* 291
          11.1.2 *Requirement for 3D printing concrete* 295
    11.2 *Printability of 3D Printing Concrete* 296
          11.2.1 *The definition of printability* 296
          11.2.2 *The test for printability* 297
          11.2.3 *Criteria to evaluate the loading bearing capacity* 299
          11.2.4 *Printable cement-based materials* 301
    11.3 *Interlayer Bonding and Rheology* 303
          11.3.1 *The characterization of interlayer bonding* 303
          11.3.2 *The effect of rheological properties on interlayer bonding* 305
          11.3.3 *The effect of rheological properties on interface durability* 308
          11.3.4 *The effect of rheological properties on interface microstructure* 309
    11.4 *Summary* 311
    References 311

Index   317

Preface

Rheology, the study of the flow and deformation of matter, is an independent and prominent branch of the natural sciences. It is an extremely useful technical tool and a scientific basis for cementitious materials. Advanced technologies such as pumping and 3D printing impose additional requirements on the rheological properties of cement-based materials. Besides, rheology for cement-based materials extends beyond the study of flow and deformation. The significance of the physical and chemical reactions underlies the growing importance of rheology. This intricate and intriguing rheology of cement-based materials attracts the interest of an increasing number of researchers. A lot of research has been done in the rheology of cement-based materials, encompassing its fundamentals, measurements, and predictions. This does not imply, however, that we have a complete grasp of the rheology of cement-based materials and can accurately predict as well as control it. Still, many scientific issues are yet to be clarified.

This book focuses on the flow and rheological behavior of fresh cement-based materials. It reviews the fundamentals, measurement techniques, and applications involved in these areas, and presents the state-of-the-art progress made recently. The authors of this book have long been involved in research pertaining to the rheology of fresh cement-based materials at Central South University, Hunan University, and Ghent University. The essential sources for this book were the graduate students who obtained their M.Sc. and Ph.D. degrees in this field of study in the past few years. Obviously, this book could not have been completed without their acquired knowledge, experience, and outstanding contributions. The authors would like to express their gratitude to everyone who contributed to the content of this book. They also gratefully recognize the contributions of reviewers and students who assisted in organizing the content of the entire book into the present layout, as well as those who drew the updated graphs and tables.

The intended audience of this book includes students, researchers, and engineers in concrete technology. Researchers may be inspired by the comprehensive overview of the rheology of cement and concrete. Practicing engineers can also benefit from this book by imparting fundamental knowledge and practical techniques.

Qiang Yuan
Caijun Shi
Dengwu Jiao

MATLAB® is a registered trademark of The MathWorks, Inc. For product information, please contact:

The MathWorks, Inc.
3 Apple Hill Drive
Natick, MA 01760-2098 USA
Tel: 508-647-7000
Fax: 508-647-7001
E-mail: info@mathworks.com
Web: www.mathworks.com

Authors

Qiang Yuan is a professor in the School of Civil Engineering at Central South University, China. He is the author of *Transport and Interactions of Chlorides in Cement-Based Materials*, also published by CRC Press.

Caijun Shi is a professor in the College of Civil Engineering at Hunan University, China. He is the author of *Alkali-Activated and Concretes* and *Transport and Interactions of Chlorides in Cement-Based Materials*, also published by CRC Press.

Dengwu Jiao is a postdoctoral researcher in the Department of Structural Engineering and Building Materials at Ghent University.

Chapter 1

Introduction to rheology

1.1 THE SUBJECT AND OBJECT OF RHEOLOGY

Rheology, first proposed by Professor Eugene Cook Bingham of Lafayette College (Bingham, 1922), is *the study of the flow and deformation of matters*. It is a branch of physics. The term rheology originates from Greek words *rheo* indicating *flow* and *logia* meaning *the study of*. This definition has been accepted by the American Society of Rheology founded in 1929. In the beginning, rheology was used to describe the properties and behavior of asphalt, lubricants, paints, plastics, rubber, etc. Over the past 100 years, rheology has been developed as an independent and important branch of natural sciences, and is widely applied in many industry sectors and research fields such as:

- Plastics, rubber, and other polymer melts and solutions;
- Food stuff such as chocolate, ketchup, and yogurt;
- Metals, alloys, etc.;
- Concrete, ceramics, glass, and rigid plastics in fresh state and at long periods of loading;
- Lubricants, greases, and sealants;
- Personal care stuff and pharmaceuticals;
- Paints and printing inks;
- Mud, coal, mineral dispersions, and pulps;
- Soils, glaciers, and other geological formations;
- Biological materials such as bones, muscles, and blood.

In a word, rheology has come of age in many ways.

Rheology is principally concerned with the following question: "How does a material response to an externally applied force?" That is, rheology describes the relationship between force, deformation, and time. Before introducing the rheology, it is necessary to give a short introduction to the elastic, viscous, and viscoelastic behaviors of solids and/or liquids from the historical perspective, which is summarized in Figure 1.1.

An elastic solid subjected to a constant force (within the elastic limit) will undergo a finite deformation, and this deformation will be fully recovered after removal of the applied force. This nature of material is commonly known as elasticity. The force F (N) applied per unit area A (m^2) is defined as stress σ (Pa), and the displacement gradient is called strain γ. In 1678, Robert Hooke established the *True Theory of Elasticity* (Barnes et al., 1989), stating that "the power of any spring is the same proportion with the tension thereof", i.e. Hooke's elastic model. A representative linear elastic solid is spring as presented in Figure 1.2, where the applied force is directly proportional to its distance within the elastic limit. The spring will return its initial shape immediately after removing the applied force. If the applied stress exceeds the elastic limit, the spring will be distorted permanently. The relationship between

Figure 1.1 Development history of rheology.

Figure 1.2 The Hooke's elastic model.

the applied stress σ (in Pa) and the strain γ (no unit) for Hooke's law follows the constitutive equation:

$$\sigma = G\gamma \qquad (1.1)$$

where G is the elastic modulus in Pa, which is a measure of the resistance of the solid to deformation. The famous applications of Hooke's law include retractable pens, manometers,

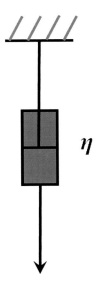

Figure 1.3 The Newton's viscous law.

spring scales, and the balance wheel of clock. It should be mentioned that Hooke's law is only applicable to describe the deformation behavior of solids on a small scale. If the strain exceeds the capacity of the material, Hooke's law will no longer be valid.

With respect to an ideal viscous liquid, the material deforms continuously under applied stress, and the deformation cannot be recovered upon removal of the load. In this context, Isaac Newton published a book *Philosophie Principia Mathematica* in 1687 and proposed Newton's viscous law, represented by a dashpot as shown in Figure 1.3, to describe this ideal viscous behavior. When applying an external stress to a dashpot, it starts to deform immediately and goes on deforming at a constant velocity v (m/s) until the stress is removed. The rate of the velocity is defined as the shear rate, represented by the symbol $\dot{\gamma}$ in s^{-1}, which can be calculated by:

$$\dot{\gamma} = v/h \tag{1.2}$$

where h is the gap height in m. A higher stress is generally required for a greater strain of the liquid. The coefficient of the applied stress and the shear rate is defined as shear viscosity or dynamic viscosity η in Pa.s, which can be described by:

$$\eta = \sigma/\dot{\gamma} \tag{1.3}$$

The viscosity, associated with the dissipation of kinetic energy of the system, is a quantitative parameter describing the internal fluid friction between elements. Newton's law is the basic flow model of most of the pure liquids, and these liquids are defined as Newtonian fluids, which will be further discussed later. The stress response to deformation for Hooke's law and Newton's law is presented in Figure 1.4.

Before the 1900s, Hooke's law and Newton's law were widely recognized to describe the behaviors of solids and liquids, respectively. In 1835, Wilhelm Weber found that although the silk thread is a solid-like material, its behavior cannot be ideally described by Hooke's

4 Rheology of Fresh Cement-Based Materials

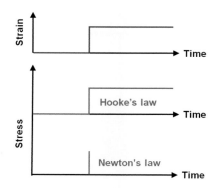

Figure 1.4 Stress response to deformation of the Hooke's law and the Newton's law.

Figure 1.5 The Maxwell model.

law alone. Indeed, there is no clear distinction between solid and liquid from the viewpoint of rheology. Elastic solid materials can undergo irrecoverable deformations under sufficiently high shear strains, and viscous liquid probably exhibits elastic behavior at extremely low shear strains. This is in agreement with the argument that all materials can flow under sufficient time (Barnes et al., 1989).

Since Newton's law and Hooke's law are inapplicable to some liquids and solids, the term viscoelasticity was introduced to describe the behavior falling in-between the ideal elastic behavior (Hooke's law) and the ideal viscous behavior (Newton's law). The Maxwell model and the Kelvin–Voigt model, as presented in Figures 1.5 and 1.6, respectively, are the most typical models for the viscoelastic materials. The Maxwell model, proposed by James Clerk Maxwell in 1867, consists of a spring and a dashpot connected in series, which is a representation of the simplest viscoelastic liquid model. Under an externally applied stress σ, the Maxwell material shows elastic behavior at very short times and is governed by elastic

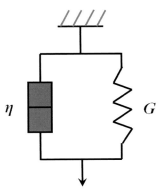

Figure 1.6 The Kelvin–Voigt model.

modulus G, while after a longer period of time, the viscous behavior becomes predominant and is governed by viscosity η. This means that the strain has two components: one is the immediate elastic component corresponding to the spring, and the other is the viscous component which grows with time. The relationship between the strain and the exerted stress for the Maxwell model can be described by:

$$\frac{d\gamma}{dt} = \frac{\sigma}{\eta} + \frac{1}{G} \cdot \frac{d\sigma}{dt} \tag{1.4}$$

If applying a sudden deformation to a Maxwell material, the stress decays on a characteristic timescale of η/G. This is also known as relaxation time, and this phenomenon is defined as stress relaxation. On this basis, a Maxwell material shows elastic solid-like behavior at the deformation time shorter than the relaxation time.

The simplest viscoelastic solid can be represented by the Kelvin–Voigt model with a spring and a dashpot connected in parallel, as shown in Figure 1.6. After applying an external stress, the dashpot retards the response of the spring, and thus a delay is required for the strain to develop. That is, the system behaves like a viscous liquid initially (shorter than the relaxation time of η/G) and then elastically over longer time scales. The evolution of strain γ and stress σ with respect to time for the Kelvin–Voigt model can be characterized by:

$$\sigma = G\gamma + \eta \frac{d\gamma}{dt} \tag{1.5}$$

The Maxwell model applies to liquids with the elastic response, while the Kelvin–Voigt model is used to describe the behavior of solids with a viscous response. However, these two models are not enough to evaluate the viscoelastic behavior of a real system. In this case, some more complicated models such as the Bingham model and the Burger model are proposed to describe the response of viscoelastic fluids. For example, the Bingham model can be regarded as a three-parameter model, which is combined by a parallel unit of a dashpot, a plastic slider, and a spring connected in series (Liingaard et al., 2004), as shown in Figure 1.7. The spring with elastic modulus G is a time-independent component, representing the elastic behavior. The parallel unit of a dashpot with a viscosity η and a slider with a threshold stress σ_y is a time-dependent component, describing the viscoplastic response. The

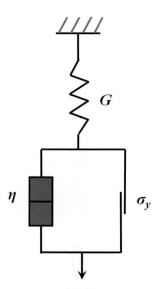

Figure 1.7 Conceptual structure of the Bingham model.

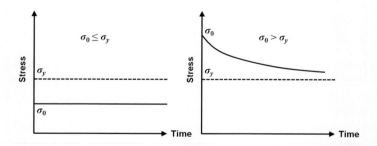

Figure 1.8 Stress response to deformation of the Bingham model.

relationship between shear rate and applied stress for the Bingham model can be described as:

$$\frac{d\gamma}{dt} = \begin{cases} \dfrac{1}{G}\dfrac{d\sigma}{dt} & \text{for } \sigma \leq \sigma_y \\ \dfrac{1}{G}\dfrac{d\sigma}{dt} + \dfrac{\sigma - \sigma_y}{\eta} & \text{for } \sigma > \sigma_y \end{cases} \qquad (1.6)$$

It can be seen that the second part of Eq. (1.6) is analogous to the constitutive equation of the Maxwell model in Eq. (1.4). The only difference in the Bingham and Maxwell models is that the stress is replaced by the difference $\sigma - \sigma_y$. Therefore, the Bingham material shows elastic behavior at $\sigma \leq \sigma_y$, while under the applied stress higher than the threshold σ_y, it exhibits viscous behavior. The stress relaxation to deformation of the Bingham model is shown in Figure 1.8.

Introduction to rheology 7

The flow behavior of pure fluids such as oil and water follows Newton's law. However, the behaviors of some fluids, e.g. paint, cheese, ketchup, paste, and suspension, do not obey the basic Newtonian flow law. The science of studying the flow behavior of fluid is called rheology. In the following parts, the basic principles of fluid rheology including definitions of viscosity, Newtonian flow, non-Newtonian flow, and the rheology in cement-based materials are briefly illustrated.

1.2 BASIC PRINCIPLES OF RHEOLOGY

1.2.1 Definition of viscosity

In general, there are two basic types of flow behavior, i.e. shear flow and extensional flow. This book mainly focuses on the shear flow, where layers of fluid slide one another with different velocities. A typical schematic of shear flow is shown in Figure 1.9. As seen, the upper layer slides at the maximum velocity v (m/s), while the bottom layer remains stationary. The external force F (N) results in a shear stress τ (Pa) in a unit area A (m²), as depicted in Eq. (1.7). As a response, the upper layer moves a given distance x (m) with the bottom layer remaining stationary. This flow behavior is also defined as laminar flow. If the distance between the upper layer and the bottom layer is denoted as h (m), the deformation gradient across the sample is defined as shear strain γ (dimensionless), and the gradient of the velocity of the material is called shear rate $\dot{\gamma}$ in s⁻¹, as calculated by *Eqs.* (1.8) and (1.9), respectively. The plot of shear stress against shear rate is also called flow curve (Banfill, 1990):

$$\tau = F/A \qquad (1.7)$$

$$\gamma = x/h \qquad (1.8)$$

$$\dot{\gamma} = v/h \qquad (1.9)$$

Viscosity, indicated by η (Pa.s), is defined as the coefficient ratio between shear stress and shear rate under a state of steady shear, as shown in Eq. (1.10), which is similar to Eq. (1.3). This definition is also called dynamic viscosity:

$$\eta = \tau/\dot{\gamma} \qquad (1.10)$$

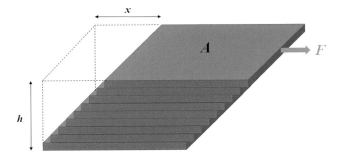

Figure 1.9 Typical diagram of shear flow.

Besides, there are a variety of viscosity terms used in practice. For example, the ratio between the dynamic viscosity and the density is defined as the kinematic viscosity:

$$v = \frac{\eta}{\rho} \tag{1.11}$$

where v is the kinematic viscosity in m²/s and ρ is the density of the fluid in m³/kg, respectively. The kinematic viscosity is a measure of the internal resistance of a fluid to flow under gravity. The apparent viscosity is the ratio of the shear stress to the shear rate of a certain point on a flow curve. For linear Newtonian flow, which will be discussed later, the value of apparent viscosity is similar to the value of dynamic viscosity. However, for nonlinear flow curves, the apparent viscosity is the slope of a line drawn from the origin to a point on a flow curve, which is dependent on the shear rate.

The differential viscosity is defined as the derivative of shear stress against shear rate:

$$\eta_{\text{diff}} = \frac{\partial \tau}{\partial \dot{\gamma}} \tag{1.12}$$

where η_{diff} is the differential viscosity in Pa.s. Plastic viscosity, which is the most popular viscosity parameter in the rheology of cement-based materials, is defined as the limit of differential viscosity with the shear rate approaching infinity:

$$\eta_{\text{pl}} = \mu = \lim_{\dot{\gamma} \to \infty} \frac{\partial \tau}{\partial \dot{\gamma}} \tag{1.13}$$

By contrast, the limit of the differential viscosity as the shear rate approaches zero is defined as zero shear viscosity, indicated by η_0:

$$\eta_0 = \lim_{\dot{\gamma} \to 0} \frac{\partial \tau}{\partial \dot{\gamma}} \tag{1.14}$$

In the case of suspensions, there are also some specific viscosity terms. For example, the ratio of the viscosity of a suspension η to that of the suspending medium η_s is called relative viscosity η_r, described by:

$$\eta_r = \frac{\eta}{\eta_s} \tag{1.15}$$

1.2.2 Newtonian flow

Fluids can be classified by the concept of viscosity. The Newtonian fluid is the simplest fluid following Newton's viscous law (see Eq. 1.10), with the viscosity independent of the shear rate or the shear stress. The constant viscosity of a Newtonian fluid is just like the statement in the book *Philosophie Principia Mathematica* that "the resistance which arises from the lack of slipperiness of the parts of the liquid, other things being equal, is proportional to the velocity with which the parts of the liquid are separated from one another". The flow curve of Newtonian fluid can be expressed by a series of plots of shear stress versus shear rate, which follows a linear line passing through the origin with the slope of η, as shown in Figure 1.10. However, no real fluid can be described by the Newtonian model perfectly.

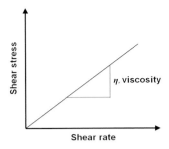

Figure 1.10 The Newtonian model.

Table 1.1 Viscosity of common substances

Substance	Viscosity (mPa.s)	Temperature (°C)	References
Air	0.01	20	Chhabra and Richardson (2011)
Water	1.0016	20	Rumble (2018)
Mercury	1.526	25	Rumble (2018)
Benzene	0.604	25	Rumble (2018)
Whole milk	2.12	20	Fellows (2009)
Coffee cream	10	25	Rumble (2018)
Olive oil	56.2	26	Fellows (2009)
Castor oil	600	20	Chhabra and Richardson (2011)
Honey	2000–10,000	20	Yanniotis et al. (2006)
Ketchup	5000–20,000	25	Koocheki et al. (2009)
Bitumen	10,000,000,000	20	Chhabra and Richardson (2011)

Instead, some common liquids and gases (e.g. water, alcohol, and air) under ordinary conditions can be assumed to be Newtonian fluid for practical calculations. The viscosity of some common Newtonian fluids is summarized in Table 1.1.

1.2.3 Non-Newtonian flow

All fluids that do not follow Newton's law of viscosity, i.e. constant viscosity independent of shear stress, are defined as non-Newtonian fluid. In this kind of fluid, the viscosity changes with the applied shear stress or shear rate history. Indeed, compared to the Newtonian flow curve passing through the origin with a constant scope, the flow curve of non-Newtonian fluids is different. For some non-Newtonian fluids with shear-independent viscosity, e.g. Bingham flow, the flow curve still exhibits a different behavior compared to Newtonian flow. In this case, the concept of viscosity is inadequate to describe the rheological behavior of non-Newtonian fluids. In this context, various models or constitutive equations have been developed to distinguish the behavior of non-Newtonian fluids and idealize the flow curves, as presented in Figure 1.11.

As mentioned earlier, all non-Newtonian fluids exhibit a distinct flow behavior compared with the Newtonian flow. In reality, some fluids possess yield stress, which is the minimum stress when flow occurs. In other words, applying an external shear stress lower than the yield stress, the fluid cannot flow and only deform elastically. When the applied shear stress

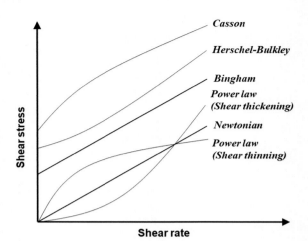

Figure 1.11 Typical common constitutive models for non-Newtonian fluids.

exceeds the yield stress, the internal network structure will be destroyed and the material starts to flow like a fluid. Materials with yield stress are also considered to be visco-plastic materials. For the material with zero yield stress but inconstant (or shear rate-dependent) viscosity, its rheological behavior can be characterized by the power law model with an exponential relationship between shear stress and shear rate (Ostwald, 1929), as expressed by:

$$\tau = a\dot{\gamma}^b \tag{1.16}$$

where a is the consistency index in Pa.s^b and b is the flow index (dimensionless). As shown in Figure 1.11, if the exponent b is higher than 1, the flow curve is concave upward, and the viscosity increases with the shear rate, indicating that the material shows shear-thickening or dilatant behavior. By contrast, if the exponent b is lower than 1, the flow curve is concave downward. It means that the material exhibits shear-thinning or pseudo-plastic behavior, i.e. the viscosity decreases with the increase of shear rate. For suspensions, the shear-induced behavior is related to the distribution of particles in the suspending medium, with particle rearrangement along the direction of applied shear for shear-thinning and clusters/jamming formation for shear thickening, as shown in Figure 1.12. The shear-induced behavior depends on the concentration and intrinsic properties of the dispersed particles such as shape, size, and density. Generally, a suspension with a particle concentration higher than 75% tends to show shear-thickening behavior.

For a material with a yield stress and a shear rate (stress) independent viscosity, its rheological behavior can be described by the Bingham model (Bingham, 1917). The constitutive equation is expressed as:

$$\tau = \tau_0 + \mu\dot{\gamma} \tag{1.17}$$

where τ_0 and μ are the yield stress in Pa and the plastic viscosity in Pa.s, respectively. The plastic viscosity in the Bingham model refers to the same physical meaning as the viscosity in a Newtonian model, i.e. the slope of the curve between shear stress and shear rate. Toothpaste is a classic example of Bingham fluid.

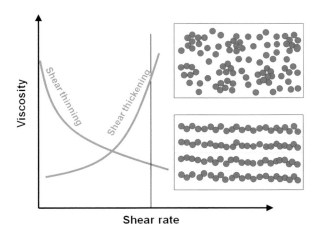

Figure 1.12 Shear-thinning and shear-thickening behaviors of suspensions.

For material with a yield stress and a shear rate-dependent viscosity, the flow behavior can be characterized by the Herschel–Bulkley model (Herschel and Bulkley, 1926), as expressed by:

$$\tau = \tau_0 + a\dot{\gamma}^b \tag{1.18}$$

where a is the consistency index in Pa.s^b and b is the flow index (dimensionless). The material exhibits shear-thinning behavior when $b<1$, while the material shows shear-thickening behavior when $b>1$. If $b=1$, this model reduces to the Bingham equation. It should be mentioned that the plastic viscosity cannot be directly obtained from the Herschel–Bulkley model. Instead, an equivalent plastic viscosity μ_e can be calculated from the consistency index and the flow index as follows:

$$\mu = \frac{3a}{b+2}\dot{\gamma}_{\max}^{b-1} \tag{1.19}$$

where $\dot{\gamma}_{\max}$ is the maximum shear rate. Another model describing the nonlinearity of the flow of fluids with yield stress is the Casson model:

$$\tau^{1/2} = \tau_0^{1/2} + \mu^{1/2}\dot{\gamma}^{1/2} \tag{1.20}$$

In addition to the above-stated models, Table 1.2 summarizes some other equations describing the rheological behavior of non-Newtonian fluids.

1.2.4 Thixotropy

The viscosity of the pseudo-plastic fluids is time-independent. However, certain fluids are showing time-dependent shear-thinning behavior. This kind of fluid is called thixotropic fluid. The concept of thixotropy was first introduced by Freundlich (Freundlich and Juliusburger, 1935), from two ancient Greek words: *thixis* meaning *touch* and *tropo* meaning

Table 1.2 Summary of rheological equations for non-Newtonian fluids

Model	Equation	References
Power	$\tau = A\dot{\gamma}^n$	Ostwald (1929)
Bingham	$\tau = \tau_0 + \mu\dot{\gamma}$	Bingham (1917)
Modified Bingham	$\tau = \tau_0 + a\dot{\gamma} + b\dot{\gamma}^2$	Feys et al. (2007), Feys et al. (2008)
Herschel–Bulkley	$\tau = \tau_0 + a\dot{\gamma}^b$	Herschel and Bulkley (1926)
Casson	$\tau^{1/2} = \tau_0^{1/2} + \mu^{1/2}\dot{\gamma}^{1/2}$	Papo and Piani (2004)
Generalized Casson	$\tau^m = \tau_0^m + [\eta_\infty \dot{\gamma}]^m$	Papo and Piani (2004)
Papo–Piani	$\tau = \tau_0 + \eta_\infty \dot{\gamma} + K\dot{\gamma}^n$	Papo and Piani (2004)
De Kee	$\tau = \tau_0 + \eta\dot{\gamma}e^{-\alpha\dot{\gamma}}$	Yahia and Khayat (2003)
Vom Berg	$\tau = \tau_0 + B\sinh^{-1}(\dot{\gamma}/C)$	Vom Berg (1979)
Eyring	$\tau = a\dot{\gamma} + B\sinh^{-1}(\dot{\gamma}/C)$	Atzeni et al. (1985)
Roberston–Stiff	$\tau = a(\dot{\gamma} + C)^b$	Yahia and Khayat (2003)
Atzeni	$\dot{\gamma} = a\tau^2 + b\tau + \delta$	Atzeni et al. (1983)
Williamson	$\tau = \eta_\infty \dot{\gamma} + \tau_f \dfrac{\dot{\gamma}}{\dot{\gamma} + \Gamma}$	Papo (1988)
Sisko–Ellis	$\tau = a\dot{\gamma} + b\dot{\gamma}^c$ (c < 1) $\eta = \eta_\infty + K\dot{\gamma}^{n-1}$	Papo (1988)
Shangraw–Grim–Mattocks	$\tau = \tau_0 + \eta_\infty\dot{\gamma} + \alpha_1[1 - \exp(-\alpha_2\dot{\gamma})]$	Papo (1988)
Yahia–Khayat	$\tau = \tau_0 + 2(\sqrt{\tau_0\eta_\infty})\sqrt{\dot{\gamma}e^{-\alpha\dot{\gamma}}}$	Yahia and Khayat (2003)

to change. Simply, thixotropy means that change occurs by touch. Technically, thixotropy is defined as *the continuous decrease in viscosity with time when applying an external shear stress to a rest suspension, and then a subsequent recovery of viscosity over time when the shear stress is removed* (Mewis and Wagner, 2009), as illustrated in Figure 1.13. Two characteristics can be recognized from this definition. First, the thixotropic material shows a time-dependent decrease of viscosity under flow state, i.e. time-dependent shear-thinning viscosity. Second, thixotropy is reversible. Thixotropy only appears in shear-thinning fluids, due to the fact that the breakdown structure cannot reform immediately after the removal of the shear stress. Daily-used products such as toothpaste, paints, ketchup, and hair gel are thixotropic fluids.

Thixotropy can be evaluated by hysteresis loop area (Jiao et al., 2019b, Roussel, 2006), structural breakdown area (Khayat et al., 2012), static yield stress (Lowke, 2018), thixotropic index (Patton, 1964), stress recovery (Qian and Kawashima, 2016), storage modulus (Jiao et al., 2019a, Yuan et al., 2017), etc. Readers are referred to Jiao et al. (2021) and Mewis and Wagner (2009) for further details. Thixotropy is important for applications that require the material to flow easily during placing, but then the structure is built quickly when the flow stops. A typical usage of thixotropy in engineering is the extrusion-based 3D printing process (De Schutter et al., 2018, Yuan et al., 2019), where the material needs to flow easily in the pump line and nozzle and builds up its mechanical strength immediately after arriving at the final position.

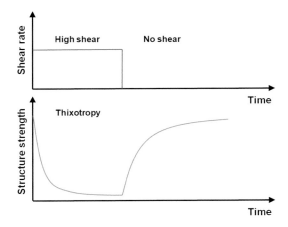

Figure 1.13 Schematic diagram of thixotropy. (Adapted from Lowke, 2018.)

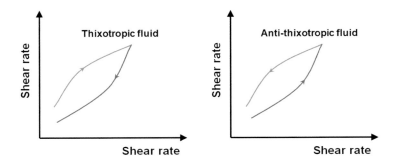

Figure 1.14 Thixotropy versus anti-thixotropy.

1.2.5 Anti-thixotropy (rheopexy)

Opposite to thixotropy, the time-dependent shear-thickening behavior is defined as rheopexy, also called anti-thixotropy or negative thixotropy. The time-dependent increase in viscosity during flow is generally caused by flow-induced aggregation. The difference between thixotropy and anti-thixotropy behaviors is presented in Figure 1.14. The anti-thixotropic materials are much less common. Typical examples of rheopectic fluids include gypsum pastes, cream, and printer inks, which become more stiffening by continuous shaking or mixing. Besides, suspensions with a solid volume fraction lower than 10% generally exhibit a negative thixotropic behavior.

Overall, a schematic representation of the flow behavior of various fluids is provided in Figure 1.15.

1.3 CEMENT-BASED MATERIALS

1.3.1 History of cement and concrete

Joseph Aspdin, an English bricklayer and builder, is generally recognized as the first inventor of Portland cement. In 1824, he patented the method of binder production from the burned

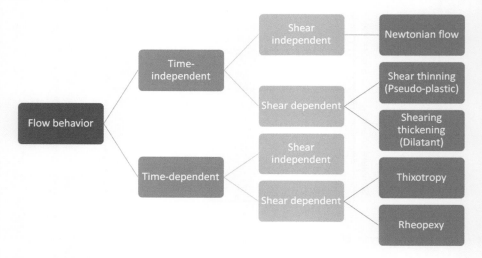

Figure 1.15 Schematic representation of flow behavior of various fluids.

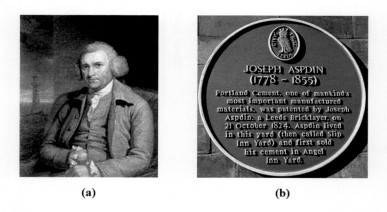

Figure 1.16 (a and b) Joseph Aspdin and a plaque in Leeds.

mixture of limestone and clay, and this powder can react with water to develop strength. This product is named Portland cement for the first time because its color resembled that of stone from the location of Portland. Of course, the cement invented by Aspdin was nothing more than a hydraulic lime, and its mineralogy was completely different from that of the toady's cement. Even so, Aspdin's patent gave him the priority for using the term Portland cement. Aspdin's cement patent is undoubtedly a pioneering work that opens a door for the development of modern cement. Consequently, Joseph Aspdin was acclaimed as the father of Portland cement, as shown in Figure 1.16. The first cement plant was built in Grodziec, Poland and started its production in 1857. However, the cement quality at that time was very low. Without the great progress in its production, it cannot be used for large-scale applications (Hewlett and Liska, 2019).

In the history of Portland cement, the production technique and quality control are the most important to the large-scale production of modern cement. The invention of the rotary kiln aided in the large-scale production of modern cement possible (Hewlett and Liska,

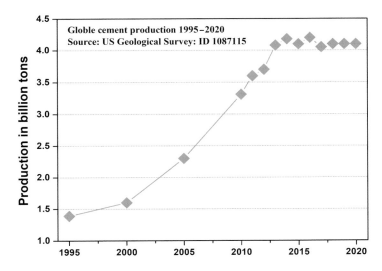

Figure 1.17 Cement production worldwide from 1995 to 2020 (in billion tons).

2019). Cement manufacture was changed from a batch process to a continuous production process, and the quality of cement was significantly improved. In 1877, Thomas Crampton first patented a kiln with a revolving cylinder lined with firebricks. In 1885, Frederick Ransome, an American, produced a 6.5 m long cylindrical kiln, but it didn't work well. In 1898, Hurry and Seaman, both Americans, successfully built the first fully operational rotary kiln supplied with energy by coal. From then on, cement has become the major construction material, and the cement industry has boomed to meet the huge demands for the construction of buildings, roads, bridges, dams, and factories all over the world. Figure 1.17 gives the worldwide output of cement from 1995 to 2020 (US Geological Survey). In 2019, China yielded 56.2% of the worldwide output of cement (CEMBUREAU), as shown in Figure 1.18, due to the massive construction of infrastructure recently.

Considering both economic and technical reasons, cement is mainly used together with sand and stone to form mortar and concrete. On the one hand, sand and stone are much cheaper than cement, and thus the cost can be reduced. On the other hand, sand and stone are more stable in terms of volume or chemistry, and their application can improve the volume and chemical stability of mortar or concrete. Mortar is a mixture of sand and a stiff paste of cement made with a proper amount of water, which is often used to bind bricks or blocks. Mortar also has some special applications such as grouting mortar and floor mortar. Concrete is a mixture of coarse aggregate (e.g. small stones or gravel), fine aggregate (e.g. sand), cement, and enough water, which is often used for the construction of primary structural members.

The vast need of infrastructure and the various requirements for construction materials have stimulated the invention and development of many benchmark concrete techniques. These technological advancements make concrete the most important and widely used construction material. The most important dates in the history of the development of contemporary concrete are the following:

1867—Joseph Monier introduced the reinforcement of concrete, which is the major composite used in infrastructure.

1870—The production of precast elements began. Precast elements are still very popular currently, since concrete elements can be manufactured in the factory. The structural

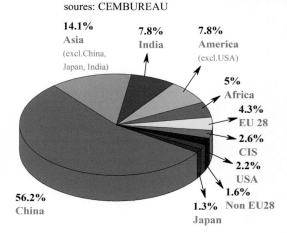

Figure 1.18 World cement production in 2019 by region and main countries.

quality, noise, and pollution can be better controlled. Many governments and authorities are trying to increase the percentage of the precast element and reduce the cast-in-site element.

1907—Koenen invented the prestressed concrete. This form of element fully uses the compressive strength of concrete and the tensile strength of steel. Prestressed concretes are used in many heavy loading structures, such as large-span bridges and beams.

1924—Bolomey proposed the formula for concrete strength calculation. The Bolomey equation is still one of the most important equations used for modern concrete 100 years later.

1950—Water-reducing agent, which was polymerized by naphthalene formaldehyde sulfonate salts, was first used to increase the flowability of concrete. The technology for chemical admixtures developed very fast, with water-reducing agents evolving into the superplasticizer. The application of superplasticizer makes high-strength concrete possible.

1980—Aïtcin first introduced high-performance concrete (HPC). HPC exceeds the properties and constructability of normal concrete. Normal and special materials are used to make these specially designed concretes that must meet a combination of performance requirements.

1988—Okamura invented self-compacting concrete (SCC). This type of concrete can flow and fill up the formwork under its own weight without vibration. Due to its economic and technical benefits, SCC has been widely used in industries, such as precast industry and heavily reinforced section.

1994—De Larrard first introduced ultra-high-performance concrete (UHPC) with low porosity, high compressive strength (above 120 MPa), high durability, and self-compactability, based on an optimized particle-packing model (de Larrard and Sedran, 1994). In 1997, UHPC was first used to construct a pedestrian bridge in Canada.

1.3.2 Fresh properties of cement-based materials

Fresh properties of concrete determine whether strength grade, elastic modulus, durability, and even color of hardened concrete meet the design or not. In the early history of concrete, workers assessed the fresh properties of concrete by subjective methods, which made quantitative assessment hard. Abrams was probably the first well-known researcher who

Introduction to rheology 17

recognized the importance of the workability of fresh concrete. He proposed the slump test to quantitatively evaluate the workability of fresh concrete. The slump test is still used up to now and universally accepted as a quality control measure of fresh concrete owing to the following reasons:

- The testing setup for the slump test is quite cheap and simple.
- The slump test is quite easy to perform.
- The test result, to some extent, not only reflects the flowability of concrete, but also roughly judges the bleeding and cohesiveness of concrete.

As pointed out by Tattersall (Tattersall and Banfill, 1983), although the slump test has been the most widely used method for the workability of concrete, it is an empirical test without a scientific base and the testing result is barely related to basic physical constants. Different operators may give different slump values for the same concrete. Moreover, two concretes with the same slump value may work quite differently. Therefore, rheology is introduced to describe the fresh properties of concrete from other disciplines for reference. Through a large number of investigations, it was found that many mature analytical and physical models in rheology can be applied to concrete, and various rheological testing techniques can also be employed to characterize the viscoelastic behavior of concrete.

Common rheometers can be directly applied to cement paste which is a suspension with micro-sized particles suspended in water. Therefore, the rheological behavior of cement paste was studied as early as the 1950s. Tattersall published his pioneering work on the structural breakdown of cement paste in the journal of *Nature* (Tattersall, 1955). In this work, a theoretical understanding of the structural breakdown of cement paste was attempted. Afterward, many other papers on the rheological behavior of cement paste were published.

Due to the large size of coarse aggregate in fresh concrete, common rheometers are not suitable for measuring concrete. By the 1970s, a rotating vane or coaxial cylinder with a wide gap was invented for concrete and used to measure the relationship between torque and rotational speed, whereas the complex mathematic transformation was not made for the testing results. At that time, researchers were just trying to find a more accurate testing method to replace the slump test. Tattersall successfully developed a practical testing setup, which was called the two-point test (Figure 1.19), that could be employed both in the lab

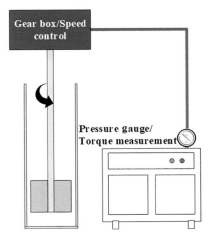

Figure 1.19 Schematic diagram of Tattersall's two-point test rheometer. (Adapted from Ferraris, 1999.)

and on-site (Tattersall and Bloomer, 1979). This test measures the shear stress under at least two shear rates, allowing to calculate the rheological parameters, e.g. yield stress and plastic viscosity. This testing setup started to bring the concept of rheology to concrete science and technology. However, due to the low content of paste, concrete still cannot flow like water. Most of the aggregates in concrete directly contact with each other when concrete flows. Thus, concrete flows with a friction and collision regime. Rheology based on continuum mechanics is not a good option for characterizing the flow behavior of this material. Tattersall argued the advantages of two-point testing of fresh concrete workability over the empirical test methods that had been employed to date (Tattersall and Banfill, 1983).

The invention of SCC in 1988 immediately buildup a close link between concrete and rheology. SCC is a type of concrete that is rich in cement paste. Most of the aggregates are suspended in the paste. Therefore, SCC can flow like water just under the self-weight, and the aggregates are no longer in direct contact during the flow process. Therefore, the rheological theory and experiment are perfectly applicable to SCC. The enthusiasm for rheology research was inspired by the invention and wide application of SCC. The rheology of SCC is completely different from ordinary concrete, bringing totally a different engineering behavior. High flowability of SCC is required to ensure self-compacting property. The density of water is $1000\,kg/m^3$, while it is 2600–2700 and $3150\,kg/m^3$ for aggregate and cement particles, respectively. On the other hand, bleeding and segregation stemming from the large density difference of constituents more possibly happen for SCC than ordinary concrete if the proper mix design of SCC is not taken. The high viscosity of the paste is required by SCC to improve the resistance of bleeding and segregation.

Followed by Tattersall's two-point rheometer, many other more sophisticated testing setups with a more accurate analytical model, computer-controlled systems, and automatic data collection were developed. For example, Wallevik developed a new coaxial cylinder rheometer for fresh concrete in the 1990s (Wallevik and Gjørv, 1990). This rheometer has been successfully commercialized as the BML, and now it upgrades to its higher version, ConTec Visco 5, as shown in Figure 1.20a. A typical dimension of the rheometer has a bob diameter of 100 mm and a container with a diameter of 145 mm, resulting in a capacity of 17 L of concrete. It is designed for concrete with slumps greater than 120 mm. In this rheometer, the outer cylinder rotates, while the inner bob registering the torque remains stationary.

In 1993, the biggest coaxial rheometer with a container radius of 1200 mm, a bob of 760 mm, and a capacity of 500 L of concrete was developed in France. It was named CEMAGREF (Coussot, 1993), as presented in Figure 1.20b. However, this rheometer has not been commercially available due to its large volume.

In 1994, the IBB rheometer (see Figure 1.20c) was developed by Beaupre in Canada (Beaupre, 1994). The vane motion of IBB was planetary and not axial. The testing sample was about 21 L of concrete. It was designed for concrete with slumps between 20 and 300 mm.

In 1996, the BTRHEOM was developed under the direction of de Larrard et al. at the Laboratoire des Ponts et Chaussées (LCPC) in Paris (Hu et al., 1996). The BTRHEOM rheometer, as shown in Figure 1.20d, was the only concrete rheometer based on parallel plate geometry. This rheometer required about 7 L of concrete that has at least 100 mm of the slump, and it was designed to be portable so that it could be used at a construction site.

In 2004, the ICAR rheometer, as depicted in Figure 1.20e, was developed at the International Center for Aggregates Research (ICAR) at the University of Texas at Austin by Koehler and his colleagues (Koehler and Fowler, 2004). This rheometer was well designed and portable for on-site testing. Different rotators and containers are available for concrete with different aggregate sizes. The ICAR rheometer is available for concrete with a slump higher than 75 mm.

Introduction to rheology 19

Figure 1.20 Concrete rheometers: (a) ConTec Visco 5, (b) CEMAGREF (adapted from Coussot, 1993), (c) IBB rheometer, (d) BTRHEOM rheometer, and (e) ICAR rheometer.

Since the behavior of concrete may range from a very stiff state to SCC and various test rheometers are used, the rheological results from different authors are hard to compare. Thus, organizations from different countries carried out several programs for the comparison of different rheometers. These test programs have been executed with several rounds. The first round-robin test was carried out in 2000 in Nante, France (Ferraris and Brower, 2000). It was found that different types of mixtures were ranked statistically in the same order by all the rheometers for both yield stress and plastic viscosity. However, obvious differences in absolute values given by the various rheometers were found. Another round-robin test carried out in 2003 in Cleveland OH, USA (Ferraris and Brower, 2003), confirmed the findings obtained in 2000. It was also pointed out that small variations in the concrete could cause large changes in the rheometer results. The RILEM TC MRP (Measuring Rheological Properties of Cement-based materials) organized a new round-robin test in Bethune, France in 2018 to evaluate the rheological properties of different mortar and concrete mixtures (Feys et al., 2019).

The rheological properties of fresh concrete evolve with time, due to the continuous hydration of cementitious materials. During this period, hydration products fill in the voids and connect in-between the particles (Huang et al., 2020a, Huang et al., 2020b), and a percolated network is progressively formed. It has been shown that the C–S–H bridge is the main reason for the rigidification of cement paste at the early age (Mostafa and Yahia, 2017, Roussel et al., 2012). Thus, the viscoelasticity and rheological properties of fresh cement paste change with continuous cement hydration, which is important to many engineering application scenarios of cement-based materials, e.g. mixing, pumping, casting, 3D concrete printing, and smart casting (Jiao et al., 2021, Roussel, 2006, Roussel, 2018).

It is well known that concrete is strong in compression and weak in tension. With the increase of compressive strength, concrete becomes more brittle, leading to catastrophic failures. Fiber-reinforced concrete (FRC) has been studied for many years to overcome this tension weakness of all types of concrete. Fiber-reinforced concrete is a promising composite, whereas the incorporation of fibers in concrete changes the rheological properties. The suspension of FRC is a non-Newtonian fluid, which can display differences in normal stress. Flow resistance arises due to the opposite movement of the particles. Thus, the rheology of fiber-reinforced concrete is also different from that of ordinary concrete.

It can be said that the rheology of cement-based materials is far more complicated than other materials for the following reasons:

- Complex constituents. Cement-based materials are inorganic ones which may include various types of particles such as cement, fly ash, slag, silica fume, aggregates, and fibers. Furthermore, many organic polymers are often used as their key admixtures.
- Evolving with time. Hydration of cement begins and continuously goes on after the contact of cement and water. The ongoing hydration results in the time-varying rheology of cement-based materials.
- Multi-scale particles ranging from nm to mm. Cement-based materials include micro-sized particles such as cement and fly ash, and sand and coarse aggregate with the size of mm.

However, the complexity of the rheology of cement-based materials has not dampen the researchers' enthusiasm. Numerous researches have been carried out, and are carried out, on the rheology of cement-based materials, including the fundamentals, measurements, and predictions of rheology. Rheology for cement-based materials is not just "a study of flow and deformation". In addition, the underlying physical and chemical reactions behind the flow and deformation become more and more crucial for the study of rheology. This brings the rheology to the molecular level, i.e. micro-rheology. It means that the key interest is devoted not only to the movements of physical points but also to the chemical and physical reactions that happen inside a point during the deformation of the medium. The developed techniques and new findings in the field of rheology play an important role in the advancement of new concrete technologies.

1.4 THE SCOPE OF THIS BOOK

As stated above, rheology is a very useful technical tool and scientific basis for cement-based materials. In particular, new technologies put forward new requirements for rheological properties. The complicated and interesting rheology of cement-based materials attracts the attention of more and more researchers, and much funding around the world has been allocated to this topic. This book focuses on the flow and rheological behavior of fresh cement-based materials. It should be mentioned that the rheology of hardened cement and concrete is beyond the scope of this book, and hence will not be discussed. Three most important parts of the rheology of cement-based materials are discussed in this book.

The first part deals with the fundamentals of the rheology of cement-based materials. The basic knowledge of rheology is briefly introduced in this part. Cement-based materials are classified into two main groups based on the particle sizes, i.e. cement paste and concrete (including mortar). Cement paste is a suspension with an interstitial solution and micro-sized particles. Colloidal interaction between particles has a great influence on the rheology of cement paste. On a larger scale, sand and aggregate are in the size of mm. Lubrication, friction, and collision may dominate the flow regime of concrete, depending on the rheological properties of interstitial fluid (i.e. cement paste) and the aggregate ratio. All these are discussed in detail in this part.

The second part discusses the measurement techniques for the rheology of cement-based materials. More than 100 techniques have been developed to evaluate the rheological properties of cement-based materials, either rheometers with sound scientific bases or empirical methods. In addition, due to the large size of aggregate in concrete, common rheometers do not apply to concrete. Therefore, many rheometers specialized in concrete have been developed. Obtaining the accurate rheological parameters of cement-based materials is crucial to this field.

The third part deals with the application of rheology in concrete technology. New technologies always come with new requirements for materials, even new materials. From SCC, new cementitious materials, to digital fabrication, they put forward new requirements on the rheological properties of concrete.

REFERENCES

Atzeni, C., et al. (1983). "New rheological model for Portland cement pastes." *Il Cemento*, 80.
Atzeni, C., et al. (1985). "Comparison between rheological models for Portland cement pastes." *Cement and Concrete Research*, 15(3), 511–519.
Banfill, P. (1990). "The rheology of cement paste: Progress since 1973", in *Properties of Fresh Concrete*: Proceedings of the International RILEM Colloquium, edited by HJ Wierig, Chapman & Hall. CRC Press, Boca Raton, FL, pp. 3–9.
Barnes, H. A., et al. (1989). *An introduction to rheology*. Elsevier, Amsterdam, the Netherlands.
Beaupre, D. (1994). *Rheology of high performance shotcrete*. Doctoral dissertation, University of British Columbia, Vancouver, Canada.
Bingham, E. C. (1917). *An investigation of the laws of plastic flow*. US Government Printing Office, Washington, United States.
Bingham, E. C. (1922). *Fluidity and plasticity*. McGraw-Hill, New York, United States.
Chhabra, R. P., and Richardson, J. F. (2011). *Non-Newtonian flow and applied rheology: Engineering applications*. Butterworth-Heinemann, Oxford, United Kingdom.
Coussot, P. (1993). *Rhéologie des boues et laves torrentielles: étude de dispersions et suspensions concentrées*. Editions Quae, Versailles, France.
de Larrard, F., and Sedran, T. (1994). "Optimization of ultra-high-performance concrete by the use of a packing model." *Cement and Concrete Research*, 24(6), 997–1009.
De Schutter, G., et al. (2018). "Vision of 3D printing with concrete - Technical, economic and environmental potentials." *Cement and Concrete Research*, 112, 25–36.
Fellows, P. J. (2009). *Food processing technology: Principles and practice*. Elsevier, Amsterdam, the Netherlands.
Ferraris, C. F. (1999). "Measurement of the rheological properties of high performance concrete: State of the art report." *Journal of Research of the National Institute of Standards and Technology*, 104(5), 461–478.
Ferraris, C. F., and Brower, L. E. (2000). Comparison of Concrete Rheometers: International test at LCPC (Nantes France) in October, 2000.
Ferraris, C. F., and Brower, L. E. (2003). Comparison of Concrete Rheometers: International tests at MB (Cleceland OH, USA) in May, 2003.
Feys, D., et al. (2007). "Evaluation of time independent rheological models applicable to fresh self-compacting concrete." *Applied Rheology*, 17(5), 56244–56241–56244–56210.
Feys, D., et al. (2008). "Fresh self compacting concrete, a shear thickening material." *Cement and Concrete Research*, 38(7), 920–929.

Feys, D., et al. (2019). "An overview of RILEM TC MRP Round–Robin testing of concrete and mortar rheology in Bethune, France, May 2018." *Proceedings 2nd International RILEM Conference Rheology and Processing of Construction Materials (RheoCon2)*, Dresden, Germany.

Freundlich, H., and Juliusburger, F. (1935). "Thixotropy, influenced by the orientation of anisometric particles in sols and suspensions." *Transactions of the Faraday Society*, 31, 920–921.

Herschel, W. H., and Bulkley, R. (1926). "Konsistenzmessungen von gummi-benzollösungen." *Kolloid-Zeitschrift*, 39(4), 291–300.

Hewlett, P., and Liska, M. (2019). *Lea's chemistry of cement and concrete*. Butterworth-Heinemann, Oxford, United Kingdom.

Hu, C., et al. (1996). "Validation of BTRHEOM, the new rheometer for soft-to-fluid concrete." *Materials and Structures*, 29(10), 620–631.

Huang, T., et al. (2020a). "Understanding the mechanisms behind the time-dependent viscoelasticity of fresh C3A–gypsum paste." *Cement and Concrete Research*, 133, 106084.

Huang, T., et al. (2020b). "Evolution of elastic behavior of alite paste at early hydration stages." *Journal of the American Ceramic Society*, 103(11), 6490–6504.

Jiao, D., et al. (2019a). "Structural build-up of cementitious paste with nano-Fe_3O_4 under time-varying magnetic fields." *Cement and Concrete Research*, 124, 105857.

Jiao, D., et al. (2019b). "Time-dependent rheological behavior of cementitious paste under continuous shear mixing." *Construction and Building Materials*, 226, 591–600.

Jiao, D., et al. (2021). "Thixotropic structural build-up of cement-based materials: A state-of-the-art review." *Cement and Concrete Composites*, 122, 104152.

Khayat, K. H., et al. (2012). "Evaluation of thixotropy of self-consolidating concrete and influence on concrete performance." *Proceedings of the 3rd Iberian Congress on Self Compacting Concrete*, Madrid, Spain, 3–16.

Koehler, E. P., and Fowler, D. W. (2004). "Development of a portable rheometer for fresh portland cement concrete." Technical Report, International Center for Aggregates Research (ICAR), The University of Texas at Austin, Austin, United States.

Koocheki, A., et al. (2009). "The rheological properties of ketchup as a function of different hydrocolloids and temperature." *International Journal of Food Science & Technology*, 44(3), 596–602.

Liingaard, M., et al. (2004). "Characterization of models for time-dependent behavior of soils." *International Journal of Geomechanics*, 4(3), 157–177.

Lowke, D. (2018). "Thixotropy of SCC—A model describing the effect of particle packing and superplasticizer adsorption on thixotropic structural build-up of the mortar phase based on interparticle interactions." *Cement and Concrete Research*, 104, 94–104.

Mewis, J., and Wagner, N. J. (2009). "Thixotropy." *Advances in Colloid and Interface Science*, 147–148, 214–227.

Mostafa, A. M., and Yahia, A. (2017). "Physico-chemical kinetics of structural build-up of neat cement-based suspensions." *Cement and Concrete Research*, 97, 11–27.

Ostwald, W. (1929). "de Waele-Ostwald equation." *Kolloid Zeitschrift*, 47(2), 176–187.

Papo, A. (1988). "Rheological models for cement pastes." *Materials and Structures*, 21(1), 41–46.

Papo, A., and Piani, L. (2004). "Flow behavior of fresh Portland cement pastes." *Particulate Science and Technology*, 22(2), 201–212.

Patton, T. C. (1964). "Paint flow and pigment dispersion." *Paint Flow and Pigment Dispersion*, pp. 479–479.

Qian, Y., and Kawashima, S. (2016). "Use of creep recovery protocol to measure static yield stress and structural rebuilding of fresh cement pastes." *Cement and Concrete Research*, 90, 73–79.

Roussel, N. (2006). "A thixotropy model for fresh fluid concretes: Theory, validation and applications." *Cement and Concrete Research*, 36(10), 1797–1806.

Roussel, N., et al. (2012). "The origins of thixotropy of fresh cement pastes." *Cement and Concrete Research*, 42(1), 148–157.

Roussel, N. (2018). "Rheological requirements for printable concretes." *Cement and Concrete Research*, 112, 76–85.

Rumble, J. R. (2018). *CRC handbook of chemistry and physics*. CRC Press, Boca Raton, FL.

Tattersall, G. (1955). "Structural breakdown of cement pastes at constant rate of shear." *Nature*, 175(4447), 166–166.

Tattersall, G. H., and Banfill, P. F. (1983). *The rheology of fresh concrete*. Pitman Books Limited, London, England.

Tattersall, G. H., and Bloomer, S. (1979). "Further development of the two-point test for workability and extension of its range." *Magazine of Concrete Research*, 31(109), 202–210.

Vom Berg, W. (1979). "Influence of specific surface and concentration of solids upon the flow behaviour of cement pastes." *Magazine of concrete research*, 31(109), 211–216.

Wallevik, O., and Gjørv, O. (1990). "Development of a coaxial cylinders viscometer for fresh concrete." *Proceedings of the Properties of Fresh Concrete: Proceedings of the International RILEM Colloquium*, Edited by H.J. Wierig, CRC Press, Hanover, Germany, pp. 213–224.

Yahia, A., and Khayat, K. H. (2003). "Applicability of rheological models to high-performance grouts containing supplementary cementitious materials and viscosity enhancing admixture." *Materials and Structures*, 36(260), 402–412.

Yanniotis, S., et al. (2006). "Effect of moisture content on the viscosity of honey at different temperatures." *Journal of Food Engineering*, 72(4), 372–377.

Yuan, Q., et al. (2017). "On the measurement of evolution of structural build-up of cement paste with time by static yield stress test vs. small amplitude oscillatory shear test." *Cement and Concrete Research*, 99, 183–189.

Yuan, Q., et al. (2019). "A feasible method for measuring the buildability of fresh 3D printing mortar." *Construction and Building Materials*, 227, 116600.

Chapter 2

Rheology for cement paste

2.1 INTERACTION BETWEEN PARTICLES IN THE PASTE

Cement paste is a suspension system containing water and cement particles, and even mineral admixture particles in some cases. Its rheological properties have universal characteristics as suspensions. The volume fraction of suspending particles, the particle shape and size distribution, and the rheological properties of the liquid medium can directly influence the rheological properties of the cement paste. In addition, hydration of cementitious particles and functions of chemical admixtures lead the rheological properties of cement paste to be more complex. The interparticle interactions can be changed due to the presence of hydrates and chemical admixtures. From a micromechanical view, the rheological properties of cement paste are determined by the interparticle interactions, including colloidal interactions, Brownian forces, hydrodynamic forces, and contact forces.

2.1.1 Colloidal interaction

Several types of non-contact interactions occur within a cementitious suspension (Flatt, 2004b). At a short distance, cement particles interact via (generally attractive) van der Waals forces (Flatt, 2004a). Also, there are electrostatic forces that result from the absorption of ions on the particle surfaces (Flatt and Bowen, 2003). Polymer additives used in many modern cementitious materials can introduce steric hindrance (Banfill, 1979, Yoshioka et al., 1997), which is believed to predominate over electrostatic repulsion. Each of these different interactions introduces non-contact forces between particles, the magnitude of which depends primarily on their separation distance.

Van der Waals interactions were shown to dominate all other colloidal interactions in the case of cement pastes and therefore dictate the interparticle distance (Flatt, 2004a, Flatt et al., 2009). The interparticle force is given by:

$$F \cong \frac{A_0 a^*}{12 H^2} \quad (2.1)$$

where a^* is the radius of curvature of the "contact" points, H is the surface-to-surface separation distance at "contact" points, and A_0 is the non-retarded Hamaker constant (Roussel et al., 2010).

Without inclusion of polymers, the value of H is estimated to be in the order of a couple of nm (Flatt, 2004b, Roussel et al., 2010). However, the steric hindrance induced by the absorption of polymers affects the value of H, and H is dictated by the conformation of the adsorbed polymer on the surface of the cement grains (Flatt et al., 2009). Orders of magnitude for typical polymers are around 5 nm (Flatt et al., 2009). Recent advances have shown that it shall be possible to correlate the molecular structure of the polymer to its surface conformation (Flatt et al., 2009).

2.1.1.1 Van der Waals force

The van der Waals force is consisted of three components (Hiemenz and Rajagopalan, 1997, Wallevik, 2003, Jiao et al., 2021a), i.e. Keesom, Debye, and London interactions, of which the London dispersion force is the most important one. This force is caused by the interaction between induced dipoles in the neighboring molecules, and it always leads to an attraction between particles. The van der Waals potential is the product of the Hamaker constant and the geometrical factor. The Hamaker constant depends on the dielectric properties of the particles and the suspending medium, while the geometrical factor depends on the size and shape of the particles and the separation distance between the particles. For cement, due to the large complexity of the material, no values for the Hamaker constant have been measured (Flatt, 2004a). Published results are based on assumptions and measurements of other minerals.

2.1.1.2 Electrostatic repulsion

Particles in suspensions are electrically charged on their surface. In the suspending medium, counter-ions, which are ions with opposite charges, are attracted to the particles, resulting in a decreasing concentration of these counter-ions with increasing distance from the particle, creating the diffuse double layer (Hiemenz and Rajagopalan, 1997). When two particles approach, their diffuse double layers will start to overlap, creating a higher concentration of counter-ions. As a result, osmotic pressure is created to neutralize this overconcentration of counter-ions and the particles are repelled from each other.

2.1.1.3 Steric hinder force

Adding a polymer layer to a particle surface can create a geometrical barrier for other particles. As a result, two particles cannot approach closer than the separation distance created by the polymers. In contrast to the van der Waals forces and the electrostatic repulsion forces, the absolute value of the potential caused by the polymers is not a monotone decreasing function of the interparticle distance. Instead, it shows a more abrupt change from its characteristic value near the particle surface, to zero at a particle distance equal to the effective length of the polymers (Hiemenz and Rajagopalan, 1997). This repulsion mechanism is called steric hindrance and is a purely geometrical phenomenon.

2.1.2 Brownian forces

Every material is subjected to thermal agitation when the temperature is higher than 0 K. The thermal energy equals kT, where k is the Boltzmann constant and T is the temperature in Kelvin. For larger particles suspended in a certain medium, the energy exerted by temperature is relatively low, compared to gravity and viscous drag forces. The thermal activation does not influence the behavior of the suspension. However, for particles smaller than 1 μm, the Brownian force (kT/a, which equals Brownian/thermal energy, divided by the particle radius a) has a similar or higher order of magnitude as gravity, meaning that it can no longer be neglected. A dimensionless parameter N_r estimating the relative magnitude of Brownian motion over interparticle force is expressed by Eq. (2.2):

$$N_r = \frac{A_0 d}{12 H \cdot kT} \tag{2.2}$$

where A_0 is the Hamaker constant (J), d is the particle diameter (m), k is the Boltzmann constant (m^2.kg/(s.K)$^{-1}$), T is the absolute temperature (K), and H is the surface–surface separation distance (m). If N_r is higher than 1, the thermal agitation is negligible compared with the van der Waals attractive forces. For conventional cementitious suspensions without polymers and ultrafine additives, Brownian motion can be neglected compared to the colloidal interactions (Jiao et al., 2021b).

The Brownian force causes colloidal particles (also called Brownian particles) to move permanently in a random pattern (Perrin, 1916). This Brownian motion can be slowed down by the high viscosity of the suspending medium through a high diffusive flux. As particle size decreases, Brownian motion gains importance relative to the other forces. A lower limit of Brownian particles is set at 1 nm, below which the size of molecules becomes significant and the homogeneity of the suspending medium can no longer be assumed.

2.1.3 Hydrodynamic force

For suspension at flow state, a particle in the suspension experiences a hydrodynamic force tending to facilitate the particle to move. Meanwhile, a viscous drag force is applied to the particle to dissipate the kinetic energy. In the case of a spherical particle in a shear field, the magnitude of the hydrodynamic force is proportional to the square of the particle diameter (d^2) (Roussel et al., 2010). The hydrodynamic force F_H exerted by the shear field on a particle can be calculated using Eq. (2.3):

$$F_H = \frac{3}{2}\pi\eta_s d^2 \dot{\gamma} \tag{2.3}$$

where η_S is the viscosity of the carrier fluid (Pa.s), d is the particle diameter (m), and $\dot{\gamma}$ is the shear rate (s^{-1}).

2.2 EFFECT OF COMPOSITIONS ON RHEOLOGY

The mixture proportions of cement-based materials vary owing to differences in design requirements. Cement paste usually consists of cementitious materials, water, and chemical admixtures. The rheology of cement paste is influenced by the ratio between the components and the characteristics of each component. In this section, the influences of volume fraction of cementitious materials, characteristics of the interstitial solution, proportion, and properties of cementitious materials, and chemical admixtures on paste rheology are reviewed.

2.2.1 Volume fraction

As a suspension, the rheology of cement paste is largely influenced by the volume fraction of the suspending particles. For suspension with a low volume fraction of particles (less than 0.03%), Einstein suggested a simple equation for the viscosity of suspension (Eq. 2.4):

$$\eta = \eta_s(1 + 2.5\phi) \tag{2.4}$$

where η_s is the viscosity of the surrounding fluid and ϕ is the volume fraction of the solid particles. The volume fraction of the cement paste is usually much higher than 0.03%, and the Krieger–Dougherty model (Eq. 2.5) suggested a better equation for the viscosity of suspension with a high-volume fraction:

$$\eta - \eta_s \left(1 + \frac{\phi}{\phi_m}\right)^{-[\eta]\phi_m} \tag{2.5}$$

where ϕ_m is the maximum volume fraction and $[\eta]$ is the intrinsic viscosity. Based on these models, the viscosity of cement paste is influenced by the concentration of cement powders. A higher water-to-cement ratio indicates a lower viscosity.

Furthermore, the packing density regarding a high volume concentration and microstructure formation also determine the viscosity of cement paste. Since cement paste is a highly concentrated suspension, the interactions between particles are important because cement particles are not spherical, and the geometry is changed by the progress of hydration. The yield stress of cementitious suspension is also influenced by the volume fraction. According to Zhou et al. (1999), the yield stress of alumina suspensions increased in power law with the growing volume fraction. A yield stress model named Yodel was developed by Flatt and Bowen (2006) to characterize the yield stress of particulate suspension as follows:

$$\tau_0 \cong m \frac{A_0 a^*}{d^2 H^2} \frac{\phi^2 (\phi - \phi_{\text{perc}})}{\phi_m (\phi_m - \phi)} \tag{2.6}$$

where m is a pre-factor, which depends on the particle size distribution, d is the particle average diameter, a^* is the radius of curvature of the contact points, H is the surface-to-surface separation distance at contact pints, A_0 is the non-retarded Hamaker constant, ϕ is the solid volume fraction, and ϕ_m is the maximum packing fraction of the powder. It shows that apart from the volume fraction of the particles, particle size, particle size distribution, maximum packing, percolation threshold, and interparticle forces together affect the yield stress of the particulate suspensions.

The solid volume fraction in cement paste not only influences its rheological parameters but also affects its flow pattern. Roussel et al. (2010) summarized the dominating interactions between particles in cement paste with different volume fractions and shear rate ranges, and the corresponding flow pattern, as shown in Figure 2.1.

There exists a percolation volume fraction ϕ_{perc} below which there are no direct contacts nor interactions between the particles and above which the suspension displays yield stress. And there exists a critical volume fraction ϕ_{div} above which the yield stress and the viscosity diverge. This critical fraction depends on the degree and/or strength of flocculation of the suspension. Besides, there exists a transition volume fraction of the order of $0.85\phi_{\text{div}}$ which separates suspensions in which the yield stress is mainly dominated by the van der Waals interactions network and concentrated suspensions where the yield stress is mainly due to the direct contacts network. The van der Waals forces dominate the hydrodynamic forces in the low strain rate regime and give rise to a shear-thinning macroscopic behavior. With the increase in strain rate, hydrodynamic forces become more dominant. At high strain rates, however, particle inertia dominates, possibly leading to shear-thickening behavior.

On the other hand, unlike many other suspensions, cement paste is thixotropic, and such character is influenced by the volume fraction of the cement paste. Thixotropy is a reversible characteristic of the material being thinned under the condition of disturbance, and the original structure will be restored over time after the disturbance is stopped (Jiao et al., 2021a). It origins from the aggregation between cement particles, which can be influenced

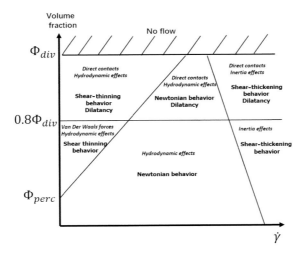

Figure 2.1 Rheo-physical classification of cement suspensions. (Adapted from Roussel et al., 2010.)

by the distance between particles and the volume fraction. According to the experimental work by Lowke (2018), the thixotropic structural build-up of cement paste increases with increasing solid volume fraction. A larger solid volume fraction leads to a shorter distance between particles, and the contributions of colloidal interactions and initial hydration reactions to thixotropy both depend on particle distance.

2.2.2 Interstitial solution

Water reducers are indispensable ingredients in flowable cement paste mixtures, enhancing the flowability by dispersing the particles. On the other hand, the non-adsorbed polymers in the interstitial solution may increase the viscosity of the paste, especially in high-performance or ultra-high-performance cement paste, in which the water-to-binder ratio is very low and a massive dosage of superplasticizer is used. As reported by Liu et al. (2017), the relationship between the viscosity of superplasticizer (SP) solution η_s and the concentration of SP in the solution can be fitted in Eq. (2.7). A larger concentration of SP corresponds to a larger viscosity of the solution:

$$\eta_s = \eta_0 (1 - \phi_{SP})^a \tag{2.7}$$

Apart from the concentration of the polymers, the ion concentration in the interstitial solution can also influence the rheology of cement paste. On the one hand, the ionic concentration influences the thickness of the electrical double layers of the particles (Hunter, 2001, Cosgrove, 2010). The Debye–Hückel length ($1/\kappa$) equation represents the thickness of the double layer (Eq. 2.8) (Overbeek, 1984, Yang et al., 1997).

$$\frac{1}{\kappa} = \sqrt{\frac{\varepsilon \varepsilon_0 RT}{2F^2 I}} \tag{2.8}$$

where ε_0 is the permittivity of the vacuum, ε is the dielectric constant (relative permittivity) of the dispersion medium, R is the gas constant, T is the absolute temperature, F

is the Faraday constant, and I is the ionic strength. As the ionic concentration increases, the thickness of electrical double layers decreases, which can result in increased/stronger agglomeration.

On the other hand, the sulfate ion concentration in cement paste affects the adsorption behavior of SP on cement particles (Yamada et al., 1998, Yamada et al., 2000). The adsorption of SP decreases in the case of high sulfate ion concentration in the aqueous phase due to the competitive adsorption of sulfate ion and SP on cement particles (Yamada et al., 1998, Yamada et al., 2001).

2.2.3 Cement

The mineral composition of ordinary cement clinker is mainly composed of C_3S, C_2S, C_3A, and C_4AF. An appropriate amount of gypsum is added to the clinker for grinding cement. The hydration rate and the water demand of reaction of each mineral composition are different. It can be expected that the rheological properties of cement paste may be affected by the mineral compositions of cement. Various ions are released into the water when cement contacts with water, such as SO_4^{2-}, OH^-, Na^+, and K^+. As mentioned in Section 2.1.5, agglomeration between particles can be enhanced by increasing ion strength, leading to larger yield stress. Besides, a higher hydration rate is expected to a quicker fluidity loss and vice versa. For paste with superplasticizers, these ions may affect the adsorption of superplasticizer onto cement particles and thus affect the rheological properties of cement paste. The fineness of cement (about 350–400 m^2/kg for ordinary cement) also has a great influence on its rheological properties, because fine cement needs more water for a given flowability, and fine cement hydrates faster than coarser one.

The effects of chemical composition and physical characteristics on rheological properties had been studied by many researchers. Hope and Rose (1990) studied the effects of cement composition on water requirement under the constant slump. They found that the required content of mixing water increased for cement with high Al_2O_3 or C_2S contents and decreased for cement with high ignition loss, high carbonate addition, or high C_3S content. Havard and Gjorv (1997) examined the influence of the gypsum to hemihydrate ratio of cement on concrete rheology. They stated that for cement with high contents of C_3A and alkalis, a reduction in the ratio of gypsum to hemihydrate resulted in a decrease in yield stress but little change in plastic viscosity. For cement with lower contents of C_3A and alkalis, the effects of the gypsum to hemihydrate ratio were less pronounced. Furthermore, a reduction in the sulfate content from 3% to 1% caused a decrease in both yield stress and plastic viscosity. Dils et al. (2013) investigated the chemical composition and fineness of cement on rheological properties of ultra-high-performance concrete for a given slump flow. They found that cement with high C_3A and specific surface, high alkali content, and a lower content of SO_3 gave the worst workability.

Chen and Kwan (2012) showed that the influence of superfine cement on rheological properties depends on the water content of cement paste. The addition of superfine cement increased yield stress and apparent viscosity at W/C ≥ 0.24, while at W/C ≤ 0.22, the addition of superfine cement decreased yield stress and apparent viscosity. At lower water content, the addition of fine particles can fill the voids, and it increases the packing density and releases the water between cement particles, and significantly increases the water films coating the particles in the cement paste, consequently improving the rheological properties of cement paste. At higher water content, the influence of increasing superfine cement contents on the water film thickness is not pronounced, and owing to the high specific surface area, superfine cement increased the yield stress and plastic viscosity. From the above analysis, the chemical composition (especially the contents of C_3A, C_3S, SO_3, and alkali), ignition loss, physical characteristics such as fineness and specific surface area, and the water content have a great influence on rheological properties.

2.2.4 Mineral admixture

Mineral admixtures are usually indispensable in modern cement-based materials as supplementary cementitious materials, due to their positive influence on the workability, strength, and durability as well as environmental benefits. This section introduces several most used mineral admixtures including fly ash, ground blast furnace slag, silica fume, and limestone powder, and their effects on the rheology are reviewed. Since the mineral admixtures modify the rheology of cement-based materials by changing the paste phase, and concrete rheology is positively correlated to the rheology of the paste phase, the results of the effects of mineral admixtures on concrete rheology are also included in this section.

2.2.4.1 Fly ash

Fly ash, high in SiO_2 and Al_2O_3, and low in CaO, is one type of pozzolanic material, which is sufficiently reactive when mixed with water and $Ca(OH)_2$ at room temperature (Taylor, 1997). Fly ash can delay the early age of hydration, lengthen the setting time, and enhance the long-term strength. Fly ash consists of crystal, vitreous, and a small amount of unburnt carbons. The vitreous includes smooth and spherical-shaped vitreous particles and irregular-shaped and low-porosity small particles. Unburnt carbon presents a loose and porous shape. The chemical and phase compositions depend on the mineral composition of coal, combustion conditions, and collector setup. Physically, fly ash presents as fine particles with particle sizes ranging from 0.4 to 100 μm, low specific gravity (2.0~2.2 g/cm³), high specific surface area (300~500 m²/kg), and light texture (Chen and Kwan, 2012). The type of fly ash determines the surface texture. For example, the surface of Class C fly ash (FAC) particles appears irregular and cellular (Figure 2.2a), while Class F fly ash (FAF) particles exhibit spherical morphology and smooth surface texture (Figure 2.2b).

The addition of fly ash has a great influence on the rheological properties of concrete, as shown in Figure 2.3. Laskar and Talukdar (2008) found that low levels of fly ash could lead to a reduction of yield stress, while a slight increase in yield stress could be observed at the high content of fly ash. Beycioğlu and Aruntaş (2014) indicated that fly ash positively

Figure 2.2 (a and b) SEM images of Class C and F fly ash particles (Ahari et al., 2015b).

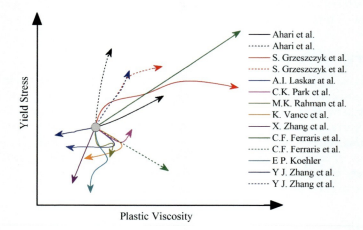

Figure 2.3 Several typical results about the effects of fly ash on rheological properties (Grzeszczyk and Lipowski, 1997, Zhang and Han, 2000, Ferraris et al., 2001, Koehler and Fowler, 2004, Park et al., 2005, Laskar and Talukdar, 2008, Vance et al., 2013, Rahman et al., 2014, Ahari et al., 2015b). (Derived from Jiao et al., 2017.)

affected the flowability, passing ability, and viscosity of self-compacting concrete due to the spherical geometry and smooth surface caused a reduction in water demand. Jalal et al. (2013) found that the ball bearing-shaped fly ash particles resulted in an improvement in the rheological properties of fresh self-compacting concrete (SCC), and led to an increase in the slump flow diameter from 800 to 870 mm and a reduction in the T_{500} values from 1.7 to 1.1 s. However, Park et al. (2005) found that the mixtures without fly ash showed slightly higher yield stress than those with fly ash, and the yield stress and plastic viscosity slightly increased with the increase of fly ash. Rahman et al. (2014) observed that fly ash significantly increased the flocculation rate and plastic viscosity of SCC.

The type of fly ash is correlated to the rheological properties of concrete. Ahari et al. (2015a, 2015b) studied the influence of various amounts of FAF and FAC on the rheological properties of SCC. They found that FAF significantly decreased plastic viscosity in comparison to mixtures with FAC. The fineness or particle size distribution of fly ash can also significantly affect workability. Ferraris et al. (2001) and Li and Wu (2005) reached the same conclusion that as the mean size of fly ash particles increased, the slump flow decreased to a certain value and then gradually increased, and the optimum size was about 3 μm. Lee et al. (2003) pointed out that the fluidity of paste increased as the particle size distribution became wider.

Fly ash has lower specific gravity than cement. The replacement of cement with the same mass of fly ash could increase the volume of paste, proportionally decrease the cement concentration in the paste, and therefore reduce the number of flocculation cement particle connections, which is called the "dilution effect" (Malhotra and Mehta, 2004, Bentz et al., 2012). The spherical geometry and smooth surface of fly ash particles promote particle sliding and reduce frictional forces among the angular particles, which is called the "ball-bearing effect". This effect can be magnified with the particle size distribution of fly ash slightly coarser than cement due to the increase in separation distance between neighboring particles (Vance et al., 2013). The addition of fine fly ash improves the flowability due to the filling effect. The packing density of paste is improved; the water retained inside the particle flocs is decreased; and consequently, the fluidity of paste is increased. In addition, fly ash can delay the early age of hydration and lengthen the setting time. It is worth mentioning that the unburnt carbons in fly ash can greatly adsorb superplasticizer molecules or water, even resulting in negative zeta potential (Felekoğlu et al., 2006), which leads to higher viscosity. Stated thus, the replacement level, surface texture, particle size distribution or specific surface area, and content of unburned carbon of fly ash affect the rheological properties of paste by dilution effect, ball-bearing effect, filling effect, and adsorption effect.

2.2.4.2 Ground blast furnace slag

Ground blast furnace slag (GBFS), finely granular and almost fully noncrystalline, is a by-product of the steel-manufacturing industry. The glass content of GBFS is predominated by the rate of quenching. The mass contents of CaO, SiO_2, and Al_2O_3 are up to 90%. GBFS has hydraulic properties but is much slower than Portland cement. The specific gravity of slag is approximately 2.90, and its bulk density varies in the range between 1200 and 1300 kg/m³. The specific surface area of GBFS measured by the Blaine method is 375–425, 450–550, and 350–450 m²/kg in the United Kingdom, the United States, and India, respectively. In China, the specific surface area of GBFS is higher than 450 m²/kg (Pal et al., 2003).

Since the replacement of cement with slag improves the workability of mixtures and reduces CO_2 emissions, GBFS has been widely applied in cement paste, mortar, and concrete. Effects of GBFS on the rheological properties of cement-based materials from literature are shown in Figure 2.4. As can be seen from Figure 2.4, most studies found that the addition of GBFS decreases plastic viscosity, while the effect of GBFS on yield stress is uncertain. Park et al. (2005) indicated that the yield stress decreased and then increased as GBFS increased, but the plastic viscosity decreased with the addition of GBFS. Ahari et al. (2015b) stated that the replacement of Portland cement with GBFS decreased the yield stress and plastic viscosity of the mixtures, regardless of the water-to-binder (*w/b*) ratio. They also found that an 18% replacement of GBFS in the mixture with a *w/b* ratio of 0.44 reduced the breakdown area approximately by 10%. In the study by Derabla and Benmalek (2014), the specific surface area of granulated slag and crystallized slags are 228 and 485 m²/kg,

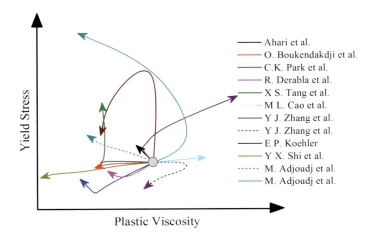

Figure 2.4 Several typical results about the effects of slag on rheological properties (Zhang and Zhang, 2002, Koehler and Fowler, 2004, Shi et al., 2004, Park et al., 2005, Boukendakdji et al., 2012, Cao et al., 2012, Derabla and Benmalek, 2014, Ezziane et al., 2014, Tang et al., 2014, Ahari et al., 2015b). (Derived from Jiao et al., 2017.)

respectively, while the activity of these two slags is near to that of cement. Even though the specific surface area of granulated slag is lower than that of cement, the plastic viscosity of the mixture containing granulated slag decreases from 150 to 121 Pa.s due to the absorption of superplasticizer particles. At the same replacement level, the plastic viscosity of the mixture containing crystallized slag decreases from 150 to 91 Pa.s. However, the addition of GBFS can increase the plastic viscosity in some cases. Tattersall (1991) reported that for the mixtures with low cement content (200 kg/m^3), the addition of slag reduced yield stress and increased plastic viscosity, which was used at replacement levels of 40% and 70%. Tang et al. (2014) studied the rheological properties of cement paste with a high volume of ground slag under different slump flow conditions by varying the dosage of superplasticizer. They found that the pastes with the ground slag had higher plastic viscosity, poorer stability, and lower velocity of flow, compared to ordinary cement paste under the same slump flow condition. These results can be attributed to the fact that a high volume of ground slag with a high specific surface area requires a larger amount of water than cement.

Generally, GBFS has a high specific surface area and a high chemical activity, which may have positive or negative effects on the rheological properties of cement-based materials. The rheological properties of cement concrete are improved for the following reasons: entrapped water within cement particles may be released by the micro-filling effect of fine GBFS particles, and a high specific area may lead to more adsorption of superplasticizer. However, high specific surface area and high chemical activity of ground blast furnace slag require a large amount of water than cement particles, and thus the rheological properties may be decreased. As a result, the influence of GBFS on rheological properties depends on the replacement level, chemical composition, specific surface area or fineness, adsorption effect, and water requirement.

2.2.4.3 Silica fume

Silica fume is the by-product of electric arc furnace-produced silica metal and silica alloys. Silica fume is an extremely fine powder. The average particle size is about 0.1–0.3 μm, and the particles below 0.1 μm are more than 80%. The specific surface area of silica fume is 20,000–28,000 m^2/kg, which is 80–100 times and 50–70 times higher than that of cement and fly ash, respectively. The particles are round and tend to be agglomerated. Therefore, superfine silica fume particles can fill the voids between other particles, which improves the gradation, increases the packing density of cementitious materials, and even has a lubrication effect. The high specific surface area of silica fume could even adsorb superplasticizer molecules with multi-layers (Nehdi et al., 1998, Park et al., 2005, Laskar and Talukdar, 2008). Silica fume, high in SiO_2 (85%–96%) with very fine vitreous particles, has higher chemical activity than fly ash. The high fineness and high chemical activity of silica fume can increase the water demand and interparticle friction (Collins and Sanjayan, 1999, Nanthagopalan et al., 2008, Benaicha et al., 2015).

Effects of silica fume on the rheological properties of cement-based materials from literature are shown in Figure 2.5. As can be seen from Figure 2.5, many studies found that the addition of silica fume increased both yield stress and plastic viscosity, and reduced the fluidity of cement-based materials. The addition of silica fume also significantly increased the flocculation rate (Rahman et al., 2014, Ahari et al., 2015b). As a result, silica fume can be used in the concrete which requires high uniformity and cohesiveness, e.g. pumping concrete, underwater concrete, and shotcrete, as an inorganic viscosity-modifying agent. However, the addition of silica fume may have different effects on the rheological properties. Zhang and Han (2000) stated that silica fume could decrease the viscosity and yield stress of cement paste. Ahari et al. (2015a, 2015b) found that the incorporation of silica

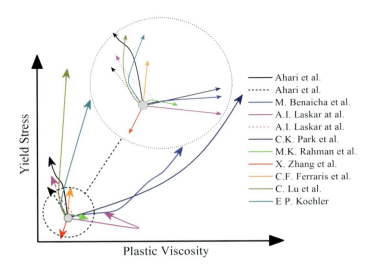

Figure 2.5 Several typical results about the effects of silica fume on rheological properties (Zhang and Han, 2000, Ferraris et al., 2001, Koehler and Fowler, 2004, Park et al., 2005, Laskar and Talukdar, 2008, Rahman et al., 2014, Ahari et al., 2015b, Lu et al., 2015). (Derived from Jiao et al., 2017.)

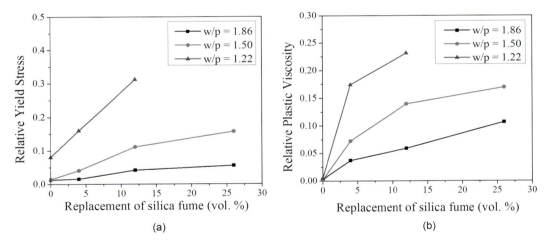

Figure 2.6 Effect of replacement of silica fume on rheological parameters (Nanthagopalan et al., 2008).

fume reduced the plastic viscosity and increased the yield stress and breakdown area. Yun et al. (2015) found that silica fume led to a remarkable increase in flow resistance, while it slightly reduced plastic viscosity. The addition of silica fume could obtain a different result with different superplasticizer types and water-to-binder ratios. Laskar and Talukdar (2008) found that silica fume increased the initial yield stress of fresh concrete in the case of polycarboxylate-based superplasticizer and decreased it in the case of sulfonated naphthalene polymer. Effects of water-to-powder ratio (w/p) and silica fume on yield stress and plastic viscosity are shown in Figure 2.6. As can be seen from Figure 2.6, both yield stress and plastic viscosity show a higher increment with the decrease of water-to-powder ratio.

Therefore, it is necessary to understand the interaction between silica fume and superplasticizer type or content when investigating the effect of silica fume on the rheology of ordinary concrete. In summary, specific surface area and chemical activity are the most important factors affecting the rheological properties because of micro-filling effect, lubricating effect or friction effect, and adsorption effect. Furthermore, the efficiency of silica fume can be influenced by the superplasticizer type or content and water-to-binder ratio.

2.2.4.4 Limestone powder

The inert limestone powder is a kind of high-quality and cheap mineral admixture. Its main component is calcium carbonate, $CaCO_3$. The surface of limestone particles is irregular and rough (Ma et al., 2013b), which increases the adhesion and friction between cement particles. The mean particle size of limestone powder used in concrete is from below 1 μm to more than several tens of μm. Limestone particles have a high adsorption capacity of superplasticizer, which increases the dispersing ability of the concrete system. Besides, limestone particles are often ground in finer powders than cement. Consequently, more water is needed for paste with the fine limestone powder.

The influences of limestone powder on rheological properties from literature are summarized in Figure 2.7. Some researchers found that the incorporation of limestone powder increased the yield stress and plastic viscosity, leading to a reduction in the workability of concrete, while others showed that the addition of limestone powder resulted in a decrease in yield stress and plastic viscosity. In addition, Rahman et al. (2014) stated that increasing amounts of limestone powder led to an increase in the flocculation rate significantly. For instance, the structuration rate of concrete with 15% limestone powder is about 1.5 times higher than that of reference concrete. The influence of limestone powder on rheological properties depends on the particle packing and water demand, which could be transferred into specific surface area or particle size distribution (Vance et al., 2013). Ma et al. (2013a) noted that yield stress and plastic viscosity increased with decrease in the particle size of limestone powder. Vance et al. (2013) found that as the particle size of limestone powder increased from 0.7 to 15 μm, there was a decrease in both yield stress and plastic viscosity. They stated that particles finer than the ordinary Portland cement usually increase the yield

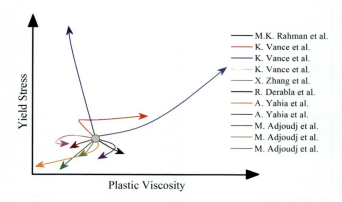

Figure 2.7 Several typical results about the effects of limestone powder on rheological properties (Yahia, 1999, Zhang and Han, 2000, Vance et al., 2013, Derabla and Benmalek, 2014, Ezziane et al., 2014, Rahman et al., 2014). (Derived from Jiao et al., 2017.)

stress and plastic viscosity, while the addition of particles coarser than cement induces an opposite effect. This is because the finer particles reduce the particle spacing and increase interparticle contacts, whereas the coarser particles could increase particle spacing and reduce the shear resistance of the suspension. Furthermore, larger limestone particles exhibited a decrease in the area of water films and provided more extra water (Uysal and Yilmaz, 2011), and thus reduced the possibility of flocculation.

Although the effect of particle size distribution is very important, the production methods of limestone powder cannot be ignored. Felekoğlu et al. (2006) considered that the finer limestone filler did not significantly change the viscosity, while the coarse limestone filler was effective in increasing the viscosity. The finer limestone filler is a filtration system by-product of crushed stone production. However, the coarse limestone powder is a special production of whitened limestone filler with lower adsorptivity. In a word, the effect of limestone powder on rheological properties can be ascribed to the morphologic effect, filling effect, and adsorption effect by particle size distribution and production methods of limestone powder.

2.2.4.5 Ternary binder system

Ternary binder system is a good way to enhance various properties of cement-based materials, including workability, rheological properties, strength, durability, cost, and CO_2 emissions. This section discusses the rheological properties of ternary blends containing fly ash and slag or silica fume or limestone. Laskar and Talukdar (2008) observed that the rheological parameters of concrete with a ternary binder system lay in between the values with the single mineral admixtures at each replacement level. Tattersall and Banfill (1983), Park et al. (2005), and Gesoğlu and Özbay (2007) observed the same phenomena. Kashani et al. (2014) investigated the rheological behavior of cement-blast furnace slag-fly ash ternary pastes. They showed that the width of the particle size distribution was the key parameter controlling the yield stress of ternary pastes. As a result, even a low volume of fly ash also has a significant effect on workability, and any other mineral additives with a broad particle size distribution may have a comparable effect. Vance et al. (2013) found that in ternary pastes containing limestone with low fly ash contents (5%), the plastic viscosity increased with the replacement of fine limestone and remained unchanged for coarse limestone. However, at higher fly ash contents (10%), the yield stress decreased even for pastes containing fine limestone. Although the particle spacing reduces and the number of interparticle contacts increases with the addition of fine limestone, the ball-shaped fly ash separates the solid grains, compensating for the positive effect of fine particle additions, thus reducing the yield stress and plastic viscosity.

The addition of ultrafine mineral admixtures such as silica fume can increase the packing density, fill the voids between cement particles, and release the water entrapped in agglomerate structure to form excess water films for lubrication. But at the same time, the water film thickness will be thinned down due to its large surface area. By adding a cementitious material whose fineness lies in between cement and silica fume, such as fly ash, the water entrapped in the agglomerate structure can be released without excessively increasing the surface area (Li and Kwan, 2014). In this way, a larger water film thickness and better flowability can be obtained. Moreover, the addition of mineral admixtures with a broader particle size distribution can also increase the particle packing and the excess water, and therefore can efficiently separate the cement particles. In this case, the intensity of colloidal interaction between particles will drop substantially, leading to the destruction of the percolated network of particles, which will then result in yield stress reduction (Kashani et al., 2014).

2.2.5 Chemical admixtures

With the wide application of high-performance concrete, admixtures have become an indispensable part of concrete. The types of chemical admixtures have a great influence on the rheological properties of fresh concrete.

2.2.5.1 Superplasticizer

Due to economic and technical benefits, superplasticizers are almost used for all modern concretes. The yield stress and viscosity of cement-based materials can be dramatically decreased by the addition of superplasticizer under the action of electrostatic or/and steric hindrance effects. The efficiency of superplasticizer depends on the adsorption of superplasticizer molecules onto cement particles and the repulsive force developed by the adsorbed molecules. For the new generation of superplasticizer, it seems that the steric hindrance effect becomes dominant over the electrostatic effect (Flatt and Bowen, 2006). With the increase of superplasticizer, the average surface-to-surface separation distance increases, the colloidal interaction between particles decreases, and therefore the yield stress of cementitious materials decreases (Perrot et al., 2012). Beyond that, the addition of superplasticizer leads to a higher surface coverage by polymers—causing an increase in effective layer thickness and a reduction in the maximum attraction between the particles, thus decreasing the number of sites available for nucleation, and increasing the bridging distance between the particles—and consequently improves the rheological properties of fresh concrete (Lowke et al., 2010, Kwan and Fung, 2013).

The efficiency of superplasticizer depends on its type and structure. Figure 2.8 demonstrates the rheological properties of SCC made with a different type of superplasticizer. It can be observed from Figure 2.8 that the concrete mixture with polycarboxylate-based superplasticizer has a higher yield stress and a lower plastic viscosity compared to those with naphthalene sulphonate-based superplasticizer at the same slag content. Papo and Piani (2004) found that modified polyacrylic was the most effective superplasticizer to improve the rheological properties of cement paste among three types of superplasticizers based on melamine resin, modified lignosulphonate, and modified polyacrylate. The naphthalene

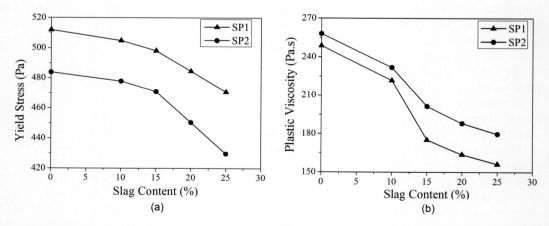

Figure 2.8 Effect of superplasticizer type on rheological properties of SCC (Boukendakdji et al., 2012): (a) yield stress and (b) plastic viscosity. SP1: a polycarboxylate-based superplasticizer; SP2: a naphthalene sulphonate-based superplasticizer.

superplasticizer is beneficial for dispersing the cement, with a low air-entraining effect and slump retention effect. The polycarboxylate superplasticizer consisting of a backbone of polyethylene, grafted chains of PEO, and carboxylic groups as adsorbing functional groups, has the potential to disperse cement particles, restrain the slump loss, and shorten the setting time (Zuo et al., 2004, Hanehara and Yamada, 2008, Boukendakdji et al., 2012). Toledano-Prados et al. (2013) found that liquid polycarboxylate proved more effective than solid polycarboxylate. Different admixtures that had the same main chain and same polymer structure but different molecular weight and different side chain density of carboxylic acid groups have a great effect on the rheological properties of SCC (Mardani-Aghabaglou et al., 2013). The yield stress values of SCC mixtures were only affected by the dosage of superplasticizer, while increasing both the molecular weight and side chain density of carboxylic acid groups of admixture increased the plastic viscosity values due to the increase in steric hindrance.

2.2.5.2 Viscosity-modifying agent

Viscosity-modifying agents (VMA) are relatively new admixtures used to improve the rheology of concrete with higher uniformity and better cohesiveness. Incorporation of VMA in concrete can reduce the risk of separation of heterogeneous concrete and obtain stable concrete for underwater repair, curtain walls, and deep foundation walls. The mode of action of VMA depends on the type and concentration of the polymer in use. The mechanism of action with welan gum and cellulose derivatives can be classified into three categories, i.e. adsorption effect, association effect, and intertwining effect (Khayat, 1995). The adsorbed cellulose ether can slow down the nucleation of calcium silicates, generate repulsive steric forces to replace van der Waals attractive forces, and then form a new interaction network that could bridge the cement grains (Brumaud et al., 2014). The inorganic VMA with a high surface area increases the content of fine particles and the water-retaining capacity of paste, thereby the thixotropy.

The concrete modified with a VMA exhibits high plastic viscosity, high yield value, and shear-thinning behavior. The efficiency of VMA depends on its type and concentration (Khayat, 1995). Schmidt et al. (2010) stated that the influence of VMA on yield stress was much stronger than the effect of polycarboxylate ether superplasticizer (PCE) at 20°C, and the mixtures with VMA based on modified potato starch show better performance than those with VMA based on diutan gum, irrespective of the PCE type. Yun et al. (2015) found that VMA based on hydroxypropyl methyl cellulose tended to reduce the flow resistance and increase the plastic viscosity of high-performance wet-mix shotcrete, worsening the shootability and pumpability. Leemann and Winnefeld (2007) showed that VMA based on polysaccharide caused the highest increase in yield stress, while VMA based on microsilica had the lowest at an equivalent plastic viscosity. Assaad et al. (2003) showed that the addition of VMA significantly increased the thixotropy for high-flowability concrete. Brumaud et al. (2014) found that the critical deformation increased with the VMA dosages. One possible explanation is that the number of bridging grains increases and therefore the probability of large interparticle relative displacement increases. Another possible explanation comes from the fact that the hydrophobic interactions of high VMA concentrations could form aggregates of large size, which could tolerate higher stretching (Bülichen and Plank, 2012, Bülichen et al., 2012). Different types of VMA may have the same effect. Benaicha et al. (2015) showed that the concrete made with 10% of silica fume and the one with 0.1% VMA had the same characteristics in resistance of sieve segregation, plastic viscosity, yield stress, and even mechanical properties. They believed that VMA could be replaced by silica fume, depending on the availability of materials.

VMA is often used in combination with superplasticizer to improve rheological properties and mechanical strength. The presence of two dissimilar chemicals can lead to a number of issues related to incompatibility, which affects the properties of concrete. Khayat (1995) stated that cellulose derivatives were always used in conjunction with melamine-based superplasticizer because of their incompatibilities with a naphthalene-based superplasticizer. Prakash and Santhanam (2006) indicated that the flow properties of pastes produced with combinations of superplasticizer based on sulfonated naphthalene formaldehyde or polycarboxylic ether and VMA based on welan gum were satisfactory. The only difference was that the polycarboxylic ether–welan gum combination showed the evidence of thixotropy, but the same was not observed for sulfonated naphthalene formaldehyde–welan gum combinations. In addition, the rheological behavior of mixtures with VMA is in general dependent upon shear stress. Bouras et al. (2012) found that at sufficiently low shear stress, the mixtures with VMA exhibited shear thinning, while at relatively high shear stress, the pastes became shear thickening. At low shear stress, the flow of paste induces the defloc-culation of solid particles and VMA polymer disentanglement and alignment. The analysis above shows that the efficiency of the viscosity-modifying agent depends on the type and concentration of VMA, the type of superplasticizer, and the applied shear stress.

2.2.5.3 Air-entraining agent

The air-entraining agent has been widely used to improve resistance to freezing and thawing damage, and to a lesser extent, the workability of concrete. The air-entraining agent is a mixture of various surfactants (Du and Folliard, 2005). The chemical nature of the air-entraining agent contains a hydrophilic head, having a strong attraction for water, and hydrophobic tails, having repulsion for water. From the rheological point of view, entrained air bubbles play a role in lubrication, and increasing paste volume, together with the character of the air-entraining agent, affects the consistency of cement-based materials.

There are many studies about the effect of the air-entraining agent on rheological properties. He et al. (2011) found that consistency of mortar showed a slightly increasing trend with the increase of air-entraining amount, although the value of consistency, in general, was not high. Yun et al. (2015) found that the use of an air-entraining agent tended to reduce both flow resistance and torque viscosity, effectively improving the pumpability of high-performance wet-mix shotcrete, while the reduction rate of torque viscosity decreased with the increased air content. Tattersall and Banfill (1983) demonstrated that flow resistance and torque viscosity continued to decrease until the air content reached 5% and then stabilized beyond that point. This result can be contributed to the lubrication effect and volume effect of air bubbles. However, Carlsward et al. (2003) and Wallevik and Wallevik (2011) revealed that the addition of an air-entraining agent strongly reduced the plastic viscosity, while the effect on yield stress was not significant. Struble and Jiang (2005) indicated that the yield stress increased and the plastic viscosity decreased with increasing air content, which was also reported by Rahman and Nehdi (2003). An explanation for the increase in yield stress is proposed that air bubbles are attracted to cement particles to form bubble bridges and thus increase bonding between particles. Once the bubble bridges are broken and paste can flow, the air bubbles reduce plastic viscosity, apparently acting as a lubricant. Thus, there is competition between these two actions: without shear, the bubbles act as flocculating particles to increase yield stress, while with shear, the bubbles act as lubricant to reduce plastic viscosity (Edmeades and Hewlett, 1998). Moreover, the rheological behavior of suspensions with bubbles also depends on the shear rate due to the fact that the bubbles are generally stiff compared to the suspending fluid and thus do not store any energy (Ducloué et al., 2015).

2.3 EFFECT OF TEMPERATURE ON RHEOLOGY

Temperature usually affects the rheological properties of cement-based materials by changing the actions of admixtures and the interaction between cementitious material particles. Moreover, temperature impacts the time-dependent rheological properties by changing the flocculation rate of cementitious material particles and the kinetics of cement hydration.

For the cement paste containing water-reducing agent, according to Fernàndez-Altable and Casanova (2006), the viscosity decreased with the increasing temperature, while the static and dynamic yield stress increased with the elevated temperature. However, the influence of temperature on yield stress was only obvious for paste with a low dosage of water reducer. For cement pastes with a high dosage of water reducer, the dependence of yield stress on temperature was found.

Nawa et al. (2000) found that the fluidity of cement paste did not change monotonically with increasing temperature. They found that the fluidity was smallest at 20°C, while the fluidity loss was larger at a higher temperature. On the one hand, the increase in temperature will increase the adsorption of polycarboxylate superplasticizer, enhance the steric resistance between cementitious material particles, and improve the fluidity of paste. On the other hand, the increase in temperature promotes the hydration process, and some polymers will be buried in the laminates of hydration products, thus reducing their performance.

On the performance of superplasticizer at different temperatures, Kong et al. (2013) studied the rheological properties of Portland cement pastes made with polycarboxylate superplasticizers. According to their work, a higher temperature can lead to a more significant amount of adsorbed polycarboxylate ester/ether on the cement surface and a lower amount of free water in fresh cement pastes, because of the higher hydration rate of cement. Yamada et al. (1999) attributed the low dispersibility of polycarboxylate superplasticizer at a lower temperature to the high sulfate ion concentration in mixing water.

Compared with the effect of temperature on the initial rheological properties, the effect of temperature on the time-varying rheological properties is more significant. Al Martini and Nehdi (2009), Al Martini (2008), and Nehdi and Al Martini (2009) studied the effect of high temperature on the rheology of cement paste and its change with time. It was found that such influences were affected by the superplasticizer (SP) dosage. When the SP dosage was below the saturation level, the increase in yield stress over time at high temperature was obviously larger, while at the dosage higher than the saturation level, the influence of temperature on yield stress evolution over time was less obvious. A similar principle was also applicable to the effect of temperature and time on plastic viscosity. In the studies of Petit et al. (2006, 2009, 2007), coupled effects of temperature and time on rheological parameters were studied, and correlations between hydration time and rheological parameters were established. They found that both yield stress and plastic viscosity vary linearly with material temperature and elapsed time for mixtures made with polymelamine or polynaphtalene superplasticizer. However, for flowable mortar with polycarboxylic superplasticizer, rheological properties can be influenced by both mixture proportioning and temperature. Besides, the effect of temperature on the variation of viscosity is more obvious in the mortar with a *w/b* of 0.42 than that of 0.53.

2.4 EFFECT OF SHEARING ON RHEOLOGY

As is well known, the operation processes of modern engineering applications of concrete include mixing, transporting, pumping, and formwork casting. At each process, fresh concrete is subjected to different shear rates. For example, the process of mixing can provide a "most deflocculated state" for the fresh concrete due to its high shear rates (Roussel, 2006). The applied shear rate during transport is associated with the rotational speed, drum volume, and rheological properties of the concrete itself, and its value varies from 1 to $8\,s^{-1}$ (Wallevik and Wallevik, 2017). For the process of pumping, the fresh concrete is subjected to very high shear stress and high pressure at the same time. As a result, the yield stress and air content are markedly increased and the viscosity of fresh concrete is decreased (Secrieru et al., 2018, Shen et al., 2021). However, the only source of applied shear stress of concrete after casting into formwork is gravimetric force, and thus the applied shear rate is usually less than $0.1\,s^{-1}$ (Papanastasiou, 1987). It is worth mentioning that the shear rate experienced by the cement pastes inside fresh concrete depends on the nature of the concrete (Roussel, 2006). The operation process, i.e. shear history, plays an important role in the evolutions of rheological properties and stability of fresh concrete over time.

The dispersion and agglomeration of particles in suspensions are significantly dependent on the applied shear rates. As mentioned before, the operations of mixing and pumping with high shear rates can almost totally break down the network structures of cement paste. Although paste in concrete suffers higher shear rates than pure paste under the same shear conditions, Helmuth et al. (1995) and Ferraris et al. (2001) pointed out that the rheological properties of cement paste prepared with a high-speed mixer were comparable with that of paste mixed in concrete. Williams et al. (1999) also mentioned that the well-mixed cement pastes contained few agglomerate structures and possessed low plastic viscosity. However, Han and Ferron (2015, 2016) found that the rheological properties of cement paste worsened once the mixing intensity was higher than a threshold value due to the agglomerates with a larger mean chord size. Recently, Mostafa et al. (2015) evaluated the effectiveness of additional shearing on the dispersion state of cement suspensions. They found that applying a rotational shearing with the shear rate corresponding to the transition shear rate could significantly improve the dispersing degree of fresh cement suspensions.

The rheological properties of concrete are relevant to the dispersion state of the paste matrix (Tregger et al., 2010), and it is influenced by the shear history (Ferron et al., 2013). Generally, a higher shear rate can break down the flocs of binders, thus improving the flowability of the paste (Jiao et al., 2018). However, there are opposite findings as well. As reported by Helmuth (1980), a very high shear rate can degrade the fluidity of cement paste. Han and Ferron (2016) found that weaker but larger cement agglomerates can form in the paste under high mixing rates, and thus the rheological properties of cement paste can increase after exceeding a threshold mixing speed. On the other hand, shear history can also influence the aggregation and breakage kinetics of the paste (Ferron et al., 2013). Ferron et al. (2013) found that the time scale needed for aggregation is longer than the breakdown of the flocculation structure. Ma et al. (2018) found that longer shearing duration can induce a more dispersed structure and decrease the static yield stress and the rebuild rate of static yield stress. Besides, increasing the pre-shearing time can increase the storage modulus value, leading to a more percolated C–S–H network (Ma et al., 2018). It is important to understand how the action of shearing influences the rheological behavior of cement-based materials to evaluate their rheology in the real situation more precisely.

2.5 EFFECT OF PRESSURE ON RHEOLOGY

Cement-based materials are pressured when they flow through pipes driven by pumps or in the formworks, and their rheological properties can be different under pressure. The influence of pressure on the rheological properties of cement pastes, which assumably represented the lubricating layer that forms along with the profile of concrete during pumping, was evaluated using a rotational rheometer with a high-pressure cell in the work of Kim et al. (2017). Cement pastes with water-to-cement ratios ranging from 0.35 to 0.6 were tested according to a protocol designed to simulate the conditions of an actual pumping process based on field tests. The shear rates, shearing durations, and pressure levels from 0 to 30 MPa were experimentally simulated. The test results indicated that below a certain water-to-cement ratio (0.40), elevated pressures lead to changes in the rheological properties, while changes were negligible when the ratio was above this threshold. Further, at low water-to-cement ratios, the thixotropy of cement pastes can reverse into rheopexy after pressurization.

2.6 SUMMARY

Cement paste can be regarded as a suspension system containing particles from microscopic to macroscopic dimensions. Interactions between particles, including colloidal interaction, Brownian forces, and hydrodynamic forces, are the origin governing the rheology of cement paste.

Cement paste is thixotropic, and its rheology is time-dependent. The rheological properties of cement paste are not only influenced by mixture proportion but also by processing procedures. Mixture proportion factors, i.e. volume fraction of cementitious materials, characteristics of the interstitial solution, proportion and properties of cementitious materials, and chemical admixtures, can change the rheological behavior of cement paste by modifying the interactions between particles. The mineral composition of cement influences the hydration rate and water demand in paste, and thus affects the paste rheology. The chemical environment in the interstitial solution can also be altered, and it can further change the agglomeration of cement particles. Mineral admixtures, such as fly ash, slag powder, silica fume, and limestone powder, mainly influence the rheological properties by changing the physical properties of particles including specific surface area, particle size distribution, and particle shape. Chemical admixtures including superplasticizer, VMA, and the air-entraining agent can change the rheology by directly influencing particle interactions.

Processing factors, i.e. temperature, shearing, and pressure, alter the rheology of cement paste by changing the flocculation and chemical hydration in cement paste. Increasing temperature can increase superplasticizer adsorption and promote cement hydration in cement paste. Shearing effect can break part of the flocs and change the dispersion state. Furthermore, additional dissolution and superplasticizer adsorption may occur on newly generated surfaces of cement particles, promoting cement hydration and SP adsorption. The knowledge of the rheology of cement paste under pressure remains pretty limited.

REFERENCES

Ahari, R. S., Erdem, T. K., and Ramyar, K. (2015a). "Effect of various supplementary cementitious materials on rheological properties of self-consolidating concrete." *Construction and Building Materials*, 75, 89–98.

Ahari, R. S., Erdem, T. K., and Ramyar, K. (2015b). "Thixotropy and structural breakdown properties of self consolidating concrete containing various supplementary cementitious materials." *Cement and Concrete Composites*, 59. doi:10.1016/j.cemconcomp.2015.03.009.

Al-Martini, S. (2008). "Investigation on rheology of cement paste and concrete at high temperature," *Doctoral dissertation, University of Western Ontario.* Available at: http://search.proquest.com.ezaccess.library.uitm.edu.my/docview/305111374?accountid=42518.

Al Martini, S., and Nehdi, M. (2009). "Coupled effects of time and high temperature on rheological properties of cement pastes incorporating various superplasticizers." *Journal of Materials in Civil Engineering*, 21(8), 392–401. 10.1061/(ASCE)0899-1561(2009)21:8(392).

Assaad, J., Khayat, K. H. and Mesbah, H. (2003). "Assessment of thixotropy of flowable and self-consolidating concrete." *ACI Materials Journal*, 100(2), 99–107. doi:10.14359/12548.

Banfill, P. F. G. (1979). "A discussion of the papers "rheological properties of cement mixes" by M. Daimon and DM Roy." *Cement and Concrete Research*, 9(6), 795–796.

Benaicha, M., et al. (2015). "Influence of silica fume and viscosity modifying agent on the mechanical and rheological behavior of self compacting concrete." *Construction and Building Materials*, 84, 103–110.

Bentz, D. P., et al. (2012). "Influence of particle size distributions on yield stress and viscosity of cement-fly ash pastes." *Cement and Concrete Research*, 42(2), 404–409. doi:10.1016/j.cemconres.2011.11.006.

Beycioğlu, A., and Aruntaş, H. Y. (2014). "Workability and mechanical properties of self-compacting concretes containing LLFA, GBFS and MC." *Construction and Building Materials*, 73, 626–635.

Boukendakdji, O., Kadri, E.H., and Kenai, S. (2012). "Effects of granulated blast furnace slag and superplasticizer type on the fresh properties and compressive strength of self-compacting concrete." *Cement and Concrete Composites*, 34(4), 583–590.

Bouras, R., Kaci, A., and Chaouche, M. (2012). "Influence of viscosity modifying admixtures on the rheological behavior of cement and mortar pastes." *Korea-Australia Rheology Journal*, 24(1), 35–44.

Brumaud, C., et al. (2014). "Cellulose ethers and yield stress of cement pastes." *Cement and Concrete Research*, 55, 14–21.

Bülichen, D., Kainz, J., and Plank, J. (2012). "Working mechanism of methyl hydroxyethyl cellulose (MHEC) as water retention agent." *Cement and Concrete Research*, 42(7), 953–959.

Bülichen, D., and Plank, J. (2012). "Mechanistic study on carboxymethyl hydroxyethyl cellulose as fluid loss control additive in oil well cement." *Journal of Applied Polymer Science*, 124(3), 2340–2347.

Cao, M., Zhang, C., and Han, L. (2012). "Experimental study on the rheological properties of fresh mineral powder concrete based on two-point method." *Hunningtu (Concrete)*, (01), 138–141.

Carlsward, J., et al. (2003). "Effect of constituents on the workability and rheology of self-compacting concrete." *Proceeding of the Third International RILEM Conference on SCC*, Reykjavik, Island, Proceedings PRO, pp. 143–153.

Chen, J. J., and Kwan, A. K. H. (2012). "Superfine cement for improving packing density, rheology and strength of cement paste." *Cement and Concrete Composites*, 34(1), 1–10.

Collins, F., and Sanjayan, J. G. (1999). "Effects of ultra-fine materials on workability and strength of concrete containing alkali-activated slag as the binder." *Cement and Concrete Research*, 29(3), 459–462.

Cosgrove, T. (2010). *Colloid science: Principles, methods and applications*. John Wiley & Sons, New Jersey, United States.

Derabla, R., and Benmalek, M. L. (2014). "Characterization of heat-treated self-compacting concrete containing mineral admixtures at early age and in the long term." *Construction and Building Materials*, 66, 787–794.

Dils, J., Boel, V., and De Schutter, G. (2013). "Influence of cement type and mixing pressure on air content, rheology and mechanical properties of UHPC." *Construction and Building Materials*, 41, 455–463.

Du, L., and Folliard, K. J. (2005). "Mechanisms of air entrainment in concrete." *Cement and Concrete Research*, 35(8), 1463–1471.

Ducloué, L., et al. (2015) "Rheological behaviour of suspensions of bubbles in yield stress fluids." *Journal of Non-Newtonian Fluid Mechanics*, 215, 31–39.

Edmeades, R. M., and Hewlett, P. C. (1998) "Cement admixtures", in *Lea's chemistry of cement and concrete,* edited by Hewlett, P. and Liska, M. Butterworth-Heinemann, Oxford, United Kingdom, pp. 841–905.

Ezziane, K., Ngo, T.-T., and Kaci, A. (2014). "Evaluation of rheological parameters of mortar containing various amounts of mineral addition with polycarboxylate superplasticizer." *Construction and Building Materials*, 70, 549–559.

Felekoğlu, B., et al. (2006). "The effect of fly ash and limestone fillers on the viscosity and compressive strength of self-compacting repair mortars." *Cement and Concrete Research*, 36(9), pp. 1719–1726.

Fernàndez-Altable, V., and Casanova, I. (2006). "Influence of mixing sequence and superplasticiser dosage on the rheological response of cement pastes at different temperatures." *Cement and Concrete Research*, 36(7), 1222–1230. doi:10.1016/j.cemconres.2006.02.016.

Ferraris, C. F., Obla, K. H., and Hill, R. (2001). "The influence of mineral admixtures on the rheology of cement paste and concrete." *Cement and Concrete Research*, 31(2), 245–255.

Ferron, R. D., et al. (2013). "Aggregation and breakage kinetics of fresh cement paste." *Cement and Concrete Research*, 50, 1–10. doi:10.1016/j.cemconres.2013.03.002.

Flatt, R. J. (2004a). "Dispersion forces in cement suspensions." *Cement and Concrete Research*, 34(3), 399–408. doi:10.1016/j.cemconres.2003.08.019.

Flatt, R. J. (2004b). "Towards a prediction of superplasticized concrete rheology." *Materials and Structures*, 37(5), 289–300.

Flatt, R. J., et al. (2009). "Conformation of adsorbed comb copolymer dispersants." *Langmuir*, 25(2), 845–855.

Flatt, R. J., and Bowen, P. (2003). "Electrostatic repulsion between particles in cement suspensions: Domain of validity of linearized Poisson–Boltzmann equation for nonideal electrolytes." *Cement and Concrete Research*, 33(6), 781–791.

Flatt, R. J., and Bowen, P. (2006). "Yodel: A yield stress model for suspensions." *Journal of the American Ceramic Society*, 89(4), 1244–1256. doi:10.1111/j.1551–2916.2005.00888.x.

Gesoğlu, M., and Özbay, E. (2007). "Effects of mineral admixtures on fresh and hardened properties of self-compacting concretes: Binary, ternary and quaternary systems." *Materials and Structures*, 40(9), 923–937.

Grzeszczyk, S., and Lipowski, G. (1997). "Effect of content and particle size distribution of high-calcium fly ash on the rheological properties of cement pastes." *Cement and Concrete Research*, 27(6), 907–916.

Han, D., and Ferron, R. D. (2015). "Effect of mixing method on microstructure and rheology of cement paste." *Construction and Building Materials*, 93. doi:10.1016/j.conbuildmat.2015.05.124.

Han, D., and Ferron, R. D. (2016). "Influence of high mixing intensity on rheology, hydration, and microstructure of fresh state cement paste." *Cement and Concrete Research*, 84, 95–106. doi:10.1016/j.cemconres.2016.03.004.

Hanehara, S., and Yamada, K. (2008). "Rheology and early age properties of cement systems." *Cement and Concrete Research*, 38(2), 175–195.

Havard, J., and Gjorv, O. E. (1997). "Effect of gypsum-hemihydrate ratio in cement on rheological properties of fresh concrete." *Materials Journal*, 94(2), 142–146.

He, Z. M., Liu, J. Z., and Wang, T. H. (2011). "Influence of air entraining agent on performance of inorganic thermal insulating mortar", in *Applied Mechanics and Materials*, edited by Sun, D.Y., Sung, W.P., and Chen, R. Trans Tech Publ, Stafa-Zurich, Switzerland, pp. 490–493.

Helmuth, R. A. (1980). "Structure and rheology of fresh cement paste." *7th International Congress on the Chemistry of Cement*.

Helmuth, R. A., et al. (1995). "Abnormal concrete performance in the presence of admixtures", Portland Cement Association. Available at: https://trid.trb.org/view/460037 (Accessed: 22 August 2018).

Hiemenz, P. C., and Rajagopalan, R. (1997). *Principles of colloid and surface chemistry*, 3rd ed. Marcel Dekker, New York.

Hope, B. B., and Rose, K. (1990). "Statistical analysis of the influence of different cements on the water demand for constant slump. Properties of fresh concrete." *Proceedings of the Coll, RILEM*, Chapman and Hall, p. 179e186.

Hunter, R. J. (2001). *Foundations of colloid science*. Oxford University Press, Oxford, United Kingdom.

Jalal, M., Fathi, M., and Farzad, M. (2013). "Effects of fly ash and TiO$_2$ nanoparticles on rheological, mechanical, microstructural and thermal properties of high strength self compacting concrete", *Mechanics of Materials*, 61, pp. 11–27.

Jiao, D., et al. (2017). "Effect of constituents on rheological properties of fresh concrete-A review." *Cement and Concrete Composites*, 83, pp. 146–159. doi:10.1016/j.cemconcomp.2017.07.016.

Jiao, D., De Schryver, R., et al. (2021a). "Thixotropic structural build-up of cement-based materials: A state-of-the-art review." *Cement and Concrete Composites*, 122(February), 104152. doi:10.1016/j.cemconcomp.2021.104152.

Jiao, D., Lesage, K., et al. (2021b). "Flow behavior of cementitious-like suspension with nano-Fe$_3$O$_4$ particles under external magnetic field." *Materials and Structures/Materiaux et Constructions*, 54(6). doi:10.1617/s11527-021-01801-y.

Jiao, D., Shi, C., and Yuan, Q. (2018). "Influences of shear-mixing rate and fly ash on rheological behavior of cement pastes under continuous mixing." *Construction and Building Materials*, 188, 170–177. doi:10.1016/j.conbuildmat.2018.08.091.

Kashani, A., et al. (2014). "Modelling the yield stress of ternary cement-slag-fly ash pastes based on particle size distribution." *Powder Technology*, 266, 203–209. doi:10.1016/j.powtec.2014.06.041.

Khayat, K. H. (1995). "Effects of antiwashout admixtures on fresh concrete properties." *Materials Journal*, 92(2), 164–171.

Kim, J. H., et al. (2017). "Rheology of cement paste under high pressure." *Cement and Concrete Composites*, 77, 60–67. doi:10.1016/j.cemconcomp.2016.11.007.

Koehler, E. P., and Fowler, D. W. (2004). Development of a portable rheometer for fresh Portland cement concrete, Technical Reports, International Center for Aggregates Research, The University of Texas at Austin, Austin, United States.

Kong, X., Zhang, Y., and Hou, S. (2013). "Study on the rheological properties of Portland cement pastes with polycarboxylate superplasticizers." *Rheologica Acta*, 52(7), 707–718. doi:10.1007/s00397-013-0713-7.

Kwan, A. K. H., and Fung, W. W. S. (2013) 'Effects of SP on flowability and cohesiveness of cement-sand mortar', *Construction and Building Materials*, 48, 1050–1057.

Laskar, A. I., and Talukdar, S. (2008). "Rheological behavior of high performance concrete with mineral admixtures and their blending." *Construction and Building Materials*, 22(12), 2345–2354. doi:10.1016/j.conbuildmat.2007.10.004.

Lee, S. H., et al. (2003). "Effect of particle size distribution of fly ash–cement system on the fluidity of cement pastes." *Cement and Concrete Research*, 33(5), 763–768.

Leemann, A., and Winnefeld, F. (2007). "The effect of viscosity modifying agents on mortar and concrete." *Cement and Concrete Composites*, 29(5), 341–349.

Li, G., and Wu, X. (2005). "Influence of fly ash and its mean particle size on certain engineering properties of cement composite mortars." *Cement and Concrete Research*, 35(6), 1128–1134.

Li, Y., and Kwan, A. K. H. (2014). "Ternary blending of cement with fly ash microsphere and condensed silica fume to improve the performance of mortar." *Cement and Concrete Composites*, 49, 26–35.

Liu, J., et al. (2017). "Influence of superplasticizer dosage on the viscosity of cement paste with low water-binder ratio." *Construction and Building Materials*, 149, pp. 359–366. doi:10.1016/j.conbuildmat.2017.05.145.

Lowke, D., et al. (2010). "Effect of cement on superplasticizer adsorption, yield stress, thixotropy and segregation resistance", in *Design, production and placement of self-consolidating concrete*, edited by Khayat, K. H. and Feys, D. Springer, Montreal, Canada, pp. 91–101.

Lowke, D. (2018). "Thixotropy of SCC—A model describing the effect of particle packing and superplasticizer adsorption on thixotropic structural build-up of the mortar phase based on interparticle interactions." *Cement and Concrete Research*, 104, 94–104. doi:10.1016/j.cemconres.2017.11.004.

Lu, C., Yang, H., and Mei, G. (2015). "Relationship between slump flow and rheological properties of self compacting concrete with silica fume and its permeability." *Construction and Building Materials*, 75, 157–162.

Ma, K., et al. (2013a). "Rheological properties of compound pastes with cement-fly ash-limestone powder." *Journal of the Chinese Ceramic Society*, 41(5), 582–587.

Ma, K., et al. (2013b). "Factors on affecting plastic viscosity of cement-fly ash-limestone compound pastes." *Journal of the Chinese Ceramic Society*, 41(11), 1481–1486.

Ma, S., Qian, Y., and Kawashima, S. (2018). "Experimental and modeling study on the non-linear structural build-up of fresh cement pastes incorporating viscosity modifying admixtures." *Cement and Concrete Research*, 108(January), 1–9. doi:10.1016/j.cemconres.2018.02.022.

Malhotra, V. M., and Mehta, P. K. (2004). *Pozzolanic and cementitious materials*. CRC Press, Boca Raton, FL.

Mardani-Aghabaglou, A., et al. (2013). "Effect of different types of superplasticizer on fresh, rheological and strength properties of self-consolidating concrete." *Construction and Building Materials*, 47, 1020–1025.

Mostafa, A. M., Diederich, P., and Yahia, A. (2015). "Effectiveness of rotational shear in dispersing concentrated cement suspensions." *Journal of Sustainable Cement-Based Materials*, 4(3–4), 205–214.

Nanthagopalan, P., et al. (2008). "Investigation on the influence of granular packing on the flow properties of cementitious suspensions." *Cement and Concrete Composites*, 30(9), 763–768.

Nawa, T., Ichiboji, H., and Kinoshita, M. (2000). "Influence of temperature on fluidity of cement paste containing superplasticizer with polyethylene oxide graft chains." *ACI Special Publication*, pp. 195–210. doi:10.14359/9912.

Nehdi, M., and Al Martini, S. (2009). "Estimating time and temperature dependent yield stress of cement paste using oscillatory rheology and genetic algorithms." *Cement and Concrete Research*, 39(11), 1007–1016. doi:10.1016/j.cemconres.2009.07.011.

Nehdi, M., Mindess, S., and Aıtcin, P. C. (1998). "Rheology of high-performance concrete: Effect of ultrafine particles." *Cement and Concrete Research*, 28(5), 687–697.

Overbeek, J. T. G. (1984). "Interparticle forces in colloid science." *Powder technology*, 37(1), 195–208.

Pal, S. C., Mukherjee, A., and Pathak, S. R. (2003). "Investigation of hydraulic activity of ground granulated blast furnace slag in concrete." *Cement and Concrete Research*, 33(9), 1481–1486.

Papanastasiou, T. C. (1987). "Flows of materials with yield." *Journal of Rheology*, 31(5), 385–404.

Papo, A., and Piani, L. (2004). "Effect of various superplasticizers on the rheological properties of Portland cement pastes." *Cement and Concrete Research*, 34(11), 2097–2101. doi:10.1016/j.cemconres.2004.03.017.

Park, C. K., Noh, M. H., and Park, T. H. (2005). "Rheological properties of cementitious materials containing mineral admixtures." *Cement and Concrete Research*, 35(5), 842–849.

Perrin, J. (1916). *Atoms (translated by Hammick D. LL.)*. D. Van Nostrand Company, New York.

Perrot, A., et al. (2012). "Yield stress and bleeding of fresh cement pastes." *Cement and Concrete Research*, 42(7), 937–944. doi:10.1016/j.cemconres.2012.03.015.

Petit, J. Y., et al. (2007). "Yield stress and viscosity equations for mortars and self-consolidating concrete." *Cement and Concrete Research*, 37(5), 655–670. doi:10.1016/j.cemconres.2007.02.009.

Petit, J. Y., Khayat, K. H., and Wirquin, E. (2006). "Coupled effect of time and temperature on variations of yield value of highly flowable mortar." *Cement and Concrete Research*, 36(5), 832–841. doi:10.1016/j.cemconres.2005.11.001.

Petit, J. Y., Khayat, K. H., and Wirquin, E. (2009). "Coupled effect of time and temperature on variations of plastic viscosity of highly flowable mortar." *Cement and Concrete Research*, 39(3), 165–170. doi:10.1016/j.cemconres.2008.12.007.

Prakash, N., and Santhanam, M. (2006). "A study of the interaction between viscosity modifying agent and high range water reducer in self compacting concrete", in *Measuring, monitoring and modeling concrete properties*, edited by Konsta-Gdoutos, M. S. Springer, Dordrecht, Netherlands, pp. 449–454.

Rahman, M. A., and Nehdi, M. (2003). "Effect of geometry, gap, and surface friction of test accessory on measured rheological properties of cement paste." *Materials Journal*, 100(4), 331–339.

Rahman, M. K., Baluch, M. H., and Malik, M. A. (2014). "Thixotropic behavior of self compacting concrete with different mineral admixtures." *Construction and Building Materials*, 50, 710–717.

Roussel, N. (2006). "A thixotropy model for fresh fluid concretes: Theory, validation and applications." *Cement and Concrete Research*, 36(10), 1797–1806. doi:10.1016/J.CEMCONRES.2006.05.025.

Roussel, N., et al. (2010). "Steady state flow of cement suspensions: A micromechanical state of the art." *Cement and Concrete Research*, 40(1), 77–84. doi:10.1016/j.cemconres.2009.08.026.

Schmidt, W., et al. (2010). "Effects of superplasticizer and viscosity-modifying agent on fresh concrete performance of SCC at varied ambient temperatures", in *Design, Production and Placement of Self-Consolidating Concrete,* edited by Khayat, K., and Feys, D. Springer, Dordrecht, pp. 65–77.

Secrieru, E., et al. (2018). "Changes in concrete properties during pumping and formation of lubricating material under pressure." *Cement and Concrete Research*, 108, 129–139. doi:10.1016/j.cemconres.2018.03.018.

Shen, W., et al. (2021). "Change in fresh properties of high-strength concrete due to pumping." *Construction and Building Materials*, 300, 1–32. doi:10.1016/j.conbuildmat.2021.124069.

Shi, Y., Matsui, I., and Guo, Y. (2004). 'A study on the effect of fine mineral powders with distinct vitreous contents on the fluidity and rheological properties of concrete." *Cement and Concrete Research*, 34(8), 1381–1387.

Struble, L. J., and Jiang, Q. (2005). "Effects of air entrainment on rheology." *ACI Materials Journal*, 101, 448–456.

Tang, X., et al. (2014). "Correlation between slump flow and rheological parameters of compound pastes with high volume of ground slag." *Journal of the Chinese Ceramic Society*, 42(5), pp. 648–652.

Tattersall, G., and Banfill, P. (1983). The rheology of fresh concrete. Pitman Books Limited, London, England.

Tattersall, G. H. (1991). *Workability and quality control of concrete.* CRC Press, Boca Raton, FL.

Taylor, H. F. W. (1997). *Cement chemistry.* Thomas Telford, London.

Toledano-Prados, M., et al. (2013). "Effect of polycarboxylate superplasticizers on large amounts of fly ash cements." *Construction and Building Materials*, 48, 628–635.

Tregger, N. A., Pakula, M. E., and Shah, S. P. (2010). "Influence of clays on the rheology of cement pastes." *Cement and Concrete Research*, 40(3), 384–391. doi:10.1016/j.cemconres.2009.11.001.

Uysal, M., and Yilmaz, K. (2011). "Effect of mineral admixtures on properties of self-compacting concrete." *Cement and Concrete Composites*, 33(7), 771–776. doi:10.1016/j.cemconcomp.2011.04.005.

Vance, K., et al. (2013). "The rheological properties of ternary binders containing Portland cement, limestone, and metakaolin or fly ash." *Cement and Concrete Research*, 52, 196–207.

Wallevik, J. E. (2003). *Rheology of particle suspensions: Fresh concrete, mortar and cement paste with various types of lignosulfonates.* Fakultet for ingeniørvitenskap og teknologi.

Wallevik, J. E., and Wallevik, O. H. (2017). "Analysis of shear rate inside a concrete truck mixer." *Cement and Concrete Research*, 95, 9–17. doi:10.1016/j.cemconres.2017.02.007.

Wallevik, O. H., and Wallevik, J. E. (2011). "Rheology as a tool in concrete science: The use of rheographs and workability boxes." *Cement and Concrete Research*, 41(12), 1279–1288.

Williams, D. A., Saak, A. W., and Jennings, H. M. (1999). "Influence of mixing on the rheology of fresh cement paste." *Cement and Concrete Research*, 29(9), 1491–1496. doi:10.1016/S0008-8846(99)00124-6.

Yahia, A. (1999). "Effect of limestone powder on rheological behavior of highly-flowable mortar". *Proceeding of the Japan Concrete Institute*, 21(2), 559–564.

Yamada, K., et al. (1999). "Influence of temperature on the dispersibility of polycarboxylate type superplasticizer for highly fluid concrete", *rilem.net*. Available at: https://www.rilem.net/gene/main.php?base=500218&id_publication=12&id_papier=1310 (Accessed: 3 April 2018).

Yamada, K., Hanehara, S., and Matsuhisa, M. (1998). "Fluidizing mechanism of cement paste added with polycarboxylate type superplasticizer analyzed from the point of adsorption behavior." *Proceedings of JCI*, 20(2), 63–78.

Yamada, K., Ogawa, S., and Hanehara, S. (2000). "Working mechanism of poly-beta-naphthalene sulfonate and polycarboxylate superplasticizer types from point of cement paste characteristics." *Special Publication*, 195, 351–366.

Yamada, K., Ogawa, S., and Hanehara, S. (2001). "Controlling of the adsorption and dispersing force of polycarboxylate-type superplasticizer by sulfate ion concentration in aqueous phase." *Cement and Concrete Research*, 31(3), 375–383. doi:10.1016/S0008-8846(00)00503-2.

Yang, M., Neubauer, C. M., and Jennings, H. M. (1997). "Interparticle potential and sedimentation behavior of cement suspensions: Review and results from paste." *Advanced Cement Based Materials*, 5(1), 1–7.

Yoshioka, K., et al. (1997). "Role of steric hindrance in the performance of superplasticizers for concrete." *Journal of the American Ceramic Society*, 80(10), 2667–2671.

Yun, K.-K., Choi, S.-Y., and Yeon, J. H. (2015). "Effects of admixtures on the rheological properties of high-performance wet-mix shotcrete mixtures." *Construction and Building Materials*, 78, 194–202.

Zhang, X., and Han, J. (2000). "The effect of ultra-fine admixture on the rheological property of cement paste." *Cement and Concrete Research*, 30(5), 827–830. doi:10.1016/S0008-8846(00)00236-2.

Zhang, Y., and Zhang, X. (2002). "Relationship between the content of slag powder and the characteristics of particle group and rheological property of cement paste." *Bulletin of the Chinese Ceramic Society*, 21(6), 63–67, 75.

Zhou, Z., et al. (1999). "The yield stress of concentrated flocculated suspensions of size distributed particles." *Journal of Rheology*, 43(3), 651–671.

Zuo, Y. F., Sui, T. B., and Wang, D. M. (2004). "Effect of superplasticizers on rheologic performance of fresh cement paste." Hunningtu (*Concrete*) (09), 38–39.

Chapter 3

Rheological properties of fresh concrete materials

3.1 GENERAL CONSIDERATIONS FOR GRANULAR MATERIALS

The term "the rheology of mortar and concrete" mainly indicates the evolution of viscosity, plasticity, and elasticity under shear stress (Barnes et al., 1989). Yield stress and plastic viscosity as two fundamental physical parameters are of great importance in cement-based materials. On the one hand, the rheological parameters are effective for characterization of the workability, prediction of the flow behavior, and evaluations of the pumpability, stability, and formwork filling of mortar and concrete (Kwon et al., 2013, Roussel 2007, Roussel et al., 2010). The rheological properties also play significant roles in describing the chemical hydration, microstructural formation, and setting process of fresh cementitious materials (Bogner et al., 2020, Iqbal Khan et al., 2016, Jiao et al., 2019a, Sant et al., 2008). On the other hand, the adjustment of rheological parameters such as static yield stress and plastic viscosity can be helpful to strike a balance between pumpability, extrudability, and printability (De Schutter et al., 2018, Roussel 2018, Yuan et al., 2019). Consequently, the rheological parameters are vital for the preparation of high-performance concrete (Jiao et al., 2018, Wallevik, 2003).

Fresh cement-based materials such as mortar and concrete can be regarded as highly concentrated suspensions with aggregate particles suspending in a cement paste phase. The particle size in a cementitious system varies from several nanometers to dozens of millimeters. Therefore, fresh cementitious materials can be viewed as a two-phase system, i.e., fluid-paste and solid-phase aggregates (Jiao et al., 2017b, Nielsen, 2001). Depending on the content of the cement utilized, fresh concrete is divided into three categories, i.e., lean concrete with cement content lower than 10%, normal concrete with cement content ranging from 10% to 15%, and rich concrete with cement content higher than 15% (Talbot et al., 1923). In the present chapter, the theoretical correlations between paste volume (or aggregate volume fraction) and rheological parameters for different types of concrete are introduced. The effect of characteristics of aggregates such as particle size, shape, and roughness on the rheological properties is discussed from theoretical and experimental perspectives. Furthermore, the influences of external factors such as mixing procedure, shear history, and measuring geometry on the rheology of fresh concrete are briefly illustrated.

3.2 FLOW REGIMES OF CONCRETE

Aggregate accounts for more than 75% of the total volume of conventional vibrated concrete and 60% of self-compacting concrete (SCC) (Jiao et al., 2017b). From the rheological point of view, aggregates with large grain size and rough surface restrict the flow of concrete and increase the yield stress and plastic viscosity of concrete than that of cement paste. Generally, the chemical compositions of aggregate have little effect on the rheological properties (Mahaut et al., 2008b). In this section, the theoretical calculations and empirical relationships between rheological properties (plastic viscosity and yield stress) and aggregate volume fraction are illustrated. The excess paste theory is also highlighted.

3.2.1 Relationships between aggregate volume fraction and concrete rheology

3.2.1.1 Viscosity vs aggregate volume fraction

The derivation of viscosity of a concentrated suspension is originated from the well-known Einstein viscosity equation, as expressed in Eq. (3.1):

$$\eta_r = 1 + [\eta]\phi \qquad (3.1)$$

where η_r is the relative viscosity of a suspension to its suspending fluid, ϕ is the volume concentration of particles, and $[\eta]$ is the intrinsic viscosity of solids, which is a measure of the effect of individual particles on the viscosity. The value of intrinsic viscosity is 2.5 for spherical particles, 3–5 for angular but presumably equant particles, and 4–10 for rods or fibers (Struble and Sun, 1995). The Einstein equation is restricted to dilute suspensions with very low particle volume fractions (lower than 0.05), supposing no particle interaction (Quemada, 1984). Subsequently, Robinson proposed a modified Einstein model for higher concentrations, by considering that the viscosity depends on both the volume fraction of particles and the volume of free liquid in the suspension (Robinson, 1949), as expressed by:

$$\eta_r = 1 + \frac{k\phi}{1 - V\phi} \qquad (3.2)$$

where k is a constant, equal to the intrinsic viscosity of particles (Ren, 2021), and V is the bulk volume per unit volume of solid, regarding as the reciprocal of maximum packing fraction which will be introduced later. Besides, considering the interaction between solid particles for medium and high concentrations, Roscoe proposed a modified viscosity model (Roscoe, 1952):

$$\eta_r = (1 - 1.35\phi)^{-[\eta]} \qquad (3.3)$$

As well known, the viscosity generally increases with the solid concentrations. However, due to the crowding effect, the suspension with a sufficiently high concentration cannot flow when solid particles become packed so tightly. This limit concentration is called maximum volume fraction ϕ_m, depending on the size distribution and shape of the particles. For mono-sized spherical particles, the maximum volume fraction is near 0.6–0.7, while it is higher for polydisperse particles (Struble and Sun, 1995). Based on the Einstein viscosity equation, Mooney stated a functional equation considering the maximum packing volume fraction of rigid spheres (Mooney, 1951), which can be expressed as follows:

$$\eta_r = \exp\left[\frac{[\eta]\phi}{1 - \frac{\phi}{\phi_m}}\right] \qquad (3.4)$$

The Mooney equation can be used to describe the viscosity of suspensions with low particle volume fractions (Struble and Sun, 1995). For the suspensions with high concentrations, the Mooney equation is no longer applicable. In this case, the Krieger–Dougherty equation, as shown in Eq. (3.5), was proposed to describe the full range of concentrations (Krieger and Dougherty, 1959):

$$\eta_r = \left(1 - \frac{\phi}{\phi_m}\right)^{-[\eta]\phi_m} \tag{3.5}$$

where ϕ_m is the maximum packing fraction, which is very sensitive to particle size distribution and particle shape. It should be mentioned that the value of $[\eta]\phi_m$ is always around 2 for various suspensions. In this case, the Krieger–Dougherty model can be simplified to the Maron–Piece model, as stated by (Ren, 2021):

$$\eta_r = \left(1 - \frac{\phi}{\phi_m}\right)^{-2} \tag{3.6}$$

Back again the Krieger–Dougherty model, if considering fresh concrete as a suspension with multi-size particles and ignoring the interaction between particles of different sizes, the viscosity of the multimodal suspensions can be derived from the unimodal viscosity of each size, namely the Farris model (Farris, 1968):

$$\eta = \eta_s \left(1 - \frac{\phi_1}{\phi_m^1}\right)^{-[\eta_1]\phi_m^1} \left(1 - \frac{\phi_2}{\phi_m^2}\right)^{-[\eta_2]\phi_m^2} \tag{3.7}$$

Typically, solid particles in fresh concrete are divided into coarse and fine aggregates. In this case, the Farris model can be modified in the following version (Noor and Uomoto, 2004):

$$\eta_C = \eta_P \left(1 - \frac{S}{S_{\lim}}\right)^{-[\eta_{FA}]S_{\lim}} \left(1 - \frac{G}{G_{\lim}}\right)^{-[\eta_{CA}]G_{\lim}} \tag{3.8}$$

where S and $S_{\lim}$ are the sand volume fraction and its maximum solid volume, respectively. G and $G_{\lim}$ are the gravel volume fraction and the maximum gravel solid volume. The $S_{\lim}$ and $G_{\lim}$ are selected as 0.643 and 0.575 in their study, respectively. η_C and η_P are the viscosity of concrete and paste (Pa.s), respectively. $[\eta_{FA}]$ and $[\eta_{CA}]$ are the intrinsic viscosity of fine and coarse aggregates with values of 1.9 and 3.2 in their study, respectively. Experimental results indicate that the multimode Farris model is reasonable and applicable for determining the viscosity of fresh mortar and concrete (Noor and Uomoto, 2004).

3.2.1.2 Yield stress vs aggregate volume fraction

Considering the fact that aggregates distribute in cement paste matrix uniformly, the shear stress of mortar and/or concrete is assumed to be the sum of the shear stresses resulting from the yield stress and flow of cement paste, the shear stress induced by the aggregate particle movement, and the interaction between cement paste and aggregates (Lu et al., 2008, Wang et al., 2020). Therefore, the aggregate volume fraction significantly affects the yield stress of the mortar and/or concrete.

If aggregates are divided into smaller subclasses i and the interactions of these subclasses are considered, a powerful semi-empirical model was established to correlate the yield stress of concrete with the volume fraction (De Larrard and Sedran, 2002), as shown in Eq. (3.9):

$$\tau_{0,C} = 2.537 + \sum_i \left(0.736 - 0.216\log(d_i)\right)K_i' + \left(0.224 + 0.910\left(1 - \frac{P}{P^*}\right)^3\right)K_C' \qquad (3.9)$$

where $\tau_{0,C}$ is the yield stress of concrete (Pa); $K_i' = \frac{\phi}{1-\phi_i^*}$; ϕ_i, ϕ_i^*, and d_i are the volume fraction, the maximum packing volume fraction, and the size of particles of class i, respectively; subscript C is the cement; and P and P^* are the superplasticizer dosages and saturation dosage, respectively. It should be noted that this model is only applied to the case of concrete only with one binder (Toutou and Roussel, 2006). Similarly, Noor and Uomoto stated that the yield stress of concrete is assumed to be a function of mortar and the volume fraction of aggregates (Noor and Uomoto, 2004), which can be expressed as follows:

$$\tau_{0,C} = \tau_{0,m} + f(\phi) \qquad (3.10)$$

where $\tau_{0,C}$ and $\tau_{0,m}$ are the yield stress of concrete and mortar (Pa), respectively. ϕ is the total aggregate volume fraction. It should be mentioned that the model in Eq. (3.10) based on experimental data of mortar and concrete is completely empirical.

Considering fresh concrete as a concentrated suspension with aggregates (i.e., sand and gravel) suspending in cement paste, the yield stress of the suspension is proportional to that of the constitutive cement paste (Yammine et al., 2008). The general form of the relation is expressed as follows:

$$\tau_{0,C} \approx \tau_{0,cp} f(\phi/\phi_m) \qquad (3.11)$$

where $\tau_{0,C}$ and $\tau_{0,cp}$ are the yield stress of the concrete and the corresponding cement paste (Pa), respectively. ϕ_m is the maximum packing fraction. A semi-empirical calculation for the maximum packing fraction can be expressed by Eq. (3.12) (Hu and de Larrard, 1996):

$$\phi_m = 1 - 0.45\left(d_{\min}/d_{\max}\right)^{0.19} \qquad (3.12)$$

where $d_{\min}$ and $d_{\max}$ are the smallest and largest grain diameters in the granular skeleton (m). Independent of the mixture proportion of cement paste, Eq. (3.11) is of great importance to study the relative yield stress of the concrete, i.e., the ratio of yield stresses between concrete and cement paste ($\tau_{0,C}/\tau_{0,cp}$). Based on Eq. (3.11), two examples are given here to illustrate the modified theoretical relationship between yield stress and volume fraction.

The first example is about the domination theory of hydrodynamic or frictional effects depending on the interparticle distance, i.e., fluid–particle or particle–particle interactions (Yammine et al., 2008). At low volume fractions, the interparticle interactions are hydrodynamic. The relative motions of aggregates imply the flow of cement paste. During the flowing of the concrete, additional energy dissipation induced by the possible analogous effect occurs, leading to the greater yield stress of the concrete than that of the cement paste alone. Consequently, the yield stress of concrete increases with the volume fraction of aggregates. At higher volume fractions, however, direct contacts between aggregate particles may

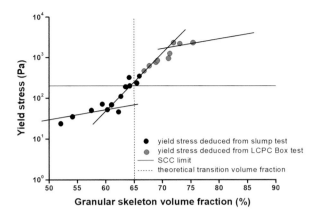

Figure 3.1 Concrete yield stress as a function of aggregate volume fraction. (Adapted from Yammine et al., 2008.)

occur. Remarkably, a "true" direct contact is hardly defined due to the particle interactions, particle roughness, and hydrodynamic effects. Instead, the physical behavior of particle contacts can be associated with the existence of a continuous network of particles in contact. In other words, this phenomenon occurs when the aggregate volume fraction is larger than a critical value ϕ_c. According to experimental and numerical results, ϕ_c should be situated around 0.5 for uniform spheres (Onoda and Liniger, 1990), and it should be increased in the case of poly-dispersed systems.

A representative relationship between concrete yield stress and aggregate volume fraction is presented in Figure 3.1, where the maximum packing of aggregates is equal to 0.84 and the transition volume fraction is in the order of 0.65. It can be seen that the direct contact between aggregate particles can be neglected below the transition value. In this regime, the effect of aggregate particles on the rheological behavior of the concrete is regarded as purely hydrodynamic. Inversely, direct frictional contacts between particles start to dominate the rheological behavior above this transition value, and their highly dissipative nature strongly increases the yield stress of the concrete. The transition volume fraction between the frictional and hydrodynamic regimes is independent of the selection of the multi-scale approach (Yammine et al., 2008).

The second example here is the quantitative relation between relative yield stress and volume fraction of aggregates. For a suspension following the Herschel–Bulkley model, the relative yield stress of concrete compared to the corresponding cement paste only depends on the solid volume fraction (Chateau et al., 2008). The relation between the yield stress and the volume fraction follows the Chateau–Ovarlez–Trung model (Chateau et al., 2008), as expressed in Eq. (3.13) with ϕ_m of 0.57:

$$\frac{\tau_c(\phi)}{\tau_c(0)} = \sqrt{\frac{1-\phi}{1-(\phi/\phi)^{2.5\phi_m}}} \tag{3.13}$$

It should be noted here that the maximum volume fraction ϕ_m is not selected as the so-called random dense packing fraction (ϕ_{RDP}), which is around 0.65 for spherical monodisperse particles. Instead, the critical volume fraction where direct contacts become important, which is the application limit of models only considering hydrodynamic interactions and also the limit between self-compacting concrete (SCC) and ordinary rheology concrete (Yammine et al., 2008), is used as the maximum volume fraction for the sharp increase of yield stress. Therefore, the parameter of ϕ_m in Eq. (3.13) is also called rheology divergence packing fraction ϕ_{div} (Ovarlez et al., 2006). Hafid et al. (2016) stated that the rheology divergence packing fraction is related to the random dense and loose packing fractions, which can be expressed as follows:

$$\phi_{\mathrm{div}} = 0.64\phi_{\mathrm{RDP}} = 0.88\phi_{\mathrm{RLP}} \tag{3.14}$$

If considering fresh concrete as a suspension with grains dispersing in mortar matrix while the fresh mortar is considered as a suspension with sand suspending in cement paste, the Chateau–Ovarlez–Trung model can be modified to the following version (Choi et al., 2013):

$$\frac{\tau_c(\phi)}{\tau_c(0)} = \sqrt{\frac{1-\phi_s}{\left(1-\phi_s/\phi_{s,m}\right)^{2.5\phi_{s,m}}}} \sqrt{\frac{1-\phi_g}{\left(1-\phi_g/\phi_{g,m}\right)^{2.5\phi_{g,m}}}} \tag{3.15}$$

where ϕ_s and ϕ_g are the volume fraction of sand and grains, respectively, and $\phi_{s,m}$ and $\phi_{g,m}$ are the maximum volume fraction of sand and grains, respectively. Recently, Kabagire et al. (2017) stated that the constant 2.5 in Eq. (3.13) can be replaced by a fitted coefficient $[\eta]^*$, representative of the modified intrinsic viscosity depending on the geometric features of particles and the shear conditions (Kabagire et al., 2019). In this case, Eq. (3.13) can be converted into:

$$\frac{\tau_c(\phi)}{\tau_c(0)} = \sqrt{\frac{1-\phi}{\left(1-\phi/\phi_m\right)^{[\eta]^*\phi_m}}} \tag{3.16}$$

where the modified intrinsic viscosity $[\eta]^*$ is determined by nonlinear regression. Furthermore, after taking the effect of paste-to-sand volume fraction into account, the relative yield stress of SCC can be calculated by (Kabagire et al., 2019):

$$\frac{\tau_c(\phi)}{\tau_c(0)} = \sqrt{\left(\frac{1-\phi}{\left(1-\phi/\phi_m\right)^{[\eta]^*\phi_m}}\right)^{\frac{d}{(\mathrm{VP/VS})^e}}} \tag{3.17}$$

where VP/VS is the paste-to-sand volume fraction, and d and e are constants with the value of 1.1 and 1.24 in their study, respectively (Kabagire et al., 2019). It is found that the relative yield stresses of SCC with crushed limestone coarse aggregate between measured and predicted by Eq. (3.17) show a good correlation coefficient of 0.75.

3.2.2 Excess paste theory

Cement paste is an essential component of concrete. A schematic diagram of the fresh concrete model is presented in Figure 3.2. It can be seen that the cement paste in concrete can be divided into two parts: one is the paste filling the voids between aggregates, and the other is the paste coating aggregates. The second part is defined as the excess paste. The excess paste layer is generally used to lubricate the solid grains and provide the flowability of the concrete mixture (Jiao et al., 2017b).

Rheological properties of fresh concrete materials 57

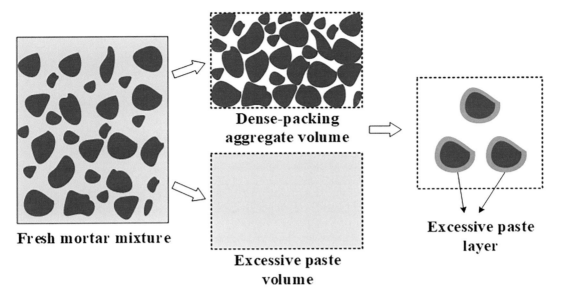

Figure 3.2 Schematic diagram of fresh concrete (SCC) model. (Adapted from Reinhardt and Wüstholz, 2006.)

There are several approaches to calculate the excess paste thickness. A rough approximation of the average interparticle distance can be expressed as follows (Yammine et al., 2008):

$$b = -d\left(1 - (\phi/\phi_m)^{-1/3}\right) \tag{3.18}$$

where b is the average interparticle distance (m), d is the particle size (m), ϕ is the particle volume fraction, and ϕ_m is the maximum packing fraction. It should be mentioned that the average distance is equal to two times the excess paste thickness. A simple approach to calculating the volume of excess paste is described by Reinhardt and Wüstholz (2006) as follows:

$$V_{paste,ex} = V_{paste} - V_{A,void} = V_{paste} - \frac{m_A}{\rho_A}\left(\frac{\rho_A}{\rho_{A,bulk}} - 1\right) \tag{3.19}$$

where $V_{paste,\,ex}$ is the volume of excess paste (m³); V_{paste} is the total paste volume (m³); $V_{A,\,void}$ is the volume of voids between aggregates (m³); m_A and ρ_A are the mass and density of aggregates (kg), respectively; and $\rho_{A,\,bulk}$ is the loose bulk density of aggregates (kg/m³). If a spherical paste layer on the aggregates is assumed and the excess paste thickness is independent of the particle size, the excess paste thickness can be calculated using Eq. (3.20):

$$V_{paste,ex} = \frac{4}{3}\pi \sum_i n_i \left((r_i + t_{paste,ex})^3 - r_i^3\right) \tag{3.20}$$

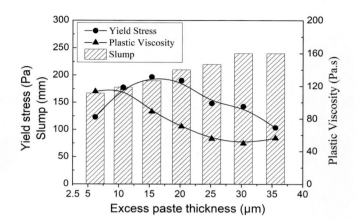

Figure 3.3 A representative relationship between rheological properties and excess paste thickness at the water-to-cement ratio of 0.4, the sand-to-aggregate ratio of 0.42, and the superplasticizer dosage of 1%. (Adapted from Jiao et al., 2017a.)

where $t_{\text{paste, ex}}$ is the excess paste thickness (m), and r_i and n_i are the radius (m) and the number of a particle of class i, respectively. Another calculation method of excess paste thickness is as follows (Oh et al., 1999):

$$t_{\text{paste,ex}} = \left(1 - 100\frac{V_S}{C_S}\right)\frac{10}{S_S V_S} \tag{3.21}$$

where $t_{\text{paste, ex}}$ is the excess paste thickness (mm), V_S is the aggregate-to-mortar volume ratio, C_S is the sand volume divided by its bulk volume (%), and S_S is the specific surface area of aggregates (cm²/cm³).

Based on the concept of excess paste thickness, the amount of excess paste plays a dominant role in determining the rheological properties of mortar or concrete than the total amount of paste. Furthermore, the influencing factors on concrete rheology such as paste volume and aggregate volume can be transferred into the parameter of excess paste thickness (Reinhardt and Wüstholz, 2006, Ren et al., 2021). A typical relationship between rheological parameters and excess paste thickness for conventional concrete is shown in Figure 3.3. At relatively low excess paste thickness, the concrete shows higher yield stress and plastic viscosity, and lower flowability. In this case, the yield stress is relatively more affected by the excess paste thickness than the plastic viscosity. With the increase of excess paste thickness, the yield stress slightly increases, while the plastic viscosity gradually decreases. The flowability and stability of the concrete are improved. At relatively high excess paste thickness, the concrete shows very high slump value, but lower yield stress and plastic viscosity. In this case, the stability of the concrete is not good enough, and a slight segregation phenomenon can be observed (Jiao et al., 2017a).

The yield stress of mortar and concrete can also be quantitatively predicted by the paste layer thickness (Lee et al., 2018). Assuming that the paste layer thickness is only dependent on the properties of cement paste, the layer thickness can be regarded as a constant value regardless of the particle volume fraction. Based on this assumption, the yield stress of SCC concerning the suspending fluid can be estimated by Eq. (3.22):

$$\frac{\tau_c}{\tau_0} = k\left(\varphi_1\left(1+\frac{b}{d_1}\right)^3 + \varphi_2\left(1+\frac{b}{d_2}\right)^3 - \varphi_c\right)^n \qquad (3.22)$$

where τ_c and τ_0 are the yield stress of concrete and the corresponding cement paste (Pa), respectively; b is the paste layer thickness (mm); k is a constant; and d_1 and d_2 are the particle size of fine and coarse aggregates (mm), respectively. φ_1 and φ_2 are the volume fraction of fine and coarse aggregates, respectively. φ_t is the percolation threshold, with the value of 0.29 for mono-sized suspension. With the experimental results of corresponding mortar with 90%-volume-fraction aggregates and wet-sieved mortar, the yield stress of concrete can be acceptably predicted by this model (Lee et al., 2018).

3.3 INFLUENCE OF AGGREGATE CHARACTERISTICS

3.3.1 Aggregate volume fraction

From the experimental point of view, a typical relationship between plastic viscosity of mortar/concrete and aggregate volume fraction is presented in Figure 3.4. It can be clearly observed that aggregates only play a significant role at high solid volume fractions. In addition, for the cement-based materials with low aggregate volume fraction, the effect of aggregate size, gradation, and surface texture is not significant (Li and Liu, 2021, Lu et al., 2008). Based on the assumption that plastic viscosity is the microscopic indication of the flow behavior of water in the voids between solid particles (De Larrard and Sedran, 2002), the results indicate that the contribution of aggregate volume fractions on the plastic viscosity of fresh lean concrete is possibly much higher than that of rich concrete. This can be explained by the increased collisions and frictions between solid particles at high aggregate volume fractions (Okamura and Ouchi, 2003).

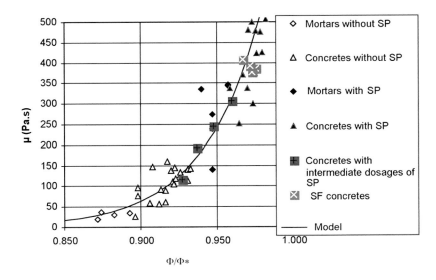

Figure 3.4 Relationship between plastic viscosity and relative particle concentration. SP is superplasticizer, and SF is silica fume. (Adapted from De Larrard and Sedran, 2002.)

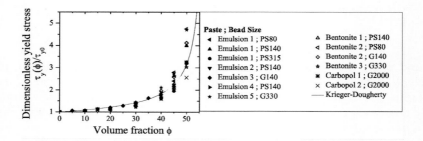

Figure 3.5 Dimensionless yield stress vs volume fraction of beads with various particle sizes in various suspensions. The solid line is based on Eq. (3.13) with ϕ_m of 0.57. (Adapted from Mahaut et al., 2008a.)

In the case of yield stress, through investigating the suspensions with rigid noncolloidal particles embedded in a yield stress fluid, Mahaut et al. (2008b) found that the relative yield stress of concrete to its cement paste only depends on the yield stress value of suspending cement paste and volume fraction of particles. However, the relative yield stress is independent of the physicochemical properties of the matrix, the bead material, and the particle size. A typical relationship between the relative yield stress and the solid volume fraction is shown in Figure 3.5. It can be observed that the relative yield stress of suspensions shows an exponential increase with the increase of particle volume fraction, with a limited increase degree at volume fraction lower than 30% and a sharp increase at volume fraction approach to 50% (Mahaut et al., 2008b).

3.3.2 Gradation and particle size

According to the excess paste theory, it can be easily concluded that for a concrete mixture with a fixed content of cement paste, the higher the aggregate packing density, the lower the fraction of voids, and thus higher the content of excess paste acting as lubrication effect. Therefore, the properties of concrete are strongly affected by the packing of aggregate particles. The existing particle packing models for mono-sized particles and poly-dispersed systems were summarized by Kumar and Santhanam (2003) and Roussel (2011). In the context of skeleton for cement-based materials, the aggregate gradation optimization process is to achieve a good aggregate packing by selecting the proper sizes and proportions of small particles to fill larger voids approaching an ideal gradation curve. An optimal aggregate gradation provides a concrete mixture with excellent properties. In this part, the most useful continuous models in cementitious materials are briefly illustrated, and the effects of particle size and gradation on the rheological properties of concrete are discussed as well.

The most well-known and applicable ideal curve of particle packing is the theoretical particle size distribution described by the Fuller–Thompson model (Fuller and Thompson, 1907), which can be expressed as follows:

$$P = \left(d_i/D_{\max}\right)^q \tag{3.23}$$

where P is the cumulative passing, d_i is the particle diameters under consideration (m), $D_{\max}$ is the nominal maximum particle size diameter (m), and q is the packing exponent of 0.5 for Fuller and Thompson (1907), and then revised to 0.45 by Talbot et al. (1923). It should be noted that the Fuller and Thompson curve is only valid for the aggregate skeleton, and it gives good results if stiff concrete mixes with low workability are used.

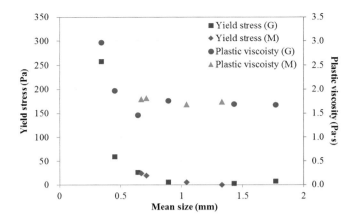

Figure 3.6 Influence of the size of fine aggregate on yield stress and plastic viscosity of mortar. (Adapted from Han et al., 2017.). G indicates mono-sized grains, and M represents mixed sands.

Considering the aggregates and cementitious materials, i.e., all the solid grains in the mixture, a modified Andreasen & Andersen model proposed by Funk and Dinger (2013) can be applied:

$$P = \frac{d_i^q - d_{min}^q}{d_{max}^q - d_{min}^q} \quad (3.24)$$

where P is the cumulative passing; d_i is the particle diameters under consideration (m); D_{min} and D_{max} are the minimum and nominal maximum particle size diameters (m), respectively; and q is the packing modulus. The results of Mueller et al. (2014) showed that the modified Andreasen & Andersen model with q of about 0.27 can be used to express the particle size distribution of all the solids in SCC with low powder content. As a result, this model is widely used in the mixture design of conventional vibrating concrete and even SCC.

From the viewpoint of experimental results, smaller aggregate particle size generally results in higher yield stress and plastic viscosity of cement-based materials, due to the high specific surface area and thus high water demand (Jiao et al., 2017b). At low aggregate volume fractions, the contribution of particle size to the rheological properties is not significant, and the influence can only be obvious at higher volume fractions (Lu et al., 2008). Hu (2005) conducted a series of experiments investigating the effect of aggregate on the rheology of cement-based materials. For mortars with a fixed water-to-cement (w/c) ratio of 0.5 and sand-to-cement ratio of 2, increasing particle size from 0.15 to 2.36 mm reduces the yield stress from 130 to 70 Pa, and decreases the plastic viscosity from 2 to 1.2 Pa.s. Han et al. (2017) also stated that increasing the fine aggregate size decreases the yield stress and plastic viscosity of fresh mortar. In addition, they also concluded that changing the size of fine aggregate when exceeding 0.70 mm will show no significant influence on the yield stress, as can be observed from Figure 3.6.

From the viewpoint of concrete, the particle size of coarse aggregate plays a more significant role in the yield stress compared to the plastic viscosity (Hu, 2005). As for the gradation, graded aggregates generally have lower uncompacted void content than single-sized aggregates (Hu and Wang, 2007, Hu and Wang, 2011). Consequently, more excess mortar is generated to lubricate the particles for graded aggregates, and thus the yield stress and plastic viscosity of concrete mixtures with graded coarse aggregates are significantly lower than that with single-sized coarse aggregates. Santos et al. (2015) stated that concrete mixtures with continuous skeleton showed higher slump flow values, whereas due to the interlocking effect, the discontinuous distribution mixtures exhibited lower flowability. In addition, the poor grading of fine aggregates can to an extent be corrected by finer sand with low fineness modulus. Lower volumetric replacement of crushed sand by dune sand, for example, decreases the yield stress and plastic viscosity of concrete, but higher volumetric replacements will exhibit an opposite influence because of the high specific surface area of dune sand (Bouziani et al., 2012, Park et al., 2018). The content of fines with a diameter less than 0.315 mm in coarse aggregate also has a significant influence on the rheological properties of fresh concrete (Aïssoun et al., 2015). It was found that increasing the content of fines from 8% to 18% could reduce the yield stress, increase the plastic viscosity, and improve the settlement resistance.

3.3.3 Particle morphology

The morphology of aggregate particles can be quantitatively assessed by shape factors with dimensionless quantities. The shape factors represent the degree of deviation of particles from an ideal shape. Common shape factors used to characterize the morphology of aggregate particles include aspect ratio (A_R), circularity (f_{circ}), and convexity (f_{conv}). A schematic diagram to define the morphological parameters is shown in Figure 3.7, where the particle area is noted as S and its perimeter is noted as P. The convex perimeter (P_{conv}) is defined as the smallest perimeter of a polygon capturing the projected particle, presented by the gray in Figure 3.7.

The aspect ratio, which is the most commonly used shape factor, is computed as the ratio of the largest (D_{max}/m) and smallest (D_{min}/m) dimensions of the projected particle, as expressed by Eq. (3.25):

$$A_R = \frac{D_{max}}{D_{min}} \tag{3.25}$$

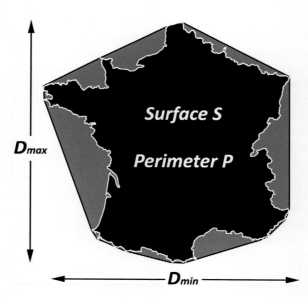

Figure 3.7 Schematic diagram to define morphological parameters. (Adapted from Hafid et al., 2016.)

Another very common shape factor is the circularity, which is defined as follows:

$$f_{circ} = \frac{4\pi S}{P^2} \quad (3.26)$$

The convexity describes the surface properties of a projected particle, which can be calculated by the ratio between convex perimeter and real perimeter, which is expressed as follows:

$$f_{conv} = \frac{P_{conv}}{P} \quad (3.27)$$

Furthermore, the shape of aggregate particles can also be evaluated by sphericity (SP) and roundness (R) (Cordeiro et al., 2016), which can be expressed by Eqs. (3.28) and (3.29), respectively:

$$SP = \frac{2P^{0.5}}{\pi^{0.5} D_{max}} \quad (3.28)$$

$$R = \frac{4P}{\pi D_{max}^2} \quad (3.29)$$

Generally, the aspect ratio describes the overall shape of a particle, whereas circularity captures the deviation of a particle from the perfect circle. The convexity is usually to capture the surface roughness of a particle. Several examples of geometrical shapes and their morphological parameters are listed in Table 3.1. The convexity and aspect ratio are supposed to be uncorrelated, whereas the circularity varies with both overall shape and surface roughness. Furthermore, a typical relationship between aspect ratio and circularity, convexity,

Table 3.1 Typical examples of geometrical shapes and their morphological parameters (Hafid, et al., 2016)

Shape	Illustration	A_R	f_{circ}	f_{conv}
Circle		100%	100%	100%
Square		100%	79%	100%
Rectangle		300%	59%	100%
Ellipsoid		300%	60%	100%
Star		100%	38%	79%

Note: Underlined value means that this parameter is constant for the specific shape.

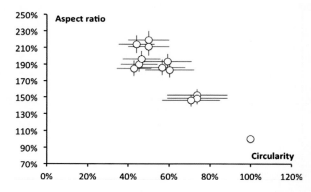

Figure 3.8 Correlations between aspect ratio and circularity. (Adapted from Hafid et al., 2016.)

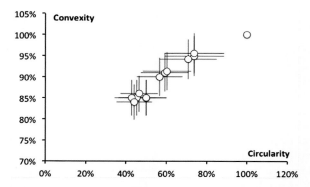

Figure 3.9 Correlations between convexity and circularity. (Adapted from Hafid et al., 2016.)

and circularity for 120 sand particles is shown in Figures 3.8 and 3.9, respectively. It can be found that convexity and circularity are well correlated for the tested particles. Therefore, aspect ratio and convexity can be used as independent morphological parameters to describe the shape of aggregate particles (Hafid et al., 2016). Moreover, the authors stated that both random dense and loose packing fractions of sand particles are strongly correlated to the aspect ratio, whereas the particle roughness seems to play a role only when the frictional contacts between particles start to dominate.

Considering the volume fraction and particle shape, Geiker et al. (2002b) used the Nielsen model (Nielsen, 2001) to estimate the relative rheological parameters of concrete. Take the plastic viscosity (μ_C) as an example, assuming particles being ellipsoids, it can be expressed by Eq. (3.30):

$$\frac{\mu_C}{\mu_0} = \frac{1 + \alpha\varphi}{1 - \varphi} \tag{3.30}$$

where μ_0 is the plastic viscosity of the corresponding cement paste (Pa.s), φ is the particle volume fraction, and α is a so-called shape function, depending on the shape functions of suspending fluid and particles. However, due to the combined effect of angularity and surface texture of aggregates, the proposed model cannot fully fit the experimental data (Geiker et al., 2002b).

Rheological properties of fresh concrete materials 65

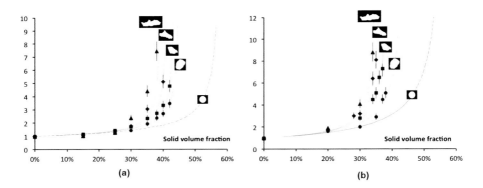

Figure 3.10 Relationship between (a) relative yield stress, (b) relative Herschel–Bulkley consistency, and solid volume fraction of suspensions with various particle shapes. Lines are fitted by the Chateau model. (Adapted from Hafid et al., 2016.)

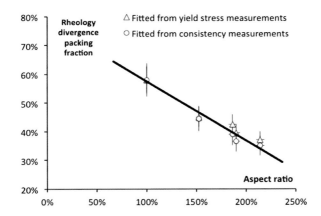

Figure 3.11 Relationship between rheology divergence packing fraction and the aspect ratio of particles. Aspect ratio of 100% indicates spherical particles. (Adapted from Hafid et al., 2016.)

Existing experimental results show that the rheological properties are affected by the shape and surface texture of aggregate particles. A representative relationship between the relative yield stress, the relative Herschel–Bulkley consistency, and the solid volume fraction for various particle shapes is presented in Figure 3.10. It can be seen that both the relative yield stress and the relative Herschel–Bulkley consistency increase with increasing solid volume fraction, with the divergence packing fraction dependent on the shape of the solids. The effect of particle shape on the relative consistency seems to be higher than that on the relative yield stress. Besides, the divergence packing fraction gradually decreases when the particle tends to be more irregular. A typical correlation between the measured divergence packing fraction and the aspect ratio for monodisperse particles is presented in Figure 3.11. It can be seen that the divergence packing fraction can decrease up to 40% for very elongated particles.

At a given solid volume fraction, spherical particles generally result in reduced yield stress and plastic viscosity of cement-based materials due to the low uncompacted void content

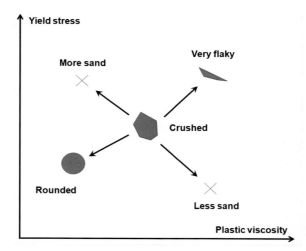

Figure 3.12 Effect of aggregate shape and sand content on rheological parameters. (Adapted from Wallevik and Wallevik, 2011.)

and low friction between particles. By contrast, irregular and elongated aggregates dramatically increase the yield stress and plastic viscosity of mortar and concrete. This can be clearly observed in Figure 3.12. It should be mentioned that the particle shape plays a more significant role in controlling the plastic viscosity and the consistency index than the yield stress (Hafid et al., 2016, Westerholm et al., 2008), and the negative effects of shaped aggregates can be reduced by increasing the paste volume (Westerholm et al., 2008). Moreover, the particle shape of finer particles (e.g., 0.125–2 mm) and the properties of filler particles with a size lower than 0.125 mm show the most important influence on the workability and rheological properties of concrete (Cepuritis et al., 2016).

From the perspective of aggregate types, mortar and concrete prepared with river sand and gravel have relatively low yield stress and plastic viscosity than the mixtures with limestone and crushed aggregates (Hu and Wang, 2007, Lu et al., 2008). As for recycled concrete aggregate (RCA) with high water absorption, experimental results suggest that it is possible to produce SCC with good flowability, excellent passing ability, and stability (Hu et al., 2013). Carro-López et al. (2015) found that replacing natural sand with 20% fine RCA has less influence on the workability and rheological properties of SCC, but further increasing fine RCA replacement significantly reduces the workability of SCC. Besides, the evolution of plastic viscosity over time increases with the replacement of natural sand with fine RCA, as shown in Figure 3.13. However, Güneyisi et al. (2016) suggested that increasing the fine RCA from 0% to 100% increased the flowability of SCC. This might be attributed to the physical properties of the utilized RCA particles. The latter authors also found that the replacement of 50% coarse aggregates by RCA improved the workability of SCC, whereas further increasing the coarse RCA utilization showed less increase in the flowability due to the angular shape of RCA. An empirical relationship between coarse RCA replacements, superplasticizer, w/c, and rheological parameters of the concrete was established by Ait Mohamed Amer et al. (2016) as follows:

$$X = A\left(1 + k_1 R_A\right)\left(1 + k_2 S_p + k_3 \frac{w}{c} + k_4 \frac{\Delta w}{c}\right) \tag{3.31}$$

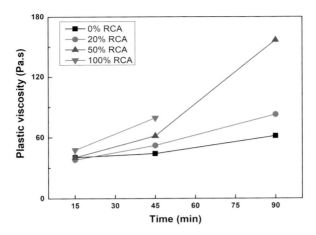

Figure 3.13 Evolution of plastic viscosity over time versus replacement of natural sand by fine RCA. (Data derived from Carro-López et al., 2015.)

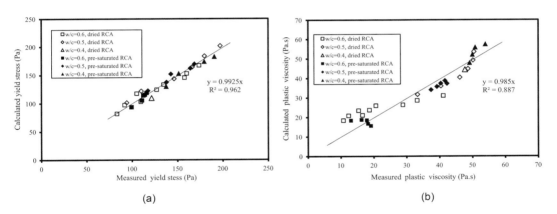

Figure 3.14 Relationship between measured rheological properties and calculated parameters by using Eq. (3.31): (a) yield stress and (b) plastic viscosity. (Adapted from Ait Mohamed Amer et al., 2016.)

where X is the yield stress (Pa) or plastic viscosity (Pa.s) of the concrete, A is a parameter depending on the control mixture, R_A is the replacement ratio of RCA (%), S_p is the superplasticizer dosage (%), $\Delta w/c$ is the pre-saturated water-to-cement ratio, and k_1–k_4 are constants. This simple empirical model provides a useful method to estimate the rheological properties of concrete containing dried or pre-saturated RCA with a high correlation coefficient (Ait Mohamed Amer et al., 2016), as can be observed in Figure 3.14.

3.4 EFFECT OF EXTERNAL FACTORS

3.4.1 Mixing process

The mixing process has a significant influence on the dispersion of suspensions, and thus the rheological behavior. In this part, the influences of mixer type, mixing time, and material addition sequence on the rheological properties of mortar and concrete are presented.

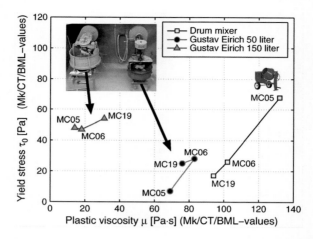

Figure 3.15 Effect of mixer type on rheological properties of concrete. (Adapted from Wallevik and Wallevik, 2011.)

Wallevik and Wallevik (2011) stated that the rheological parameters of concrete could be influenced by mixer type. Three different mixers are used, i.e., 50 and 150 l mixers from Maschinenfabrik Gustav Eirich and a typical drum mixer, as shown in Figure 3.15. It can be observed that the drum mixer gave concrete mixtures with higher plastic viscosity than real mixers. With the increase of mixer size, the yield stress of concrete mixtures decreased, but the plastic viscosity showed an opposite behavior. Moreover, they also analyzed the shear rate inside a concrete truck (Wallevik and Wallevik, 2017). They found that the shear rate is dependent on the rotational speed and charge volume of the drum, as well as the rheological parameters of the concrete. Increasing rotational speed increased the shear rate of the drum. If the charge volume is considered alone, the shear rate decreases with the volume exponentially.

Struble and Chen (2005) examined the influence of continuous agitation for up to 100 min on the rheological properties of concrete. The yield stress and plastic viscosity increased with time, and after experiencing continuous agitation, the concrete mixture showed lower yield stress and plastic viscosity than the similar mixture without agitation. By theoretical calculations, Li et al. (2004) found that the increase of yield stress and plastic viscosity with time and temperature was more remarkable in stationary state compared to an agitated state. However, overmixing could exert a negative effect on the workability and rheological properties of cement paste and concrete (Dils et al., 2012, Han and Ferron, 2016). In addition, the shear-thinning or thickening intensity of concrete can be altered by subjecting it to continuous agitation. Nehdi and Al Martini (2009) found that concrete mixtures containing superplasticizer exhibited shear-thinning behavior under prolonged mixing and elevated temperature. For the concrete mixtures beyond the saturation dosage of naphthalene sulfonate-based admixture, the rheological behavior shifted from shear-thinning to shear-thickening after subjecting to agitation at a high temperature. Recently, Jiao et al. (2019b) found that fresh mortars with glass fibers tended to show shear-thickening behavior after experiencing a rotational shear mixing.

The rheological properties of fresh cementitious materials can be affected by the material addition sequence. Van Der Vurst et al. (2014) evaluated the effects of the addition sequence of materials on rheological properties of SCC. It was found that premixing aggregates with water increased the plastic viscosity and dynamic yield stress due to the water adsorption effect of aggregates. In contrast, the workability of concretes could be significantly improved by adding aggregates in cement paste. This is in good agreement with the results of França et al. (2013). Furthermore, França et al. (2016) also examined the influence of polyvinyl alcohol (PVA) fiber and water addition sequences on the mixing and rheological properties of mortar. They found that first mixing water with dry solid parties and then introducing the fibers could save the mixing energy and thus improve the rheological properties of mortar significantly.

3.4.2 Shear history

Cementitious suspension is a kind of thixotropic and history-dependent suspension. Lack of steady state during rheological measurements of concrete tends to cause shear-thickening behavior, especially for self-compacting concrete (Feys et al., 2008, Feys et al., 2009). The relaxation period of each rotating velocity influences the implementation of steady state, and thus the measured rheological parameters. When measuring the rheological properties of concrete, non-steady state results in an overestimation of plastic viscosity and underestimation of yield stress (Geiker et al., 2002a). However, a longer rotating time has the potential to lead to the segregation of concrete mixture. To strike a balance between reducing the possible impacts of non-steady state and avoiding segregation, Geiker et al. (2002a) recommended that the rotational speed for unknown concrete should be within 0.05–0.57 rps, and the relaxation time of each rotating speed was 10 s.

3.4.3 Measuring geometry

The widely used geometry in cementitious paste includes concentric cylinder, vane, and parallel plate. The measured rheological properties depend on the type and surface of geometry. The potential artifacts, merits, and drawbacks of each geometry have been summarized by Yahia et al. (2017). The measured rheological properties of fresh concrete are also dependent on the rheometer used. The National Institute of Standards and Technology (NIST) carried out a series of experiments to compare the effect of different concrete rheometers such as IBB, BML, and two-point rheometer on the rheological parameters in France (Ferraris and Brower, 2000) and Cleveland (USA) (Ferraris and Brower, 2003). It can be concluded that the absolute yield stress and plastic viscosity are related to the rheometer used, but the rank of rheological parameters is independent of the rheometer. Hočevar et al. (2013) compared the rheological parameters of 26 concrete mixtures measured by ICAR rheometer and ConTec Visco 5. They found that the yield stress and plastic viscosity obtained using the ICAR rheometer was 42% higher and 42% lower than that by ConTec Visco 5, respectively. More details will be provided in Chapter 6.

3.5 SUMMARY

Aggregates with larger grain size and rough surface restrict the flow of concrete and increase the yield stress and plastic viscosity compared to cement paste. From a theoretical point of view, the relationship between aggregate volume fraction and plastic viscosity can be predicted by using the Krieger–Dougherty model, while its relationship with yield stress can be estimated by the Chateau–Ovarlez–Trung model. The underlying mechanisms of influence of aggregates on rheological properties can be explained by the excess paste theory.

At relatively low aggregate particle volume fractions, the effect of aggregate size, gradation, and surface texture on the rheological properties of cement-based materials is not significant. At high particle volume fractions, the yield stress and plastic viscosity generally show an exponential increase with the increase of particle volume fraction, which can be explained by the increased collisions and frictions between solid particles at high aggregate volume fractions.

For a concrete mixture with a fixed content of cement paste, the higher the aggregate packing density, the lower the fraction of voids, and thus the higher the content of excess paste acting as a lubrication effect. Therefore, the rheological properties of concrete are strongly affected by the packing of aggregate particles. Generally, the particle size of coarse aggregate plays a more significant role in the yield stress compared to the plastic viscosity, and the yield stress and plastic viscosity of concrete mixtures with graded coarse aggregates are significantly lower than that with single-sized coarse aggregates.

With regard to particle morphology, aspect ratio and convexity can be used as independent morphological parameters to describe the shape of aggregate particles. Spherical particles generally result in reduced yield stress and plastic viscosity of cement-based materials due to the low uncompacted void content and low friction between particles. The effect of particle shape on the plastic viscosity and consistency index is higher than that on the relative yield stress, and the divergence packing fraction gradually decreases when the particle tends to be more irregular.

REFERENCES

Aïssoun, B. M., et al. (2015). "Influence of aggregate characteristics on workability of superworkable concrete." *Materials and Structures*, 49(1–2), 597–609.

Ait Mohamed Amer, A., et al. (2016). "Rheological and mechanical behavior of concrete made with pre-saturated and dried recycled concrete aggregates." *Construction and Building Materials*, 123, 300–308.

Barnes, H. A., et al. (1989). *An introduction to rheology*, Elsevier, Amsterdam, the Netherlands.

Bogner, A., et al. (2020). "Early hydration and microstructure formation of Portland cement paste studied by oscillation rheology, isothermal calorimetry, 1H NMR relaxometry, conductance and SAXS." *Cement and Concrete Research*, 130, 105977.

Bouziani, T., et al. (2012). "Effect of dune sand on the properties of flowing sand-concrete (FSC)." *International Journal of Concrete Structures and Materials*, 6(1), 59–64.

Carro-López, D., et al. (2015). "Study of the rheology of self-compacting concrete with fine recycled concrete aggregates." *Construction and Building Materials*, 96, 491–501.

Cepuritis, R., et al. (2016). "Crushed sand in concrete – Effect of particle shape in different fractions and filler properties on rheology." *Cement and Concrete Composites*, 71, 26–41.

Chateau, X., et al. (2008). "Homogenization approach to the behavior of suspensions of noncolloidal particles in yield stress fluids." *Journal of Rheology*, 52(2), 489–506.

Choi, M. S., et al. (2013). "Prediction on pipe flow of pumped concrete based on shear-induced particle migration." *Cement and Concrete Research*, 52, 216–224.

Cordeiro, G. C., et al. (2016). "Rheological and mechanical properties of concrete containing crushed granite fine aggregate." *Construction and Building Materials*, 111, 766–773.

de Larrard, F., and Sedran, T. (2002). "Mixture-proportioning of high-performance concrete." *Cement and Concrete Research*, 32, 1699–1704.

De Schutter, G., et al. (2018). "Vision of 3D printing with concrete - Technical, economic and environmental potentials." *Cement and Concrete Research*, 112, 25–36.

Dils, J., et al. (2012). "Influence of mixing procedure and mixer type on fresh and hardened properties of concrete: A review." *Materials and Structures*, 45(11), 1673–1683.

Farris, R. J. (1968). "Prediction of the viscosity of multimodal suspensions from unimodal viscosity data." *Transactions of the Society of Rheology*, 12(2), 281–301.

Ferraris, C. F., and Brower, L. E. (2000). Comparison of Concrete Rheometers: International Test at LCPC (Nantes France) in October, 2000.

Ferraris, C. F., and Brower, L. E. (2003). Comparison of Concrete Rheometers: International tests at MB (Cleceland OH, USA) in May, 2003.

Feys, D., et al. (2008). "Fresh self compacting concrete, a shear thickening material." *Cement and Concrete Research*, 38(7), 920–929.

Feys, D., et al. (2009). "Why is fresh self-compacting concrete shear thickening?" *Cement and Concrete Research*, 39(6), 510–523.

França, M. S. D., et al. (2013). "Influence of laboratory mixing procedure on the properties of mortars." *Ambiente Construído*, 13(2), 111–124.

França, M. S. D., et al. (2016). "Influence of the addition sequence of PVA-fibers and water on mixing and rheological behavior of mortars." *Revista IBRACON de Estruturas e Materiais*, 9(2), 226–243.

Fuller, W. B., and Thompson, S. E. (1907). "The laws of proportioning concrete." *Transactions of the American Society of Civil Engineers*, 59(2), pp. 67–143.

Funk, J. E., and Dinger, D. R. (2013). *Predictive process control of crowded particulate suspensions: Applied to ceramic manufacturing*. Springer Science & Business Media.

Geiker, M. R., et al. (2002a). "The effect of measuring procedure on the apparent rheological properties of self-compacting concrete." *Cement and Concrete Research*, 32, 1791–1795.

Geiker, M. R., et al. (2002b). "On the effect of coarse aggregate fraction and shape on the rheological properties of self-compacting concrete." *Cement, Concrete and Aggregates*, 24(1), 3–6.

Güneyisi, E., et al. (2016). "Rheological and fresh properties of self-compacting concretes containing coarse and fine recycled concrete aggregates." *Construction and Building Materials*, 113, 622–630.

Hafid, H., et al. (2016). "Effect of particle morphological parameters on sand grains packing properties and rheology of model mortars." *Cement and Concrete Research*, 80, 44–51.

Han, D., et al. (2017). "Critical grain size of fine aggregates in the view of the rheology of mortar." *International Journal of Concrete Structures and Materials*, 11(4), 627–635.

Han, D., and Ferron, R. D. (2016). "Influence of high mixing intensity on rheology, hydration, and microstructure of fresh state cement paste." *Cement and Concrete Research*, 84, 95–106.

Hočevar, A., et al. (2013). "Rheological parameters of fresh concrete – Comparison of rheometers." *Građevinar*, 65(02), 99–109.

Hu, C., and de Larrard, F. (1996). "The rheology of fresh high-performance concrete." *Cement and Concrete Research*, 26(2), 283–294.

Hu, J. (2005). "A study of effects of aggregate on concrete rheology", Doctoral Thesis, Iowa State University.

Hu, J., et al. (2013). "Feasibility study of using fine recycled concrete aggregate in producing self-consolidation concrete." *Journal of Sustainable Cement-Based Materials*, 2(1), 20–34.

Hu, J., and Wang, K. (2011). "Effect of coarse aggregate characteristics on concrete rheology." *Construction and Building Materials*, 25(3), 1196–1204.

Hu, J., and Wang, K. J. (2007). "Effects of size and uncompacted voids of aggregate on mortar flow ability." *Journal of Advanced Concrete Technology*, 5(1), 75–85.

Iqbal Khan, M., et al. (2016). "Utilization of supplementary cementitious materials in HPC: From rheology to pore structure." *KSCE Journal of Civil Engineering*, 21(3), 889–899.

Jiao, D., et al. (2017a). "Effects of paste thickness on coated aggregates on rheological properties of concrete." *Journal of the Chinese Ceramic Society*, 45(9), 1360–1366.

Jiao, D., et al. (2017b). "Effect of constituents on rheological properties of fresh concrete-A review." *Cement and Concrete Composites*, 83, 146–159.

Jiao, D., et al. (2018). "Mixture design of concrete using simplex centroid design method." *Cement and Concrete Composites*, 89, 76–88.

Jiao, D., et al. (2019a). "Structural build-up of cementitious paste with nano-Fe_3O_4 under time-varying magnetic fields." *Cement and Concrete Research*, 124, 105857.

Jiao, D., et al. (2019b). "Effects of rotational shearing on rheological behavior of fresh mortar with short glass fiber." *Construction and Building Materials*, 203, 314–321.

Kabagire, K. D., et al. (2017). "Experimental assessment of the effect of particle characteristics on rheological properties of model mortar." *Construction and Building Materials*, 151, 615–624.

Kabagire, K. D., et al. (2019). "Toward the prediction of rheological properties of self-consolidating concrete as diphasic material." *Construction and Building Materials*, 195, 600–612.

Krieger, I. M., and Dougherty, T. J. (1959). "A mechanism for non-Newtonian flow in suspensions of rigid spheres." *Transactions of the Society of Rheology*, 3(1), 137–152.

Kumar, S., and Santhanam, M. (2003). "Particle packing theories and their application in concrete mixture proportioning: A review." *Indian Concrete Journal*, 77(9), 1324–1331.

Kwon, S. H., et al. (2013). "Prediction of concrete pumping: Part II—Analytical prediction and experimental verification." *ACI Materials Journal*, 110(6), 657–667.

Lee, J. H., et al. (2018). "Prediction of the yield stress of concrete considering the thickness of excess paste layer." *Construction and Building Materials*, 173, 411–418.

Li, T., and Liu, J. (2021). "Effect of aggregate size on the yield stress of mortar." *Construction and Building Materials*, 305, 124739.

Li, Z., et al. (2004). "Theoretical analysis of time-dependence and thixotropy of fluidity for high fluidity concrete." *Journal of Materials in Civil Engineering*, 16(3), 247–256.

Lu, G., et al. (2008). "Modeling rheological behavior of highly flowable mortar using concepts of particle and fluid mechanics." *Cement and Concrete Composites*, 30(1), 1–12.

Mahaut, F., et al. (2008a). "Yield stress and elastic modulus of suspensions of noncolloidal particles in yield stress fluids." *Journal of Rheology*, 52(1), 287–313.

Mahaut, F., et al. (2008b). "Effect of coarse particle volume fraction on the yield stress and thixotropy of cementitious materials." *Cement and Concrete Research*, 38(11), 1276–1285.

Mooney, M. (1951). "The viscosity of a concentrated suspension of spherical particles." *Journal of Colloid Science*, 6(5), 162–170.

Mueller, F. V., et al. (2014). "Linking solid particle packing of Eco-SCC to material performance." *Cement and Concrete Composites*, 54, 117–125.

Nehdi, M., and Al Martini, S. (2009). "Coupled effects of high temperature, prolonged mixing time, and chemical admixtures on rheology of fresh concrete." *ACI Materials Journal*, 106(3), 231–240.

Nielsen, L. F. (2001). "Rheology of some fluid extreme composites: Such as fresh self-compacting concrete." *Nordic Concrete Research*, 27, 83–94.

Noor, M. A., and Uomoto, T. (2004). "Rheology of high flowing mortar and concrete." *Materials and Structures*, 37, 512–521.

Oh, S., et al. (1999). "Toward mix design for rheology of self-compacting concrete." *Proceedings of Self-Compacting Concrete*, Stockholm, 13–14 September 1999, 361–372.

Okamura, H., and Ouchi, M. (2003). "Self-compacting concrete." *Journal of Advanced Concrete Technology*, 1(1), 5–15.

Onoda, G. Y., and Liniger, E. G. (1990). "Random loose packings of uniform spheres and the dilatancy onset." *Physical Review Letters*, 64(22), 2727–2730.

Ovarlez, G., et al. (2006). "Local determination of the constitutive law of a dense suspension of non-colloidal particles through MRI." *Journal of Rheology*, 50(3), 259–292.

Park, S., et al. (2018). "Rheological properties of concrete using dune sand." *Construction and Building Materials*, 172, 685–695.

Quemada, D. (1984). "Models for rheological behavior of concentrated disperse media under shear." *Advances in rheology*, 2, 571–582.

Reinhardt, H. W., and Wüstholz, T. (2006). "About the influence of the content and composition of the aggregates on the rheological behaviour of self-compacting concrete." *Materials and Structures*, 39(7), 683–693.

Ren, Q. (2021). "Rheology of cement mortar and concrete as influenced by the geometric features of manufactured sand", Ghent University.

Ren, Q., et al. (2021). "Plastic viscosity of cement mortar with manufactured sand as influenced by geometric features and particle size." *Cement and Concrete Composites*, 122, 104163.

Robinson, J. V. (1949). "The viscosity of suspensions of spheres." *The Journal of Physical Chemistry*, 53(7), 1042–1056.

Roscoe, R. (1952). "The viscosity of suspensions of rigid spheres." *British Journal of Applied Physics*, 3, 267.

Roussel, N. (2007). "Rheology of fresh concrete: From measurements to predictions of casting processes." *Materials and Structures*, 40(10), 1001–1012.

Roussel, N., et al. (2010). "Steady state flow of cement suspensions: A micromechanical state of the art." *Cement and Concrete Research*, 40(1), 77–84.

Roussel, N. (2011). *Understanding the rheology of concrete*. Elsevier.

Roussel, N. (2018). "Rheological requirements for printable concretes." *Cement and Concrete Research*, 112, 76–85.

Sant, G., et al. (2008). "Rheological properties of cement pastes: A discussion of structure formation and mechanical property development." *Cement and Concrete Research*, 38(11), 1286–1296.

Santos, A. C. P., et al. (2015). "Experimental study about the effects of granular skeleton distribution on the mechanical properties of self-compacting concrete (SCC)." *Construction and Building Materials*, 78, 40–49.

Struble, L., and Sun, G.-K. (1995). "Viscosity of Portland cement paste as a function of concentration." *Advanced Cement Based Materials*, 2(2), 62–69.

Struble, L. J., and Chen, C.-T. (2005). "Effect of continuous agitation on concrete rheology." *Journal of ASTM International*, 2(9), 1–19.

Talbot, A. N., et al. (1923). *The strength of concrete: Its relation to the cement aggregates and water*. University of Illinois.

Toutou, Z., and Roussel, N. (2006). "Multi scale experimental study of concrete rheology: From water scale to gravel scale." *Materials and Structures*, 39, 189–199.

Van Der Vurst, F., et al. (2014). "Influence of addition sequence of materials on rheological properties of self-compacting concrete." *The 23rd Nordic Concrete Research Symposium*, Reykjavik, Iceland, 399–402.

Wallevik, J. E., and Wallevik, O. H. (2017). "Analysis of shear rate inside a concrete truck mixer." *Cement and Concrete Research*, 95, 9–17.

Wallevik, O. H. (2003). "Rheology—A scientific approach to develop self-compacting concrete." *Proceedings of the 3rd International Symposium on Self-Compacting Concrete*, Reykjavik, Iceland, 23–31.

Wallevik, O. H., and Wallevik, J. E. (2011). "Rheology as a tool in concrete science: The use of rheographs and workability boxes." *Cement and Concrete Research*, 41(12), 1279–1288.

Wang, X., et al. (2020). "Effect of interparticle action on shear thickening behavior of cementitious composites: Modeling and experimental validation." *Journal of Sustainable Cement-Based Materials*, 9(2), 78–93.

Westerholm, M., et al. (2008). "Influence of fine aggregate characteristics on the rheological properties of mortars." *Cement and Concrete Composites*, 30(4), 274–282.

Yahia, A., et al. (2017). "Measuring rheological properties of cement pastes: Most common techniques, procedures and challenges." *RILEM Technical Letters*, 2, 129–135.

Yammine, J., et al. (2008). "From ordinary rheology concrete to self compacting concrete: A transition between frictional and hydrodynamic interactions." *Cement and Concrete Research*, 38(7), 890–896.

Yuan, Q., et al. (2019). "A feasible method for measuring the buildability of fresh 3D printing mortar." *Construction and Building Materials*, 227, 116600.

Chapter 4

Empirical techniques evaluating concrete rheology

4.1 INTRODUCTION

Fresh cement-based materials behave as fluids with a yield stress, which is the minimum shear stress for flow to occur (Tattersall and Banfill, 1983, Tattersall, 1991). The fundamental rheological parameters, i.e., yield stress and plastic viscosity, are usually measured using rheological tools such as viscometer and rheometer (Banfill, 2006, Barnes et al., 1989). In situ, the application of rheological apparatus is usually more expensive, difficult, and time-consuming (Wallevik and Wallevik, 2011). For example, the height of ConTec Viscometer 5, a rheometer that can measure the rheological properties of concrete with slump higher than 120 mm and a maximum aggregate particle size of 22 mm (Wallevik, 2003), is around 2 m, and thus it cannot be used for on-site quality control. For the portable ICAR rheometer, its price is still higher than $20000. In this context, simpler and cheaper conventional workability tests such as the slump test and V-funnel test are more preferred. Although they cannot give the absolute values similar to the rheological parameters, the single-point tests can still be used as effective tools to classify different materials in terms of their ability to be cast (Bouziani and Benmounah, 2013, Gram et al., 2014, Roussel, 2006).

For fresh concrete, the yield stress can be correlated to the slump or slump flow, while the plastic viscosity can be predicted by the flow time (Jiao et al., 2017, Lu et al., 2015, Wallevik, 2006). This provides theoretical foundations for evaluating rheological properties by using traditional workability tests. In this chapter, the empirical techniques for evaluating the rheological properties of cement-based materials, such as slump or slump flow, flow time (T50 or V-funnel), and L-box, are presented. The geometry, testing procedure, and data interpretation of the conventional workability tests are illustrated. The mathematical relationships between empirical parameters and rheological properties are also demonstrated. It should be mentioned that the empirical techniques for assessing the rheological properties of cement paste and mortar are beyond the scope of this chapter, with the main emphasis on concrete.

4.2 SLUMP: ASTM ABRAMS CONE

The concrete slump test is being used widely by jobsite engineers since a long time ago. Although the construction industry has faced many changes with time, this test is still performed on its old procedure and no changes have been made to this method because of its simple and easily adoptable procedure. It can be easily applied on job site as well as in the laboratory. The basic principle of concrete slump value is a gravity-induced flow of the concrete surface. The slump test is conducted based on the ASTM standard (ASTM-C143/C143M-12, 2010) or BS EN standard (12350-2, 2009), similar to the Chinese standard (GB/

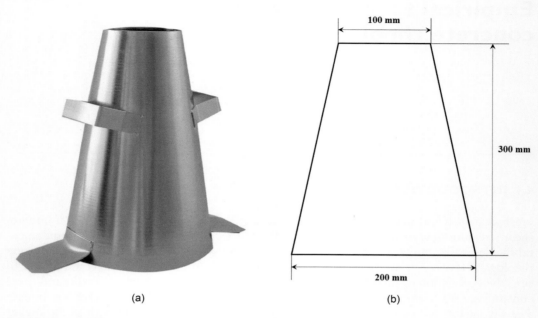

Figure 4.1 ASTM Abrams cone (a) appearance and (b) geometries.

T50080-2002, 2002). This test method is applicable to concrete with a maximum aggregate size of 37.5 mm. The slump of the tested concrete should be within the range of 15–230 mm.

4.2.1 Geometry

The apparatus needed for the slump test includes a slump cone, a tamping rod, a scoop, and a measuring device. The appearance of the slump cone and the corresponding geometries are shown in Figure 4.1a and b, respectively. The thickness of the metal cone should be more than 1.5 mm. The base and the top shall be open and parallel to each other and at right angles to the axis of the cone. The diameter of the base and top of the cone is 200 and 100 mm, respectively, and the height is 300 mm. All diameters must be within 3 mm of these diameters. The inside of the cone needs to be smooth and clean, and the cone must be free of dents, adhered mortar, and deformations. For more detailed information about the cone, refer to ASTM-C143/C143M-12 (2010).

The tamping rod is a round and straight steel rod with a diameter of 16 mm and a length of approximately 600 mm. A hemispherical tip with a diameter of 16 mm should be attached at the tamping end or both ends. The scoop must be large enough to get a representative sample of concrete and small enough so that it is not spilled during placement into the cone. The measuring device shall have markings in increments of 5 mm or smaller, and the length shall be at least 300 mm.

4.2.2 Testing procedure and parameters

The sample of the concrete for the slump test shall be representative of the entire batch, and the selection of the sample is in accordance with ASTM-C172. The testing procedure, as shown in Figure 4.2, shall be as follows:

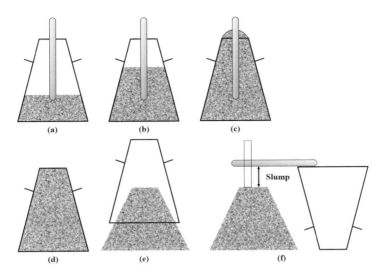

Figure 4.2 (a–f) Main procedures of slump test.

1. Moisten the slump cone to keep concrete from sticking to it. Place the slump cone on a flat, moist, and rigid surface. During concrete filling, the slump cone shall be held firmly by the operator standing on the two-foot pieces. Do not step off until the cone is full and ready to be lifted.
2. Fill the first layer of the concrete up to 70 mm of the cone, 1/3 of the cone by volume. Make sure the concrete is even inside the cone.
3. Rod the layer 25 times throughout its depth by using the tamping rod. Uniformly distribute the strokes over the cross section. Make sure to cover all the surface area inside the cone, slightly angling the rod to get the edges. Do not tap the side of the cone.
4. Fill the second layer to 160 mm, 2/3 of the cone by volume.
5. Rod the layer 25 times, making sure to penetrate the first layer by 25 mm. Again, do not tap the side of the cone.
6. Fill the last layer up to the top, where the concrete is slightly overflowing. Rod the last layer 25 times, and keep the top layer always full.
7. Strike off the excess concrete while keeping the cone steady. Clean around the rim of the cone, making sure it is full.
8. Lift the cone from the concrete carefully in a vertical direction. The lifting process should be within 3–7 s, without lateral or torsional motion.
9. Measure the slump by determining the vertical difference between the top of the mold and the displaced original center of the top surface of the specimen. The slump value is recorded in terms of mm to the nearest 5 mm of subsidence of the specimen during the test. The entire slump procedure needs to be finished within 2.5 min.
10. Clean the equipment thoroughly, and discard the used concrete.

4.2.3 Data interpretation

Slumps can be categorized in basic four types with respect to their collapsed shape, i.e., zero slump, true slump, collapsed slump, and shear slump, as shown in Figure 4.3. In zero slump, the fresh concrete does not change its shape during testing. It indicates that a small

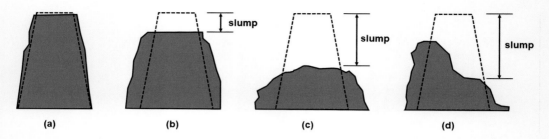

Figure 4.3 Types of the slump: (a) zero slump, (b) true slump, (c) collapsed slump, and (d) shear slump.

Table 4.1 Classification of workability based on the concrete slump

No.	Concrete slump (mm)	Degree of workability
1	0–25	Very low workability
2	25–50	Low workability
3	50–100	Medium workability
4	100–175	High workability

Table 4.2 Relation between consistency and slump values

Slump (mm)	0–20	20–40	40–120	120–200	200–220
Consistency	Dry	Stiff	Plastic	Wet	Sloppy

amount of water is used in the mixture design. The true slump is the only type which is allowed under the above-mentioned procedure. The collapse slump where concrete collapses completely indicates that a high amount of water is used to prepare the concrete or the concrete is too wet. Zero slumps and collapse slumps cannot be evaluated by normal tests for slumps because these two do not fall into the workability range, i.e., 15–230 mm. In the case of the shear slump, the top portion of the concrete slump slips away from the side, and ASTM-C143/C143M-12 (2010) advises to repeat the tests with different samples. If the shear slump occurs again, then the concrete mix shall be avoided as it is not sufficiently cohesive. The classification of workability based on concrete slumps is summarized in Table 4.1, and the relation between consistency and slump values is shown in Table 4.2.

The slump value can be used to predict the yield stress of fresh concrete, while there is no specific correlation between slump and plastic viscosity (Wallevik, 2006, Zerbino et al., 2008). In the following paragraphs, some representative relationships between slump and yield stress of fresh concrete are presented.

By establishing the relationship between yield stress and slump of wet-consistency concrete with a maximum aggregate of 20 mm and slumps from 125 to 260 mm, Murata and Kikukawa (1992) proposed an empirical equation to estimate the yield stress:

$$\tau_0 = 714 - 473\log(S_L/10) \tag{4.1}$$

where τ_0 is the yield stress of concrete in Pa, measured by a coaxial cylinder viscometer, S_L is the concrete slump in mm, and A and B are experimental constants. From the theoretical

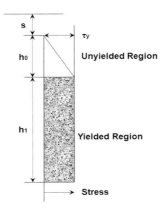

Figure 4.4 Representative stress of deformed cone after slump. (Adapted from Schowalter and Christensen, 1998.)

viewpoint, concrete in the cone can be divided into two parts, as shown in Figure 4.4. The shear stress induced by the self-weight of the material is higher than the yield stress of the material, and thus flow occurs in the lower part. However, there is no flow in the upper part. The height of the lower part gradually decreases until the shear stress in this zone reaches the yield stress, where the flow stops. Based on this assumption, Schowalter and Christensen (1998) established a relation between the final total height of the cone and the yield stress, as described by Eq. (4.2), which is independent of the slump cone geometry:

$$S_L = 1 - h_0 - 2\tau_0 \ln\left(\frac{7}{(1+h_0)^3 - 1}\right) \tag{4.2}$$

where h_0 is the height of the deformed cone. This relation had been successfully validated by Clayton et al. (2003) and Saak et al. (2004) using cylindrical molds. For the ASTM Abrams cone, Hu et al. (1996) established a relationship between the slump and the yield stress by considering the density of the concrete based on a number of numerical simulations, as described by:

$$S_L = 300 - 347\frac{(\tau_0 - 212)}{\rho} \tag{4.3}$$

where S_L is the slump in mm, τ_0 is the yield stress in Pa, and ρ is the density of the fresh concrete in kg/m³. Besides, through numerical simulating, Roussel (2006) proposed a simple linear relationship between slump and yield stress for traditional concrete with slump values varying from 50 to 250 mm:

$$S_L = 25.5 - 17.6\frac{\tau_0}{\rho} \tag{4.4}$$

where S_L is the slump in cm, τ_0 is the yield stress in Pa, and ρ is the density of the fresh concrete in kg/m³. There is a good agreement between the BTRHEOM measured yield stress and the theoretical predictions by Eq. (4.4) (Roussel, 2006).

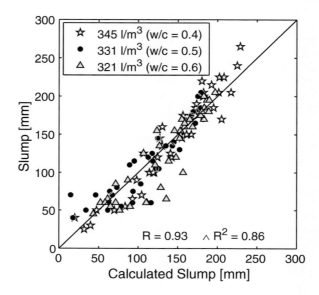

Figure 4.5 Typical relationship between a calculated slump and a measured slump based on Eq. (4.5), where α is 0.0077 mm/(Pa.l), V_m^{ref} is 345 l/m³, and τ_0^{ref} is 200 Pa. (Adapted from Wallevik, 2006.)

During the slump flow, the suspended particles must bypass one another. At the stop-flow of the slump, the suspended particles can no longer bypass one another. For a given yield stress, it is found that the stop condition occurs sooner at a relatively low volume fraction of the matrix, due to the reduced distance between particles (Wallevik, 2006). Besides, the relation between slump and yield stress becomes less dependent on the matrix volume fraction when the concrete becomes more workable because of the reduced influence of granular properties by the increased matrix lubrication. After considering the matrix lubrication effect and the granular properties, Wallevik (2006) proposed an empirical relationship between slump and yield stress of concrete, as characterized by:

$$S_L = 300 - 416 \frac{(\tau_0 + 394)}{\rho} + \alpha \left(\tau_0 - \tau_0^{ref}\right)\left(V_m - V_m^{ref}\right) \qquad (4.5)$$

where S_L is the slump in mm, τ_0 is the yield stress in Pa, ρ is the density of the fresh concrete in kg/m³, α is a constant, $\left(\tau_0 - \tau_0^{ref}\right)$ is the lubrication effect, and $\left(V_m - V_m^{ref}\right)$ is the effect of matric volume fraction. A typical relationship between the calculated slump and the measured slump, with α of 0.0077 mm/(Pa.l), V_m^{ref} of 345 l/m³ and τ_0^{ref} of 200 Pa, is shown in Figure 4.5. It can be concluded that Eq. (4.5) provides an excellent prediction of slump for a given yield stress.

4.3 SLUMP FLOW AND T_{50}

The slump flow test is generally used to evaluate the workability of high flowable concrete (HFC) and self-compacting concrete (SCC), where the standard slump test is not applicable

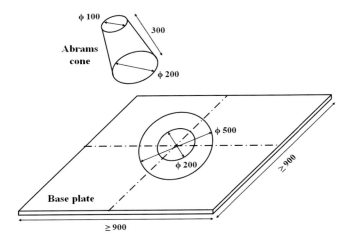

Figure 4.6 Baseplate and Abrams cone for slump flow test in mm.

anymore, since this kind of concrete cannot maintain its shape when the cone is removed. The slump flow test evaluates the horizontal free flow of the concrete in the absence of obstructions. It is a measure of the filling ability of the concrete, and it can be conducted in the laboratory as well as on-site easily.

4.3.1 Geometry and testing procedure

The apparatus for the slump flow test is similar to that of the slump test. The only difference is that a steel base plate marked with circles of 200 and 500 mm at the center is highly recommended, as shown in Figure 4.6. The testing procedure can be found in ASTM-C1611/C1611M-21 (2009) and EN-BS (2010), according to the following details:

1. Prepare at least 6 L of concrete.
2. Moisten the steel plate and the inside of the slump cone. Put the slump cone on the center of the steel plate, and hold it down firmly.
3. Fill the cone with concrete, without layering, and without any compaction. This process shall be finished within 90 s.
4. Strike off the concrete at the top of the cone with the scraper. Remove the surplus concrete around the base of the cone.
5. Raise the cone vertically within 3 s by a steady upward lift without interruption, and allow the concrete to flow out freely.
6. Measure the time from the beginning of lifting the slump cone to the moment when the maximum diameter reaches 500 mm by using a stopwatch to the nearest 0.1 s, i.e., T_{50}.
7. When the concrete flow stops, measure the largest diameter and the corresponding perpendicular diameter to the nearest 5 mm. The duration from lifting the slump cone to the end of measuring the diameters should be within 40 s.
8. Take the average value of the two diameters as the slump flow to the nearest 5 mm.
9. If the two diameters deviate more than 50 mm, an additional test shall be conducted by using a different sample from the same batch.
10. Clean the steel plate and the slump cone after testing.

Table 4.3 SCC classes for different applications (originated from Concrete (2005)) summarized by Zerbino et al. (2008)

Class	Range (mm)	Conformity criteria (mm)	Applications
SF1	550–650	520–700	Housing slabs, tunnel linings, piles and deep foundations
SF2	660–750	640–800	Waals, columns
SF3	760–850	740–900	Very congested structures

4.3.2 Data interpretation

The slump flow S_F is recorded as the average of the two diameters to the nearest 5 mm, and T_{50} is the time when the slump flow reaches 500 mm. Note that this flow time can also be designated as T_{500}, and it is lack of precision in the determination due to human eye estimation (Zerbino et al., 2008). The status of the concrete (i.e., segregation or bleeding or not) after flowing should be checked. If most coarse aggregates remain in the center and mortar/cement pastes gather at the concrete periphery, the concrete can be regarded as severe segregation. For the minor segregation, a border of mortar without coarse aggregate can occur at the edge of the pool of concrete. It should be mentioned that none of the above-mentioned phenomena does not necessarily mean that segregation will not occur due to the time-dependent evolution of workability. Based on the range of slump spread flow, SCC can be divided into several classes for different applications, as summarized in Table 4.3.

Similar to the slump, the slump spread flow is correlated to the yield stress of the concrete, while the flow time T_{50} has an apparent relationship with the plastic viscosity (Bouziani and Benmounah, 2013, Cu et al., 2020). Taking the density of concrete into account, Sedran and De Larrard (1999) connected the yield stress to the slump flow from the Abrams cone by using the following empirical equation:

$$\tau_0 = (808 - S_F)\frac{\rho g}{11,740} \tag{4.6}$$

where S_F is the slump flow in mm, g is the gravity acceleration (9.8 m/s²), and ρ is the density of concrete in kg/m³. A typical relationship between the flow time T_{50} and the plastic viscosity is expressed as follows:

$$\mu = \frac{\rho g}{10,000}(0.026 S_F - 2.39) T_{50} \tag{4.7}$$

where T_{50} is the flow time in s. It should be mentioned that the constants in Eqs. (4.6) and (4.7) depend on the concrete components or types, and the used rheometer. Similar empirical equations have been obtained by Zerbino et al. (2008), as shown in Figure 4.7, which are independent of the concrete temperature, the mixing energy, the environmental conditions, or the resting time.

4.4 V-FUNNEL TEST FLOW TIME

V-funnel test flow time is used to evaluate the filling ability of fresh SCC with a maximum coarse aggregate size of 20 mm through a narrow opening. This test provides a qualitative assessment of the relative viscosity of SCC and indicates the blocking effect caused by

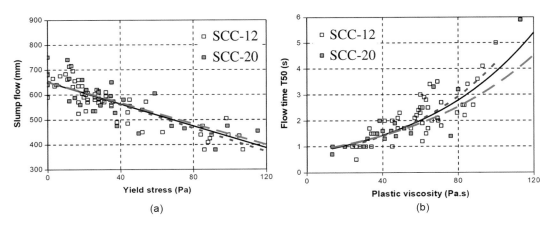

Figure 4.7 Typical relationships between rheological properties and slump flow test: (a) yield stress versus slump spread flow and (b) plastic viscosity versus T_{50}. (Adapted from Zerbino et al., 2008.)

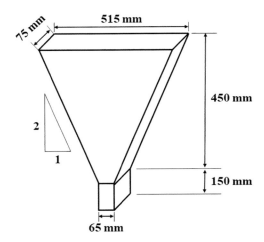

Figure 4.8 Geometric parameters of V-funnel.

segregation. A shorter flow time suggests a lower viscosity, whereas a prolonged flow time indicates the blocking susceptibility of the mix.

4.4.1 Geometry

The equipment for the V-funnel flow test includes a V-funnel, a stopwatch with an accuracy of 0.1 s, a straightedge, a bucket with a capacity of 12–14 L, and a moist towel. The V-funnel is made of steel, with geometric parameters shown in Figure 4.8. The top of the V-funnel is flat and horizontal, and a momentary releasable and watertight opening gate is installed at the bottom. The V-funnel is placed on vertical support. It should be mentioned that an O-shaped funnel, called O-funnel, can also be used as alternative equipment for the funnel flow time test.

4.4.2 Testing procedure

The testing procedure of the V-funnel test is referred to EFNARC (Concrete, 2005), according to the following details:

1. Place the V-funnel on the ground vertically. Make sure the opening top is horizontally positioned.
2. Wet the inner side of the V-funnel using the moist towel, and remove the surplus of water.
3. Close the gate, and place the bucket under the V-funnel to collect the concrete.
4. Fill the V-funnel with a representative SCC completely, without any compaction or vibration.
5. Remove any surplus of the concrete at the top of the funnel using the straightedge.
6. After a resting period of around 10 s, open the gate to let the concrete flow freely. Start to record the stopwatch when the gate opens.
7. Stop the time at the moment when clear space is visible through the opening of the funnel. The reading of the stopwatch is the V-funnel flow time, denoted as T_V. Do not move the V-funnel before it is empty.
8. Clean the V-funnel after the test.

4.4.3 Data interpretation

The V-funnel flow time T_V is the period from opening the gate until the first light enters the gate. It is recorded to the nearest 0.1 s. The V-funnel flow time for SCC should be less than 10 s. To measure the segregation resistance of concrete, the V-funnel is refilled with concrete and allowed to sit for 5 min, and then perform the V-funnel flow test. The segregation resistance is quantified by the flow-through index S_f, as calculated by:

$$S_f = \frac{T_5 - T_0}{T_0} \tag{4.8}$$

where T_0 is the initial flow-through time (s) and T_5 is the flow-through time after resting for 5 min.

There is an acceptable linear relationship between slump flow time T_{50} and V-funnel flow time T_V (Afshoon and Sharifi, 2014, Mohammed et al., 2021, Ouldkhaoua et al., 2019). Compared with T_{50}, the V-funnel flow time is affected by the plastic viscosity, the maximum aggregate size, and the geometric parameter of the bottom opening of the V-funnel (Zerbino et al., 2008). Higher plastic viscosity could possibly be estimated for SCC with coarser aggregates. Nevertheless, the V-funnel flow time can also to a certain degree be correlated to the plastic viscosity of SCC. A representative empirical relationship established by Zerbino et al. (2008) is shown in Figure 4.9, which follows the equation:

$$\mu = \frac{1}{0.013} \ln\left(\frac{T_V}{3.04}\right) \tag{4.9}$$

where μ is the plastic viscosity (Pa.s) and T_V is the V-funnel flow time (s). Besides, a linear relationship between plastic viscosity and V-funnel flow time had also been established (Boukhelkhal et al., 2016, Ouldkhaoua et al., 2019).

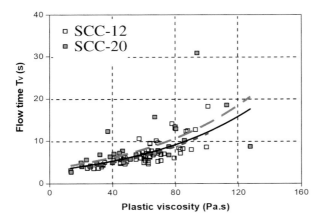

Figure 4.9 Typical relationship between plastic viscosity and V-funnel flow time. (Adapted from Zerbino et al., 2008.)

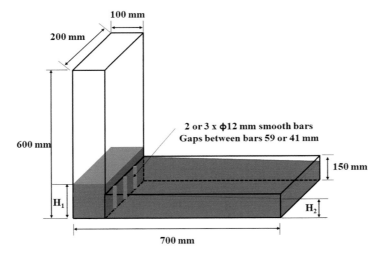

Figure 4.10 Geometric parameters of standard L-box.

4.5 OTHER METHODS

In addition to the above-mentioned tests, there are some other available empirical workability tests to evaluate the rheological properties of fresh concrete. The apparatus, testing procedure, and data interpretation of some typical empirical methods, including L-box, LCPC box, V-funnel coupled with a horizontal channel, and J-ring, are illustrated.

4.5.1 L-box

L-box is the reference method for evaluating the passing ability of SCC and its passing profile along with reinforcement. The geometric parameters of the L-box are shown in Figure 4.10. Two types of gates, one with 3 smooth bars and the other with 2 smooth bars, can be used. The gaps between neighboring bars are 41 and 59 mm, respectively.

The testing procedure of the L-box test is as follows (Nguyen et al., 2006, Skarendahl et al., 2000):

1. Wet the internal surface of the L-box, and then place it in a stable and level position.
2. Fill the vertical part of the L-box with 12.7 L representative fresh SCC.
3. Wait for 1 min to check whether the concrete is stable or not.
4. Lift the gate slowly, and let the concrete flow into the horizontal part freely.
5. When the concrete stops moving, measure the average distance of the top of the concrete in the vertical part (h_1) as well as the one at the end of the box (h_2). Three positions with one at the center and two at each side shall be recorded.
6. The L-box value is calculated as h_2/h_1.

If the L-box value is equal to 1, the concrete is perfectly flowable. Conversely, if the concrete is too stiff, then the L-box value is equal to 0. For SCC, the L-box value varies from one country to another in the range of 0.60–1. The L-box value can be linked to the rheological properties of SCC. Given that the fluid is homogeneous and the gate is lifted slowly, the relationship between the L-box measured parameters and the yield stress/density ratio of the material can be expressed as follows (Nguyen et al., 2006):

$$h_1 - h_2 = \frac{\tau_0 L_0}{\rho g}\left(\frac{l_0 L_0}{V} + \frac{2}{l_0}\right) + B\frac{\tau_0}{\rho g} \qquad (4.10)$$

where L_0 and l_0 are the length and width of the L-box channel, respectively; V is the total volume of the sample; τ_0 is the yield stress; g is the gravity acceleration; ρ is the density of concrete; and B is a dimensionless constant depending on the geometry of the L-box. Note that Eq. (4.10) works well for limestone powder suspensions, but it cannot be validated for concrete due to the lack of absolute value of the yield stress. Nevertheless, this correlation has a potential practical application in predicting the yield stress of SCC. For example, a plot of L-box value and yield stress/density ratio, as shown in Figure 4.11, is a practical tool for evaluating the yield stress of SCC in on-site works (Chamani et al., 2014). Alternatively, the relationship can be expressed by:

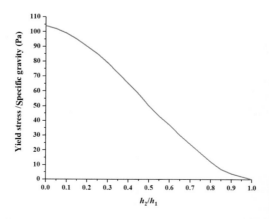

Figure 4.11 Relationship between L-box value and yield stress/density ratio. (Adapted from Chamani et al., 2014.)

Empirical techniques evaluating concrete rheology 87

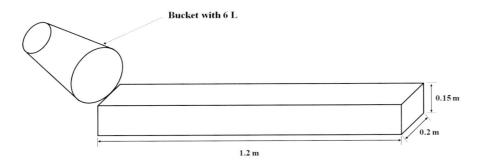

Figure 4.12 LCPC box for measuring the yield stress of SCC.

$$\frac{\tau_0}{SG} = 340.5 \begin{cases} 91.3\left(\frac{h_2}{h_1}\right)^3 - 175.3\left(\frac{h_2}{h_1}\right)^3 - 39.6\left(\frac{h_2}{h_1}\right)^3 + 104.8, & 0 \le \frac{h_2}{h_1} \le 0.5 \\ 340.5\left(\frac{h_2}{h_1}\right)^3 - 609.8\left(\frac{h_2}{h_1}\right)^3 + 214.1\left(\frac{h_2}{h_1}\right)^3 + 55.2, & 0.5 < \frac{h_2}{h_1} \le 1 \end{cases} \quad (4.11)$$

where SG is the specific gravity of the sample.

4.5.2 LCPC box

Roussel (2007) proposed a cheap and simple technique, the LCPC box, to measure the yield stress of SCC. The LCPC box test for SCC is shown in Figure 4.12. The length and the width of the LCPC box are 1.2 and 0.2 m, respectively. The height of the LCPC box is 0.15 m. The testing procedure of the LCPC box test is as follows:

1. Fill a pre-wetted bucket with 6 L tested SCC.
2. Slowly pour the SCC at the end of the LCPC box (see Figure 4.12). The pouring process shall be within 30 s.
3. Wait until the concrete flow stops.
4. Measure the spread length L, the initial height H_i, and the final height H_f of the concrete in the LCPC box.
5. Calculate the yield stress based on the following equation:

$$\tau_0 = \frac{\rho g l_0}{2L}\left[(H_i - H_f) + \frac{l_0}{2}\ln\left(\frac{l_0 + 2H_f}{l_0 + 2H_i}\right)\right] \quad (4.12)$$

where ρ is the density of the concrete, l_0 is the width of the L-box, L is the length of the flow, g is the acceleration gravity, H_i and H_f are the initial and final height, respectively. A representative correlation between the spread length and the ratio of yield stress/specific gravity is shown in Figure 4.13. The LCPC box test is a cheap, simple, and precise measurement of the yield stress of any SCC. However, it is highly dependent on the operator, and the use of a bucket can possibly influence the concrete flow (Benaicha et al., 2015).

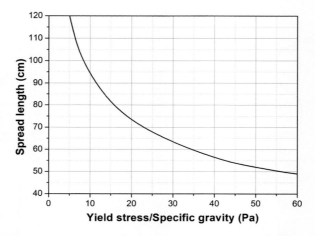

Figure 4.13 Representative correlation between the spread length from LCPC box and the ratio of yield stress/specific gravity. (Adapted from Roussel, 2007.)

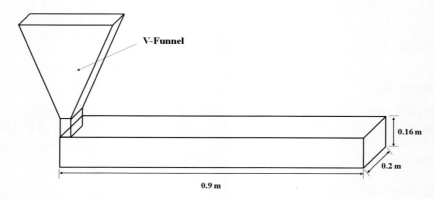

Figure 4.14 V-funnel coupled with a horizontal channel. (Adapted from Benaicha et al., 2015.)

4.5.3 V-funnel coupled with a horizontal channel

Based on the LCPC box test, Benaicha (2013) and Benaicha et al. (2015) modified the traditional V-funnel coupled with a horizontal channel to predict the plastic viscosity of concrete. The modified device is shown in Figure 4.14. The length of the channel is 0.90 m, the same length as the slump flow table. The channel has the same width as L-box, i.e., 0.20 m. The height of the channel is 0.16 m. The flow time and the flow profile of concrete in the channel can be determined at any time, which can be used to calculate the plastic viscosity of concrete, as described by:

$$\mu_p = \frac{\left(-\dfrac{8H\tau_0}{3}\left(\dfrac{2(z\tan\alpha+d)+e}{(z\tan\alpha+d)\cdot e}\right) + \rho g z\right)\dfrac{dt}{dz} + \rho\left(1-\left(\dfrac{z\tan\alpha+d}{d}\right)^2 - \xi\right)\dfrac{dt}{2dz}}{\dfrac{16H}{\pi}\left(\dfrac{[2(z\tan\alpha+d)+e]^4}{[(z\tan\alpha+d)\cdot e]^3}\right)} \qquad (4.13)$$

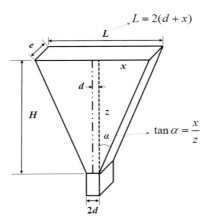

Figure 4.15 Definition of geometric parameters of the V-funnel. (Adapted from Benaicha et al., 2015.)

where d, H, e, and α are the geometric parameters of the V-funnel, as shown in Figure 4.15. dz/dt is the flow velocity of concrete, μ_p is the plastic viscosity, τ_0 is the yield stress, ρ is the concrete density, and g is the gravity. By using MATLAB® program with the Runge-Kutta method, the plastic viscosity of concrete can be obtained from the V-funnel flow time and length. A simple expression of Eq. (4.13) was described by Benaicha et al. (2017):

$$\mu_p = \frac{1}{3}(\rho - \rho_{\text{air}})g\left(\frac{Q}{e}\right)^3 \left(\frac{L}{0.95}t^{-4/5}\right)^{-5} \tag{4.14}$$

where ρ and ρ_{air} are the density of the concrete and air, respectively; Q is the volumetric flow rate; e is the channel width; L is the length of the flow; and t is the flow time. Experimental results with more than 100 different compositions showed that the calculated plastic viscosity had an excellent correlation with the measured ones using a rheometer, with a coefficient of determination of 0.9224 (Benaicha et al., 2015). The V-funnel coupled with a horizontal channel is an efficient, simple, and economical tool to characterize the plastic viscosity of concrete.

4.5.4 J-ring

J-ring test is an alternative method for evaluating the filling ability and passing ability of high flowability concrete, especially SCC. The apparatus of the J-ring test includes all the apparatuses for slump flow test and a J-ring with a rectangular section of 30 mm×50 mm and a diameter of 300 mm open steel ring drilled vertically with holes to accept threaded sections of reinforcing bars generally at a spacing of 48±2 mm, as shown in Figure 4.16.

The testing procedure of the J-ring test is referred to De Schutter (2005), according to the following steps:

1. Moisten the inside of the slump cone and base plate.
2. Place the J-ring centrally on the base plate and the slump cone centrally inside the J-ring.

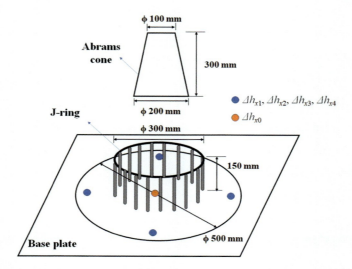

Figure 4.16 J-ring test apparatus. (Adapted from De Schutter, 2005.)

3. Fill the slump cone with about 6 L concrete by using a scoop without any external compacting action. Simply strike off the concrete level with a trowel, and remove all surplus concrete.
4. After a short rest (no more than 30 s for cleaning), raise the cone vertically and allow the concrete to flow out through the J-ring.
5. Record the time when the front of the concrete first touches the circle of a diameter of 500 mm, i.e., T_{50J}. The test is completed when the concrete flow is ceased.
6. Measure the largest diameter (d_{max}) and the one perpendicular to it (d_{perp}), to the nearest 5 mm. Calculate the average diameter.
7. Measure the difference in height between the concrete surface at the central position (Δh_0) and the four positions outside the J-ring, Δh_{x1}, Δh_{x2}, Δh_{y1}, and Δh_{y2}, as shown in Figure 4.16. Calculate the average of the difference in height at four locations in mm.
8. Clean the base plate, the J-ring, and the cone after testing.

Three parameters can be obtained from the J-ring flow test, including J-ring flow spread (S_J), J-ring flow time (T_{50J}), and J-ring blocking step (B_J), which can be expressed by:

$$S_J = \frac{d_{max} + d_{perp}}{2} \tag{4.15}$$

$$B_J = \frac{\Delta h_{x1} + \Delta h_{x2} + \Delta h_{y1} + \Delta h_{y2}}{4} - \Delta h_0 \tag{4.16}$$

S_J is expressed in mm to the nearest 5 mm, indicating the restricted deformability of SCC due to the blocking effect. T_{50J} is recorded in s to the nearest 0.1 s, which indicates the rate of deformation. B_J is expressed in mm to the nearest 1 mm, representative of the blocking effect. The acceptable difference in height between inside and outside should be between 0 and 10 mm.

Although no quantitative equation is established between the J-ring measured parameters and rheological properties, the J-ring flow time and J-ring spread flow can be empirically

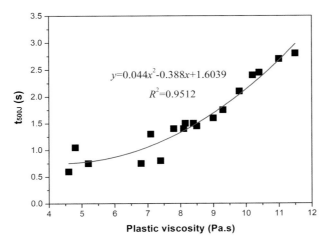

Figure 4.17 J-ring flow time T_{50J} versus plastic viscosity of SCC. (Adapted from Abo Dhaheer et al., 2015.)

correlated to the plastic viscosity and yield stress, respectively. A typical relationship between J-ring flow time T_{50J} and plastic viscosity of SCC is shown in Figure 4.17. It can be seen that a quadratic relationship between T_{50J} and plastic viscosity is observed. In the case of J-ring spread flow, it can be correlated to the spreading flow measured from the slump flow test, as shown in Figure 4.17. More details can be found in the works of Barroqueiro et al. (2019), Daoud and Kabashi (2015), and Pradoto et al. (2016).

4.6 SUMMARY

The fundamental rheological parameters including yield stress and plastic viscosity are usually measured using rheological tools. In situ, however, the application of rheological apparatus is usually more expensive, difficult, and time-consuming. Therefore, simpler and cheaper conventional workability tests such as the slump test and V-funnel tests can be used as effective tools to classify different materials in terms of their ability to be cast. The geometry, testing procedure, and data interpretation of the conventional workability tests (including slump, slump spread flow and flow time, V-funnel, L-box, and J-ring) are illustrated. The mathematical relationships between empirical parameters and rheological properties are also demonstrated. For fresh concrete, the yield stress can be correlated to the slump or slump flow, while the plastic viscosity can be predicted by the flow time (T_{50}, V-funnel flow time, or J-ring flow time).

REFERENCES

Abo Dhaheer, M. S., et al. (2015). "Proportioning of self-compacting concrete mixes based on target plastic viscosity and compressive strength: Part II - experimental validation." *Journal of Sustainable Cement-Based Materials*, 5(4), 217–232.

Afshoon, I., and Sharifi, Y. (2014). "Ground copper slag as a supplementary cementing material and its influence on the fresh properties of self-consolidating concrete." *The IES Journal Part A: Civil & Structural Engineering*, 7(4), 229–242.

ASTM-C143/C143M-12 (2010). "Standard test method for slump of hydraulic-cement concrete." Annual Book of ASTM Standards, American Society for Testing and Materials, Philadelphia, USA.

ASTM-C172. "Practice for sampling freshly mixed concrete." Annual Book of ASTM Standards, American Society for Testing and Materials, Philadelphia, USA.

ASTM-C1611/C1611M-21 (2009). "Standard test method for slump flow of self-consolidating concrete", Annual Book of ASTM Standards, American Society for Testing and Materials, Philadelphia, USA.

Banfill, P. (2006). "The rheology of fresh cement and concrete-rheology review." *British Society of Rheology*, 61, 130.

Barnes, H. A., et al. (1989). *An introduction to rheology*. Elsevier, Amsterdam, the Netherlands.

Barroqueiro, T., et al. (2019). "Fresh-state and mechanical properties of high-performance self-compacting concrete with recycled aggregates from the precast industry." *Materials (Basel)*, 12(21), 3565.

Benaicha, M. (2013). *Rheological and mechanical characterization of concrete: New approach*. LAP Lambert Academic Publishing, Chisinau, Moldova.

Benaicha, M., et al. (2015). "New approach to determine the plastic viscosity of self-compacting concrete." *Frontiers of Structural and Civil Engineering*, 10(2), 198–208.

Benaicha, M., et al. (2017). "Theoretical calculation of self-compacting concrete plastic viscosity." *Structural Concrete*, 18(5), 710–719.

Boukhelkhal, A., et al. (2016). "Effects of marble powder as a partial replacement of cement on some engineering properties of self-compacting concrete." *Journal of Adhesion Science and Technology*, 30(22), 2405–2419.

Bouziani, T., and Benmounah, A. (2013). "Correlation between v-funnel and mini-slump test results with viscosity." *KSCE Journal of Civil Engineering*, 17(1), 173–178.

BS EN standard 12350-2. (2009). "Testing fresh concrete: Slump-test." *Part 2: British Standards*.

Chamani, M. R., et al. (2014). "Evaluation of SCC yield stress from L-box test using the dam break model." *Magazine of Concrete Research*, 66(4), 175–185.

Clayton, S., et al. (2003). "Analysis of the slump test for on-site yield stress measurement of mineral suspensions." *International Journal of Mineral Processing*, 70(1–4), 3–21.

Cu, Y. T. H., et al. (2020). "Relationship between workability and rheological parameters of self-compacting concrete used for vertical pump up to supertall buildings." *Journal of Building Engineering*, 32, 101786.

Daoud, O., and Kabashi, T. (2015). "Production and fresh properties of powder type self—compacting concrete in Sudan", in *Concrete repair, rehabilitation and retrofitting IV.*, Edited by Dehn, Beushausen, Alexander and Moyo. CRC Press, Leipzig, Germany, pp. 99–99.

De Schutter, G. (2005). "Guidelines for testing fresh self-compacting concrete." *European Research Project*.

Concrete, S.C., (2005). "The European guidelines for self-compacting concrete." BIBM, et al., pp. 50–52.

EN-BS (2010). 12350-8 (2010). *Testing fresh concrete self-compacting concrete. Slump-flow test*. British Standards Institute, London, United Kingdom.

Concrete, S.C., (2005). "The European guidelines for self-compacting concrete." *BIBM*, et al., pp. 47–59.

GB/T50080-2002 (2002). "Standard for test method of performance on ordinary fresh concrete." Chinese Standard, China Academy of Building Research, Guangdong, China.

Gram, A., et al. (2014). "Obtaining rheological parameters from flow test — Analytical, computational and lab test approach." *Cement and Concrete Research*, 63, 29–34.

Hu, C., et al. (1996). "Validation of BTRHEOM, the new rheometer for soft-to-fluid concrete." *Materials and Structures*, 29(10), 620–631.

Jiao, D., et al. (2017). "Effect of constituents on rheological properties of fresh concrete-A review." *Cement and Concrete Composites*, 83, 146–159.

Lu, C., et al. (2015). "Relationship between slump flow and rheological properties of self compacting concrete with silica fume and its permeability." *Construction and Building Materials*, 75, 157–162.

Mohammed, A. M., et al. (2021). "Experimental and statistical evaluation of rheological properties of self-compacting concrete containing fly ash and ground granulated blast furnace slag." *Journal of King Saud University - Engineering Sciences*.

Murata, J., and Kikukawa, H. (1992). "Viscosity equation for fresh concrete." *ACI Materials Journal*, 89(3), 230–237.

Nguyen, T. L. H., et al. (2006). "Correlation between L-box test and rheological parameters of a homogeneous yield stress fluid." *Cement and Concrete Research*, 36(10), 1789–1796.

Ouldkhaoua, Y., et al. (2019). "Rheological properties of blended metakaolin self-compacting concrete containing recycled CRT funnel glass aggregate." *Epitoanyag-Journal of Silicate Based & Composite Materials*, 71(5):154–161.

Pradoto, R., et al. (2016). "Fly ash-nano SiO_2 blends for effective application in self-consolidating concrete." *Proceedings of 8th International RILEM Symposium on Self-Compacting Concrete*, Washington DC, 299–308.

Rezania, M., et al. (2019). "Experimental study of the simultaneous effect of nano-silica and nano-carbon black on permeability and mechanical properties of the concrete." *Theoretical and Applied Fracture Mechanics*, 104: 102391.

Roussel, N. (2006). "Correlation between yield stress and slump: Comparison between numerical simulations and concrete rheometers results." *Materials and Structures*, 39(4), 501–509.

Roussel, N. (2007). "The LCPC BOX: A cheap and simple technique for yield stress measurements of SCC." *Materials and Structures*, 40(9), 889–896.

Saak, A. W., et al. (2004). "A generalized approach for the determination of yield stress by slump and slump flow." *Cement and Concrete Research*, 34(3), 363–371.

Schowalter, W. R., and Christensen, G. (1998). "Toward a rationalization of the slump test for fresh concrete: Comparisons of calculations and experiments." *Journal of Rheology*, 42(4), 865–870.

Sedran, T., and De Larrard, F. (1999). "Optimization of self compacting concrete thanks to packing model." *RILEM Proceedings*, 7, 321–332.

Skarendahl, A., et al. (2000). "Self-compacting concrete-state-of-the-art report of RILEM TC 174-SCC." RILEM report, 23.

Suliman, M. O., et al. (2017). "Effects of stone cutting powder (Al-Khamkha) on the properties of self-compacting concrete." *World Journal of Engineering and Technology*, 05(04), 613–625.

Tattersall, G. H. (1991). *Workability and quality control of concrete*. CRC Press, Boca Raton, FL.

Tattersall, G. H., and Banfill, P. F. (1983). *The rheology of fresh concrete*. Pitman Books Limited, London, England.

Wallevik, J. E. (2006). "Relationship between the Bingham parameters and slump." *Cement and Concrete Research*, 36(7), 1214–1221.

Wallevik, O. H. (2003). "Rheology—A scientific approach to develop self-compacting concrete." *Proceedings of the 3rd International Symposium on Self-Compacting Concrete*, Reykjavik, 23–31.

Wallevik, O. H., and Wallevik, J. E. (2011). "Rheology as a tool in concrete science: The use of rheographs and workability boxes." *Cement and Concrete Research*, 41(12), 1279–1288.

Zerbino, R., et al. (2008). "Workability tests and rheological parameters in self-compacting concrete." *Materials and Structures*, 42(7), 947–960.

Chapter 5

Paste rheometers

5.1 INTRODUCTION TO THE RHEOLOGY OF CEMENT PASTE

Rheology is an important branch of natural science. It is widely used in suspension, polymer, food, coating, and cosmetic industries to evaluate the rheological properties of fluids. Principles of rheology applied to cement-based materials will provide us with an important means to accurately characterize the rheological properties of concrete. The good rheology of concrete is beneficial to pouring, forming, and transmission under complex conditions and even to the development of later strength and durability, especially in the field of pumped and sprayed concrete (Morinaga, 1973, Glab et al., 2020). At the same time, for 3D-printed concrete, to ensure its extrudability, uniformity, and buildability, the rheological properties of the concrete need to be accurately controlled within an appropriate range (Mca et al., 2020). In addition, rheological measurements have become a means of monitoring changes in the microstructure of cement-based materials (Wallevik, 2009). Cement paste is the most important component of concrete. Therefore, the rheological behavior of paste is the basis of concrete rheological properties (Ferraris and Gaidis, 1992).

Fresh cement paste is a heterogeneous material that can be regarded as a suspension of cement particles in water. The particle size ranges from nanometers (e.g., hydration products and polymers) to micrometers (e.g., cement and minerals). The mechanical interaction between solids and the chemical hydration between cementitious particles and water together affect their rheological properties (Liu et al., 2017). For a long time, most cement pastes have been considered as Bingham fluids, exhibiting yield stress (Cepuritis et al., 2019). Because of the friction between cement particles, it also has viscosity. The Bingham model describes the linear relationship between shear stress and shear rate through two rheological parameters. However, the modification of chemical admixtures and mineral additives makes the rheological behavior of cement paste more complicated (Ferraris et al., 2001, Alonso et al., 2007, Hanehara and Yamada, 1999, Khayat, 1998). The stress–strain relationship of the slurry will appear in a nonlinear form, including shear thinning and shear thickening (Ma et al., 2016, Deng et al., 2013, Maybury et al., 2017, Yang et al., 2018). Shear thinning and shear thickening may occur simultaneously, depending on which one is dominant (Liu et al., 2017). In particular, cement paste exhibits thixotropy, which usually shows that under the action of shear stress, the viscosity decreases with time, and it recovers after the removal of the shear stress. The thixotropy of cement paste increases over time due to the process of hydration (Roussel et al., 2012, Erdem et al., 2015). Therefore, thixotropy may cause an overestimation of rheological parameters. In addition, the measurement results of the rheometer are also affected by other factors, including system friction, thermal expansion, temperature, humidity (Schüller and Salas-Bringas, 2007, Nehdi and Rahman, 2004, Lewis et al., 2000), and pressure (Hyun et al., 2002, Proske et al., 2020). Therefore, under the most suitable rheological model, obtaining accurate rheological parameters of cement slurry and controlling and adjusting the number of mineral admixtures used in concrete will have good application prospects (Chen and Rothstein, 2004, Sun et al., 2007).

5.2 RHEOMETERS FOR CEMENT PASTE

5.2.1 Narrow gap coaxial cylinder rheometer

5.2.1.1 Geometry

The coaxial cylinder rheometer usually includes an inner cylinder and an outer cylinder. During the measurement process of torque, the inner cylinder or the outer cylinder maintains rotation, while the other cylinder remains stationary. For cement paste, ASTM C 1749 (ASTM C1749-17a) gives procedures for measuring rheological parameters under narrow and wide gap conditions, and it points out that for a narrow gap, the ratio between inner and outer diameters is greater than or equal to 0.92. The gap should be 10 times greater than the diameter of the cementitious particles, approximately 0.4 mm. Generally, the gap of the rheometer is slightly larger than this value by approximately 1 mm. To calculate the shear stress and shear rate from the known torque and speed, we assume that the flow of the slurry between the inner and outer cylinders is simple. First, there is no slippage between the surface of the cylinder and the slurry. Second, the flow of the slurry along the radius is laminar, no flow occurs in the vertical direction, and there is only a velocity component along the angle. This is the same assumption as the concrete rheometer. Under narrow gap conditions, the most important thing is that the shear rate can be considered constant. Therefore, the rheological parameters can be directly obtained from the shear stress and shear rate at the rotating drum. This assumption is not applicable under wide gaps because of the more complicated flow behavior (Liu et al., 2020).

To reduce the influence of wall slippage, the traditional cylindrical rotor can be replaced by a vane rotor. The gap between the blades will be filled by the slurry, which is theoretically the same as a cylindrical rotor. The effectiveness of the blade rotor instead of the cylindrical rotor is verified by numerical simulation (Zhu et al., 2010). There are two main advantages of using blades, which are determined by their geometry. First, when the four-blade rotor rotates, the outer end is not a completely smooth round surface, so there is no wall slip. Second, due to the influence of its volume and geometry, the structure of the sample is much less damaged by inserting the blade (Alderman et al., 1991). However, some researchers have found that the position where the shear surface occurs when measuring the Bingham fluid using a four-blade rotor is slightly larger than the radius of the blade rotor (Keentok et al., 1985).

5.2.1.2 Measurement principle

For a narrow gap, the shear rate at the inner cylinder can be obtained from the gap and angular velocity between the inner and outer cylinders, which can be expressed as Eq. (5.1):

$$\dot{\gamma} = \frac{\omega R_1}{R_2 - R_1} \tag{5.1}$$

where $\dot{\gamma}$ is the shear rate, ω is the angular velocity of rotation, R_1 is the radius of the inner cylinder, and R_2 is the radius of the outer cylinder.

The shear stress at the inner cylinder can be calculated by Eq. (5.2):

$$\tau = \frac{T}{2\pi R_1^2 h} \tag{5.2}$$

where τ is the shear stress at the inner cylinder, T is the torque value, and h is the rotor height.

Therefore, the rheological parameters can be obtained by the intercept and slope of the linear relationship between the shear stress at the inner cylinder and the shear rate. The unit of yield stress is Pa, and the unit of plastic viscosity is Pa.s. For wide gap conditions, rheological parameters can be obtained by the Reiner–Riwlin equation.

The shear rate is calculated from Eq. (5.3):

$$\dot{\gamma} = r \frac{d\omega}{dr} \tag{5.3}$$

The shear stress at radius r is given by Eq. (5.4):

$$\tau = \frac{T}{2\pi r^2 h} \tag{5.4}$$

For the Bingham fluids, the linear relationship between shear stress and shear rate can be expressed as Eq. (5.5):

$$\tau = \tau_0 + \mu \dot{\gamma} \tag{5.5}$$

Substitute Eqs. (5.4) and (5.5) into Eq. (5.3) as Eq. (5.6):

$$r \frac{d\omega}{dr} = \frac{T}{2\pi r^2 h \mu} - \frac{\tau_0}{\mu} \tag{5.6}$$

Integrating Eq. (5.7) is shown in Eq. (5.8):

$$\int_\Omega^0 d\omega = \int_{R_1}^{R_2} \left(\frac{T}{2\pi r^3 h \mu} - \frac{\tau_0}{\mu r} \right) dr \tag{5.7}$$

$$\Omega = \frac{T}{4\pi h \mu} \left(\frac{1}{R_1^2} - \frac{1}{R_2^2} \right) - \frac{\tau_0}{\mu} \ln\left(\frac{R_2}{R_1} \right) \tag{5.8}$$

However, at a certain speed of the inner cylinder, the slurry has an unsheared area, that is, a dead zone. At this time, the radius of the sheared area is smaller than the radius of the outer cylinder, and it is often necessary to correct the torque or speed during the calculation. Therefore, the abovementioned Reiner–Riwlin equation can be rewritten as Eq. (5.9):

$$\Omega = \frac{T}{4\pi h \mu} \left(\frac{1}{R_1^2} - \frac{1}{R_{plug}^2} \right) - \frac{\tau_0}{\mu} \ln\left(\frac{R_{plug}}{R_1} \right) \tag{5.9}$$

The radius of the plug flow area R_{plug} can be expressed as Eq. (5.10):

$$R_{plug} = \sqrt{\frac{T}{2\pi h \tau_0}} \tag{5.10}$$

Koehler and Fowler (2004) compared three methods of dead zone correction with existing concrete rheometers, namely Point Elimination Method, Independent Yield Stress Method, and Effective Annulus Method. These methods can also be used in slurry rheometers.

5.2.1.3 Measuring errors and artifacts

In the measurement of a coaxial cylindrical rheometer, the wall slip effect of a smooth cylindrical rotor has been widely accepted. This is due to a thin layer with a lower particle concentration produced near the wall. Accordingly, the actual shear rate of the slurry decreases, and the test produces a large error (Harboe et al., 2012). The most common way to reduce wall slip is to increase the roughness of the rotor surface. The thickness of the thin slip layer is also affected by the physical properties of the test fluid. For example, the particles of suspension in contact with the wall of the rheometer are affected by their radius (Haimoni and Hannant, 1988). Figure 5.1a shows the remarkable feature that the stress changes in a zigzag shape when wall slip occurs (Saak et al., 2001). Figure 5.2 shows some surface-treated rotors and the inner wall of the outer cylinder, including surface protrusions or grooves and blade rotors, sandblasting, and jagged surfaces.

The wall slip velocity is considered to be the difference between the wall velocity and the fluid velocity near the wall. When using a roughened rotor or even a bladed rotor instead of a cylindrical rotor, the results of rheological measurements may be different due to the wall slip effect (Nehdi and Rahman, 2004, Haimoni and Hannant, 1988). In terms of experiments, the coaxial cylindrical rheometer uses the roughened-surface rotor to obtain higher torque or shear stress than the cylindrical rotor (Saak et al., 2001, Harboe et al., 2011, Haimoni and Hannant, 1988). Figure 5.1b (Saak et al., 2001) shows the stress–speed relationship between the blade rotor and the cylindrical rotor. Figure 5.3 shows representative shear stress–shear rate curves under smooth, slotted, and bladed rotors. In addition, researchers have used numerical simulation methods to prove that the roughening of the surface makes the impact of wall slip no longer significant, and higher accuracy is obtained (Wang et al., 2010). Zhu et al. (2010) compared the rheological parameters obtained at different shear rates with the experimental results and reported that the blade rotor can approximately replace the cylindrical rotor to measure the rheology of the Bingham fluid. The wall slip of the cylindrical rotor underestimates the viscosity when slip occurs. This is the same as the simulation result of Barnes (1990). However, the blade may not be completely equivalent to a cylinder, and the secondary flow between the blade gaps affects the viscosity measurement. In the measurement of wall slip velocity, Ortega-Avila et al. (2016) used particle image velocimetry to measure the velocity of gel flowing in the annular gap between two concentric cylinders. The theoretical calculation of wall slip velocity has a certain mathematical foundation, and its correctness has been verified through experiments (Yoshimura and Prud'homme, 1988). Therefore, in most cases, surface roughening is effective in reducing the effect of wall slip, but how to define the most appropriate roughness conditions remains to be studied. ASTM (ASTM C 1749-17a) provides a standardized method for measuring cement slurries under rough surface conditions.

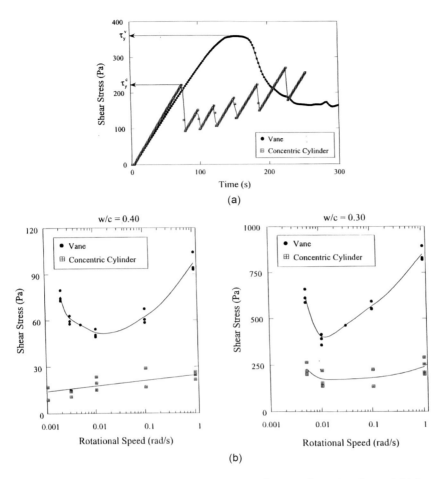

Figure 5.1 Representative stress growth curves (a) and the influence of rotational speed (b) for concentric cylinders and vanes. The rotational speed of the stress growth test is 0.01 rad/s, and the w/c ratio is 0.3 (Saak et al., 2001).

5.2.2 Plate–plate rheometer

5.2.2.1 Geometry

The parallel-plate rheometer usually includes a rotating upper plate and a stationary lower plate, and the gap between the two parallel plates can be adjusted as needed. It will be filled with slurry, and the torque applied to the slurry by the upper plate will be measured. When measuring cement paste, the gap between the two plates is not always guaranteed to be a small gap, so that the shear rate is not constant. The largest at the upper board, and the smallest at the lower board. Based on the assumption of laminar flow, the shear rate varies linearly between the upper and lower plates. In addition, whether in a smooth or nonsmooth plate, the speed of the upper plate will also cause a change in rheological parameters when the upper plate is lowered because the speed of the squeezing fluid will cause different degrees of water migration (Cardoso et al., 2015). The upper plate can be replaced with a conical plate. The greatest advantage compared to a flat plate is that the shear rate at different heights is constant, and the cone plate reduces the volume of the sample. However,

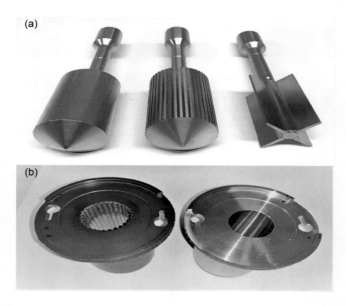

Figure 5.2 Concentric cylinder test geometry: (a) from left: smooth rotor, grooved rotor, 4-blade vane and (b) from left: grooved cup, smooth cup (Shamu and Hkansson, 2019).

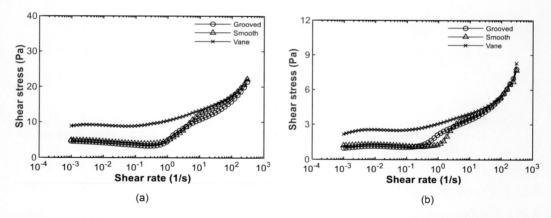

Figure 5.3 Comparison of flow curves for cement grout with w/c of 0.6 (a) and 0.8 (b) obtained in grooved, smooth, and vane geometries (Shamu and Hkansson, 2019).

the centrifugal effect at the conical plate makes the slurry unable to adhere well to the wall. Besides, it is more sensitive to the particle size, usually greater than ten times the maximum particle size.

Compared with the coaxial cylinder rheometer, the parallel-plate rheometer is more suitable for high-viscosity slurries. On the one hand, it is easier to add high-viscosity samples between the parallel plates and will not be affected by geometry. On the other hand, the parallel-plate measurement requires fewer samples, and the maximum shear rate can be changed by changing the diameter or gap. However, in the coaxial cylinder rheometer, the low-viscosity fluid can fill the gap between the inner and outer cylinders well and at the same time can better control the temperature of the sample during the measurement. There is a

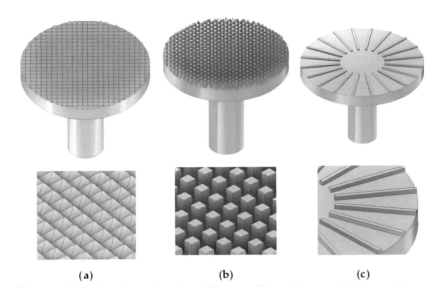

Figure 5.4 (a–c) Parallel-plate test geometry (Pawelczyk et al., 2020).

temperature difference between the samples in the gap between the parallel plates, and it is often positively correlated with the gap. Cone-plate conditions require the least amount of sample.

ASTM (C 1749-17a) provides a test procedure for parallel plates. To reduce the influence of wall slippage, methods such as sandpaper bonding, sandblasting, serrating, and grooving are also used. Figure 5.4 shows some parallel-plate test geometries. The position of the parallel plate should be first determined at the beginning of the test. The size of the gap has a significant impact on the rheological parameters (Carotenuto and Minale, 2013, Nickerson and Kornfield, 2005). It is determined by the maximum particle size of the particles in the tested slurry. When the gap is narrowed, the friction between the particles and the plate will increase, resulting in a higher normal force and the changing of the flow state. If the gap is too large, it cannot be effectively filled by slurry (Ferraris et al., 2014). To ensure that the initial measurement is at the zero position, it is more difficult to determine under a rough surface because of its geometric height limitation, which will have a greater impact on the calculation of rheological parameters.

5.2.2.2 Measurement principle

The torque and speed data are obtained by measurement, and the linear relationship between torque and speed is obtained by fitting. Take the most common Bingham model as an example, the relationship between torque and shear stress is as follows:

$$T = \int \tau \, dA \, r = \int_0^R 2\pi r^2 \left(\tau_0 + \mu \gamma\right) dr \qquad (5.11)$$

We can obtain the shear rate at the edge of the upper plate by Eq. (5.12):

$$\dot{\gamma} = \frac{\omega R}{h} \qquad (5.12)$$

Afterward, we can convert the variable in the integral into a shear rate:

$$T = \frac{2}{3}\pi\tau_0 R^3 + \int_0^{\frac{\omega R}{h}} 2\pi\mu \frac{h^3}{\omega^3}\gamma^3 \, d\gamma = A + B\gamma \tag{5.13}$$

Therefore, we can obtain the rheological parameters through the intercept and slope of the T–γ linear relationship, which can be obtained by Eq. (5.14):

$$\tau_0 = \frac{3A}{2\pi R^3} \quad \mu = \frac{2B}{\pi R^3} \tag{5.14}$$

where T is the torque, γ is the shear rate, ω is the rotational angular velocity, τ_0 is the yield stress, and μ is the plastic viscosity. A is the T–γ line intercept, and B is the T–γ line slope. The units of yield stress and plastic viscosity are Pa and Pa.s, respectively. Of course, similar rheological parameters can also be obtained through the T–ω relation. However, in some documents (Mendes et al., 2014, Rosquoët et al., 2003, Shamu and Hkansson, 2019), the inhomogeneity of the shear flow at the edge of the parallel plate was reported. Based on the Weissenberg–Rabinowitsch equation, a method to correct the shear stress at the outer edge was obtained. Similarly, it can be corrected by radius, which is $r = 2/3R$ or $r = 3/4R$. In ASTM C 1749, the viscosity can be expressed by Eq. (5.15):

$$\eta = \frac{3hT}{2\pi R^4 \omega \left(1 + \frac{1 d \ln T}{3 d \ln \omega}\right)} \tag{5.15}$$

where η is the viscosity in Pa.s, and T is the torque in N.m.

5.2.2.3 Measuring errors and artifacts

Similar to the coaxial cylindrical rheometer, the problem of wall slippage is also common in parallel-plate rheometers. Usually, a thin layer appears near the wall of the upper plate and the fluid. Hartman et al. (2002) used ATR spectroscopy to monitor the drop in particle concentration generated in the non-Newtonian fluid near the plate area, and this phenomenon was related to the particle concentration. To reduce the influence of the wall slip phenomenon, many researchers have adopted the surface-roughening method (Alonso et al., 2007, Kalyon and Malik, 2012, Nickerson and Kornfield, 2005, Nehdi and Rahman, 2004). Pawelczyk et al. (2020) and Carotenuto and Minale (2013) believed that the unevenness of the surface would change the actual gap, and they tried to establish a certain correction relationship by expanding the gap. Although the wall slip is not completely eliminated, the corrected viscosity of the Newtonian fluid with rough surface treatment is almost the same as the actual viscosity. Table 5.1 shows the expansion gap under some structural features. Although it is necessary to establish this relationship, its viscosity is unknown for cement paste, and obtaining the correction function will be more complicated due to the shear thickening or thinning behavior. If the measurement results of the rotating rheometer are used as a reference, it is possible to establish the corresponding correction relationship (Ferraris et al., 2001, 2004). The viscosity correction can be obtained by Eq. (5.16) (Pawelczyk et al., 2020, Carotenuto et al., 2012):

$$\eta_c = \frac{\eta_m H}{H + \delta} \tag{5.16}$$

Table 5.1 Values of δ for different measuring systems at $\dot{\gamma}=5.05\,\text{s}^{-1}$ (Pawelczyk et al., 2020)

Measuring system	Effective gap extension δ/mm		
	Silicon oil AK 5000	Silicon oil AK 12,500	25 wt.% suspension (AK 5000)
1 (pyramids)	0.23±0.006	0.25±0.004	022±0.011
4 (pyramids)	0.91±0.040	0.83±0.032	0.86±0.071
5 (pyramids)	0.44±0.023	0.49±0.016	0.48±0.029
8 (columns)	0.20±0.003	0.22±0.009	0.19±0.014
12 (bars)	0.35±0.009	0.37±0.004	035±0.009
17 (bars)	0.73±0.068	0.71±0.072	0.70±0.089

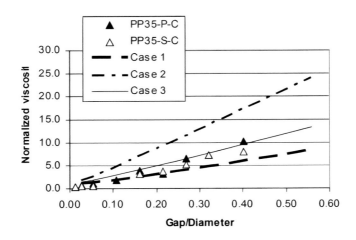

Figure 5.5 Simulation and data for the PP35-confined geometry. The three lines are for analog data, and the symbols represent experimental data (Ferraris et al., 2007).

where η_c is the corrected viscosity, η_m is the measured viscosity, H is the gap, and δ is the gap expansion value.

Some parallel-plate rheometers are tested under the confinement wall, and this may be more advantageous for higher fluidity slurries. In general, the rheological test result under the fence condition is larger compared to that of the plane plate, which may be caused by the friction between the slurry and the wall. Ferraris et al. (2007) improved the parallel-plate rheometer so that it can be used to test cement pastes with larger particle sizes. The experimental data were compared with the numerical simulation results, as shown in Figure 5.5. It can be found that the simulated data under a certain gap has a good correlation with the experimental data. In general, the smaller the gap, the higher the simulation result than the measurement, which may be due to the fact that the fluid outside the upper edge contributes more torque. With Eq. (5.17), the corrected viscosity can be obtained from the measured viscosity:

$$\eta_r = \frac{\eta_m}{\frac{h}{D}f + 1} \qquad (5.17)$$

Furthermore, when the gap is much smaller than the diameter of the confinement wall, the corrected viscosity will be closer to the measured viscosity (Ferraris et al., 2007).

Compared with the parallel-plate rheometer, the narrow gap coaxial cylindrical rheometer may be more suitable for the rheological measurement of cement slurry, which has some main advantages. For example, it is easier to add samples to the cylinder. Moreover, the outer cylinder can wrap the cement paste, making temperature control effective. By contrast, it is difficult to control the sample under the condition of a parallel plate or cone plate for the slurry with a very strong flowability. Although it can be achieved by setting a boundary wall, this will cause an overestimation of the shear stress. In addition, independent of parallel-plate or cone-plate rheometer, the slurry is susceptible to normal stress, especially for slurries with high viscoelasticity and high thixotropy. Besides, the physical properties of the material can change in the plate rheometer, such as the radial migration and evaporation of water. At present, although many studies are using rotary rheometers to test cement slurries, the rheological parameters of slurries composed of different materials are difficult to compare.

5.2.3 Other rheometers

5.2.3.1 Capillary viscometer

5.2.3.1.1 Geometry

Many types of capillary viscometers can be used to measure the viscosity of fluids, especially Newtonian fluids or fluids with Newtonian characteristics. Measurements for cement-based materials usually include pressure and gravity capillary viscometers. The pressure capillary viscometer makes the fluid flow along the capillary tube under constant pressure and obtains the pressure change over a certain distance. The shear stress and shear rate can be determined by the pressure difference and the flow rate to obtain the fluid viscosity. The gravity viscometer records the time it takes for the fluid to pass a certain distance through the capillary tube to determine the fluid viscosity (Demko, 1989). When the liquid flows along the wet pipe wall, its viscosity is proportional to the flow time.

The hydrodynamic theory of fluid flowing through a capillary viscometer is very simple and makes necessary assumptions. That is, the fluid flow in the tube is stable, incompressible, and laminar, and hence no slippage occurs at the tube wall (Rosa et al., 2020). The same assumption is also applied to other rheological measurements of cement paste. Capillary rheometers have a high aspect ratio. Normally, the ratio of capillary length to diameter is 30 or even greater. Figure 5.6 shows a commonly used Cannon–Fenske viscometer.

The measurement using the capillary viscometer usually contains only one rheological parameter. The viscosity of Newtonian fluids or some cement pastes with properties similar to Newtonian fluids can be obtained through the Poiseuille equation, while the yield stress is still unknown. For pressure capillary viscometers, when the fluid flows in the capillary without plug flow, the pressure drop and flow relationship curve may be used to describe the yield stress and viscosity of the Bingham fluid (Rosa et al., 2020).

5.2.3.1.2 Measurement principle

Before measuring, the viscometer is calibrated first. The well-stirred cement slurry was added to the calibrated viscometer and placed in a water bath to keep the temperature constant. Then, the amount of the sample was adjusted until the meniscus was placed 7 mm above the first graduation line. The sample flows freely under its weight, and the time for the meniscus

Figure 5.6 Cannon–Fenske viscometer (Rosa et al., 2020).

to flow through the first and second graduation lines along the capillary will be measured, with the time accurate to 0.1 s. After that, the capillary was changed to a smaller diameter, and the measurement was repeated. Calculate the average value twice and cannot exceed the required accuracy. The kinematic viscosity of the slurry is determined two times, and the dynamic viscosity can be obtained according to the density. The measurement is conducted concerning ASTM D445-17a. The pressure capillary viscometer needs to measure the pressure and distance at two test points.

For gravity capillary viscometers, the viscosity is calculated by Eq. (5.18):

$$v = C \times t \tag{5.18}$$

where v is the kinematic viscosity of the fluid in mm²/s, C is the instrument-related constant in mm²/s², and t is the average flow time in s.

The dynamic viscosity is calculated by Eq. (5.19):

$$\eta = v \times \rho \tag{5.19}$$

where η is the dynamic viscosity in mPa.s, and ρ is the fluid density in kg/m³.

For pressure capillary viscometer, the shear stress is defined as Eq. (5.20):

$$\tau = \frac{Pr}{8L} \tag{5.20}$$

where τ is the shear stress, r is the radius from the center of the capillary, L is the length between two points, and P is the pressure difference.

The shear rate at the wall is expressed as Eq. (5.21):

$$\dot{\gamma} = \frac{4Q}{\pi R^3} \tag{5.21}$$

where $\dot{\gamma}$ is the shear rate, and Q is the flow rate.

Dynamic viscosity is obtained by the ratio between shear stress and shear rate as Eq. (5.22):

$$\eta = \frac{\pi R^4 P}{8QL} \tag{5.22}$$

The relationship between the pressure drop and flow rate is as follows:

$$Q = \frac{\pi R_4}{8\mu L}\left(P - \frac{8\tau_0 L}{3R} + \frac{16\tau_0^4 L^4}{3P^3 R^4}\right) \tag{5.23}$$

The shear rate can be corrected by the Rabinowitsch–Mooney equation:

$$\dot{\gamma} = \frac{Q}{\pi R^3}\left(3 + \frac{d \ln Q}{d \ln \tau}\right) \tag{5.24}$$

5.2.3.1.3 Application to cement paste

Capillary viscometers are mainly used in the chemical field, but rarely used for cement slurry. Rosa et al. (2020) used a capillary viscometer for the first time to measure cement slurries with different material compositions and additives, and compared them with the results of a rotary rheometer. The viscosities obtained by the two methods are very close. The test results are shown in Table 5.2. For non-Newtonian fluid suspensions such as cement paste, there are certain restrictions (Mooney, 1931). On the one hand, the diameter of the capillary will be limited by the size of the suspending particles. If the diameter is too small, the capillary will be blocked. If the diameter is too large, the fluid will pass through the pipe quickly, resulting in large errors in readings (Liu et al. 2017). On the other hand, as time passes, the continuous hydration of cement will generate heat, which has a greater influence on the measurement results of the capillary viscometer (Rushing and Hester, 2003, Patterson and Rabouin, 1958). Therefore, a reasonable choice of capillary diameter is a very important prerequisite. ASTM D445-18 provides a standardized method for determining the dynamic and kinematic viscosity of transparent and opaque fluids (Demko, 1989). However, the test method needs to be optimized according to the complexity of cement paste.

In summary, capillary viscometer can only provide viscosity parameters related to rheological properties. Throughout the test, the viscosity is time-dependent and only a fixed value can be obtained. If the time and distance of mud movement can be continuously obtained, the change of viscosity can be characterized.

Table 5.2 Comparison of plastic viscosity between rotational rheometer (RR) and capillary viscometer (C-F)

Cementitious material	w/c	SP/c	ϕ_c	η_p [Pa.s] RR	C-F 400
CEM I 52.5-SR	0.35	0.4	0.477	0.123	0.214
	0.35	0.8	0.477	0.086	0.117
	0.47	0.4	0.404	0.058	0.087
	0.47	0.8	0.404	0.047	0.037
	053	0.4	0.376	0.037	0.042
	0.53	0.8	0.376	0.030	0.022
	0.63	0.4	0.336	0.019	0.027
	0.63	0.8	0.336	0.020	0.012
CEM II 32.5 BL-II	0.35	0.4	0.487	0.091	0.114
	0.35	0.8	0.487	0.055	0.073
	0.47	0.4	0.414	0.037	0.030
	0.47	0.8	0.414	0.032	0.029
	053	0.4	0.386	0.025	0.020
	0.63	0.4	0.345	0.013	0.020
	0.63	0.8	0.345	0.014	0.013
75% CEM I 52.5-SR+25% GGBS	0.35	0.4	0.471	0.117	0.169
	0.35	0.8	0.471	0.087	0.158
	0.47	0.4	0.399	0.049	0.058
	0.47	0.8	0.399	0.039	0.033
	0.53	0.4	0.371	0.034	0.041
	0.53	0.8	0.371	0.029	0.026
	0.63	0.4	0.332	0.019	0.016
	0.63	0.8	0.332	0.019	0.016
75% CEM II 32.5 BL-II+25% GGBS	0.35	0.4	0.476	0.074	0.087
	0.35	0.8	0.476	0.067	0.074
	0.47	0.4	0.405	0.035	0.036
	0.47	0.8	0.405	0.032	0.028
	0.53	0.4	0.377	0.027	0.026
	0.53	0.8	0.377	0.019	0.021
	0.63	0.4	0.338	0.012	0.022
	0.63	0.8	0.338	0.009	0.011

5.2.3.2 Falling sphere viscometer

5.2.3.2.1 Geometry

The principle of the falling sphere viscometer is that the speed of the ball falling in the liquid is inversely proportional to the viscosity of the liquid. A typical falling-ball viscometer is shown in Figure 5.7. It usually contains a test tube containing the test liquid and an external tube that controls the temperature. The test tube is slightly inclined, approximately 10°. The viscometer is equipped with spheres of many sizes, ranging from 11.00 to 15.81 mm in diameter. The sphere falls through the liquid in the test tube, and its viscosity is evaluated

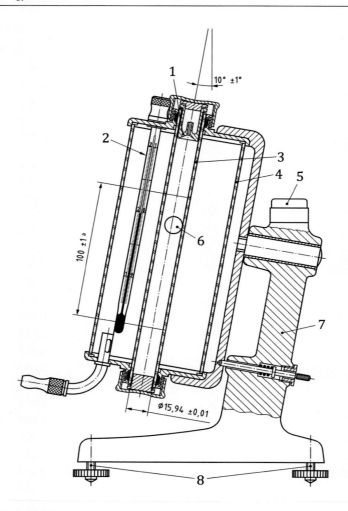

Figure 5.7 Typical falling sphere viscometer.

by the time of falling. It is necessary to obtain the calibration coefficient of the instrument through a liquid of known viscosity. The calibration coefficients for different materials and diameters can be found in DIN EN ISO 12058-1. Similar to the capillary viscometer, the falling sphere viscometer also provides only viscosity information that characterizes the rheological behavior of the slurry (Fulmer and Williams, 2002).

5.2.3.2.2 Measurement principle

Put the mixed cement paste into the measuring tube, and then put the sphere. The measuring tube is closed by the stopper. Before each series of measurements, the sphere needs to be rolled once along the length of the pipe to make the sphere surface to have complete contact with the cement paste. Turn the viscometer so that the sphere falls from the upper end and passes through the two circular graduation lines in sequence. The time for the upper or lower end of the sphere to pass through the two-scale lines will be measured, and the temperature must be kept constant during the measurement. To reduce the error, at least three measurements are taken. The above test is carried out according to DIN EN ISO 12058-1.

The dynamic viscosity of the slurry can be obtained by:

$$\eta = K(\rho_1 - \rho_2)t \tag{5.25}$$

where η is the dynamic viscosity in mPa.s, K is the instrument calibration coefficient in mm^2/s^2, ρ_1 is the density of the sphere in kg/m^3, ρ_2 is the density of the slurry in kg/m^3, and t is the average falling time in s.

5.2.3.2.3 Application to cement paste

A falling sphere viscometer is commonly used to test Newtonian fluids and some polymers and resins with low viscosity. For non-Newtonian fluids, there will be more complex flow fields near the sphere, the tube wall and the end region (Munro et al., 1979, Yoshimura, 2000). This results in the measurement of cement slurry being more restricted, and it is difficult to evaluate the shear flow of the sphere in the pipeline. Nicolò et al. (2020) showed that mixtures with different particle sizes lead to the falling process of spheres in fluid showing different shapes, including changes in the trajectory and state of motion. The same effect may also occur in the cement slurry. The measurement of fresh mortar uses a measurement method similar to a falling sphere viscometer, and the viscosity value is obtained according to the Stokes equation (Ferraris, 1999). However, it is difficult to find a similar cement slurry method. In addition, cement slurry is an opaque fluid. Therefore, the falling time of the sphere may be difficult to be accurately measured. Thompson (1949) provided a method to test opaque liquids by the falling sphere method. The realization of induction may have a positive influence on the measurement of cement slurry. The falling sphere viscometer does not have a very solid theoretical basis in the measurement of non-Newtonian fluids, which limits its application in the cement slurry. Besides, it can be affected by the complex rheological properties of cement pastes, such as thixotropy, shear thickening, or shear thinning. Therefore, the relationship between viscosity and time is difficult to define. In addition, the variation of the drop motion trajectory and shear rate may be difficult to assess and may require some theoretical support.

5.3 MEASURING PROCEDURES

5.3.1 Flow curves test

The well-mixed cement paste is first poured into the outer cylinder of the rheometer. The rotor should be turned gently to place it in the slurry so that the bottom of the rotor is in full contact, avoiding incomplete contact and compaction. Figure 5.8 shows a typical protocol for a flow curve test. During the measurement, it is necessary to maintain the temperature unchanged. By setting the number of shearing steps or the shearing time, the shear rate can be distributed stepwise or continuously. Before the formal measurement, a pre-shearing at the maximum shear rate is required. The shear time usually ranges from 30 to 60s. Afterward, the sample was set standstill for half a minute to stabilize the torque. The speed quickly reaches the initial speed, and the rising and falling parts will be measured. The rheological parameters are determined by the shear stress and shear rate of the descending section. In addition, the area enclosed by the ascending and the descending curves is positively related to the thixotropy of the material. It can be repeated several times to obtain the average value to improve the accuracy of rheological parameters.

110 Rheology of Fresh Cement-Based Materials

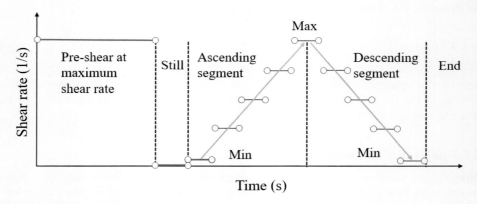

Figure 5.8 The process diagram of flow curve test.

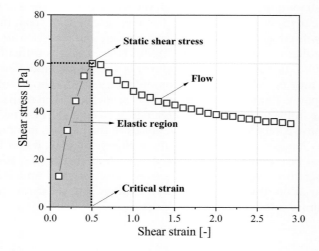

Figure 5.9 Determination of static yield stress in a shear strain test.

For parallel-plate geometry, the added sample should exceed 10%–20% to ensure that the gap is evenly filled by the slurry. Afterward, adjust the gap, and trim off the excess part. Then, the rheometer is run based on the testing protocol.

5.3.2 Static yield stress test

The increase in static yield stress depends on the coupling of chemical hydration and interactions between colloidal particles, van der Waals forces, etc. With elapsed time, a denser network structure is formed between the particles, which increases the yield stress. The slurry is first pre-sheared to reduce the effects of thixotropy. After that, the slurry was applied with a shear rate (or shear strain) of 10^{-1} to $10^{-3} s^{-1}$, and the shear stress at the critical state was the static yield stress (Feys et al., 2018). This critical state corresponds to the transition from rising to falling shear stress during shearing. Figure 5.9 shows the determination of the static yield stress and critical strain.

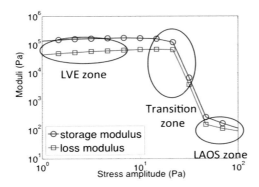

Figure 5.10 Evolution of the storage and loss moduli as a function of the stress amplitude.

5.3.3 Oscillatory shear test

5.3.3.1 Description of SAOS and LAOS

Small-amplitude oscillatory shear (SAOS) is a technical method to evaluate the deformation ability of fluid based on the classic Hooke's law. In SAOS, a strain smaller than the critical strain of slurry is applied for a continuous sinusoidal excitation to obtain a stress response, such that the investigated slurry is deformed only within its elastic region. However, the stress response usually has a certain degree of lag (Théau et al., 2016). Once the strain exceeds a critical value, the material structure is destroyed, and the stress–strain relationship is no longer linear (Betioli et al., 2009), as shown in Figure 5.10. For polymers and suspensions, the critical value of structural transformation can be determined (Yuan et al., 2017, Betioli et al., 2009, Nachbaur et al., 2001, Kallus et al., 2001). Therefore, the SAOS test method identifies the process of materials from construction to destruction.

The association of large-amplitude oscillatory shear (LAOS) data with rheological behavior provides a method for evaluating fluid rheological properties (Cho et al., 2005). Compared with SAOS, the slurry is subjected to a strain amplitude higher than the critical strain. The stress–strain relationship is no longer a pure linear relationship, and the stress cannot be expressed by a single trigonometric function. In the case of nonlinearity, the modulus is no longer independent of the strain amplitude, so there will be periodic deviations (Hyun et al., 2011). High-order harmonics have a certain contribution to affecting the behavior of non-Newtonian fluids (Giacomin et al., 2011).

Fourier transform is considered to be the most commonly used technical means to describe the properties of nonlinear behavior fluids (Hyun et al., 2002). Hyun et al. (2011) developed Fourier transform based on nuclear magnetic resonance spectroscopy and developed the Fourier transform–rheology method. The method has been applied to rheological tests of various complicated fluids. Hyun et al. (2002) proposed the concepts of generalized storage modulus and generalized loss modulus, and used them as a measure to evaluate the nonlinear behavior of viscoelasticity. Corresponding to the Fourier transform method, a higher accuracy is obtained. In a later report, the theoretical development and application of LAOS data interpretation were reviewed (Hyun et al., 2011). In addition, through Lissajous curves, the complicated high-order harmonic problem is transformed into a mathematical geometric problem, and the slope of a certain point of the stress–strain curve is used as the physical interpretation of the LAOS data (Simon, 2018).

For SAOS, the ratio of shear stress to strain can be expressed by the complex modulus, including the storage modulus and loss modulus. Among them, the storage modulus represents the amount of energy stored in the slurry due to deformation under elastic behavior, and the loss modulus represents the amount of heat or other losses that need to be overcome in the process, such as resistance between particles. LAOS also includes two parameters representing elastic and viscous behaviors, and the dynamic modulus related to the material is obtained through Fourier transformation. If the contribution of higher harmonics is ignored, the storage modulus and loss modulus under LAOS conditions may lose their meanings. However, it may still provide some information about microstructure changes to distinguish complex fluids (Hyun et al., 2002). LAOS is mainly carried out under strain or stress control through the Fourier transform method; that is, the output stress obtained is decomposed into a Fourier series. Currently, the most commonly used method for interpreting LAOS data is the geometric method of Lissajous curves.

5.3.3.2 Measurement principle

For SAOS, the strain can be obtained by Eq. (5.26):

$$\gamma(t) = \gamma_0 \times \sin(\omega t) \tag{5.26}$$

where $\gamma(t)$ is the strain at time t, γ_0 is the maximum amplitude, and t is the time.

Then, the complex modulus can be obtained:

$$G^* = \tau / \gamma \tag{5.27}$$

$$G^* = G' + iG'' \tag{5.28}$$

The relationship among the rheological parameters can be expressed in Eqs. (5.29–5.31):

$$G' = \frac{\tau_0}{\gamma_0} \times \sin \delta \tag{5.29}$$

$$G'' = \frac{\tau_0}{\gamma_0} \times \cos \delta \tag{5.30}$$

$$\tan \delta = \frac{G''}{G'} \tag{5.31}$$

where G' is the storage modulus, G'' is the loss modulus, τ_0 is the maximum stress, γ_0 is the maximum strain, and δ is the phase angle. For purely elastic materials, the phase angle is 0°, and the response is completely flexible. For pure Newtonian fluids, the phase angle is 90°, and the reaction is completely vicious. For cement-based materials, the response is somewhere in between, showing viscoelasticity that corresponds to the elastic and viscous behaviors of the material.

SAOS is mainly performed using the abovementioned rotational rheometer. When it is a strain-controlled rheometer, SAOS technique usually includes the following three processes:

Strain sweep. In this test, the frequency needs to be kept at a certain value (generally 1~2 Hz for cement paste). The material receives a continuous sinusoidal strain, and the amplitude gradually increases, usually from 10^{-5}% to 1%. When the strain exceeds a certain critical value, the internal structure breaks down, and the modulus decreases as the

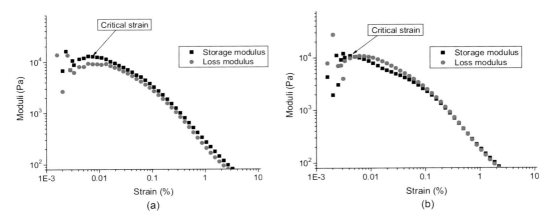

Figure 5.11 Typical oscillatory strain sweep of fresh cement pastes at 1 Hz. (a) Critical value can be easily recognized, and (b) critical value is difficult to recognize.

strain increases. Through this process, the linear viscoelastic region (LVER) and the critical strain of the material can be obtained. In some cases, the critical strain is easy to identify (see Figure 5.11a). However, the critical strain may not be easily determined in other cases (see Figure 5.11b).

Oscillatory frequency sweep: Oscillatory frequency sweep was used to evaluate the stability of the material over a wide frequency range. The amplitude was kept constant, while the frequency was increased incrementally from 0.01 to 100 Hz. Both storage modulus and loss modulus were monitored during the tests.

Oscillatory time sweep: Oscillatory time sweep test was used to investigate the evolution of rheological properties due to structural changes. In this test procedure, the amplitude was kept constant at a magnitude lower than the critical strain, and the frequency was also kept constant.

Generally, the applicable frequency of cement paste is 1 Hz for LAOS. The process carried out is very similar to the abovementioned SAOS process. Taking strain input as an example, it includes three processes: strain oscillation sweep, frequency oscillation sweep, and time oscillation sweep. Compared with SAOS, LAOS imposes larger stress or strain amplitude without a large difference in frequency. The main purpose is to analyze the transition of the stress–strain curve from the linear region to the nonlinear region. If the frequency is too high or too low, it will affect the rheological test. At higher frequencies, it may be affected by the inertia of the instrument and fluid. At the same time, for time-dependent fluids, the development and evolution of the internal structure of the material make the lower LAOS parameter at frequency relatively greatly affected.

When using a strain-controlled rheometer, the input stress is as Eq. (5.32) (Läuger and Stettin, 2010, Théau et al., 2016):

$$\tau(t) = \tau_1 \times \sin(\omega t) \tag{5.32}$$

Due to symmetry, the Fourier series of the output strain is expressed as Eq. (5.33):

$$\gamma(t) = \sum_{n=\text{odd}} \gamma_n \sin(n\omega t - \delta_n) \tag{5.33}$$

Strain can be divided into elastic and viscous parts, where J_n' and J_n'' are the storage modulus and loss modulus, respectively:

$$\gamma(t) = \tau_1 \sum_{n=\text{odd}} \left[J_n' \sin(n\omega t) - J_n'' \cos(n\omega t) \right] \tag{5.34}$$

The strain rate can also be expressed by material-related fluid properties ψ_n' and ψ_n'':

$$\dot{\gamma}(t) = \tau_1 \sum_{n=\text{odd}} \left[n\psi_n'' \cos(n\omega t) + n\psi_n' \sin(n\omega t) \right] \tag{5.35}$$

Furthermore, small-strain compliance and large-strain compliance are defined, corresponding to the slope of the stress–strain curve in Lissajous curves:

$$J_M' = \frac{d\gamma}{d\tau}\bigg|_{\tau=0} = J_M'(\omega:\tau_1) \tag{5.36}$$

$$J_M' = \sum_{n=\text{odd}} n J_n' \tag{5.37}$$

$$J_L' = \frac{\gamma}{\tau}\bigg|_{\tau=\pm\tau_1} = J_L'(\omega:\tau_1) \tag{5.38}$$

$$J_L' = \sum_{n=\text{odd}} J_n' \tag{5.39}$$

Therefore, the stress-softening ratio R can be represented by Eq. (5.40):

$$R(\omega:\tau_1) = \frac{J_L' - J_M'}{J_L'} \tag{5.40}$$

Similarly, the definition of small rate fluidities and large rate fluidities corresponds to the slope of the stress–strain rate curve in Lissajous curves.

$$\psi_M' = \frac{d\dot{\gamma}}{d\tau}\bigg|_{\tau=0} = J_M'(\omega:\tau_1) \tag{5.41}$$

$$\psi_M' = \omega \sum_{n=\text{odd}} n^2 J_n'' \tag{5.42}$$

$$\psi_L' = \frac{\dot{\gamma}}{\tau}\bigg|_{\tau=\pm\tau_1} = \psi_L'(\omega:\tau_1) \tag{5.43}$$

$$\psi_L' = \omega \sum_{n=\text{odd}} n J_n'' \tag{5.44}$$

The shear-thinning ratio Q is defined as Eq. (5.39):

$$Q(\omega, \tau_1) = \frac{\psi'_L - \psi'_M}{\psi'_L} \quad (5.45)$$

In particular, when $R=0$, it should correspond to the critical strain of SAOS. At the same time, different R and Q values correspond to different shapes of Lissajous curves. Therefore, the LAOS data can be explained by changes in the geometric relationship between the stress and strain or strain rate curves.

5.3.3.3 Application to cement paste

SAOS is widely used in high-molecular-weight polymers, suspensions, emulsions, greases, etc., and it can also be used in cement paste to evaluate the viscoelasticity and the evolution of internal structure (Yuan et al., 2017, Betioli et al., 2009, Nachbaur et al., 2001, (Papo and Caufin, 1991)). Schultz and Struble (1993) pointed out that the storage modulus of cement slurry is approximately 14–24 kPa and the critical strain is approximately 10^{-4}. As w/c increases, the storage modulus and critical strain decrease slightly. However, Yuan et al. (2017) believed that the critical strain is between 10^{-5} and 10^{-4}. The value is significantly affected by the use of a water-reducing agent, while it shows a weak correlation with w/c. Researchers have gradually connected the rheological parameters of SAOS with the structural development of cement paste and even the formation of hydration products on a microscopic scale. Betioli et al. (2009) found that SAOS parameters can correspond to the heat of the hydration process. Huang et al. (2020) established a linear relationship between the storage modulus and ettringite (AFt) content. Figure 5.12 shows the correlation between ettringite content and storage modulus. These results also prove that SAOS may have a strong correlation between early hydration progress and structure establishment. In addition, Yuan et al. (2017) found that the storage modulus was in good agreement with the static yield stress growth, as shown in Figure 5.13.

The yield stress can be expressed as the product of the critical strain and the corresponding modulus (Betioli et al., 2009). Ukrainczyk et al. (2020) compared the yield stress calculated

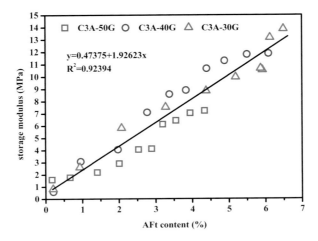

Figure 5.12 Correlation of the ettringite content (wt.%) and the storage modulus of C_3A-gypsum paste (Huang et al., 2020).

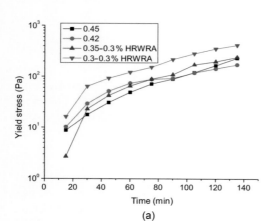 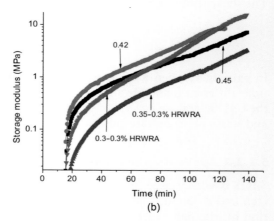

Figure 5.13 Comparison of yield stress (a) and storage modulus (b) over time (Yuan et al., 2017).

by SAOS and the mechanical model, and verified the relationship between the yield stress and SAOS parameters. Sun et al. (2006) used a shear wave reflection technique to monitor the early behavior of cement slurries with different water–cement ratios. The results have a good correlation with the storage modulus obtained directly using SAOS.

The small deformation under SAOS may not fully characterize the processing and molding of some materials. In this context, LAOS corresponding to the study of rheology under large deformation or deformation rates, mainly for the measurement of polymers (Hyun et al., 2002), is used to measure the structural behavior of some fluids with high viscosity or deformation ability. However, the research of LAOS on the measurement of the rheological properties of cement pastes is limited. Théau et al. (2016) first used LAOS to study the influence of amplitude and frequency on the rheological parameters of cement slurries, and the underlying mechanisms were explained by a Lissajous–Bowditch curve. Stress–strain or stress–strain rate curves or geometric properties of stress, such as shape, slope, enclosed area, and evolution, were linked to the thixotropy and viscoelastic behavior of the material (Qian et al., 2019). Figure 5.14 shows the LAOS parameters of cement paste.

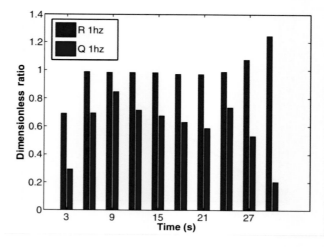

Figure 5.14 Age evolution of the nonlinear viscoelastic parameters of the NC cement paste at 1 Hz and 40 Pa; R: stress-softening ratio; Q: shear-thinning ratio (Théau et al., 2016).

5.4 SUMMARY

This chapter reviews the measurement of the rheological properties of cement paste, mainly including rheological equipment, measuring principles, testing methods, and data processing. The following conclusions can be reached.

Surface-roughening methods include sandblasting, grooving, recessing or protruding, and even blade rotors, which have a certain restrictive effect on the occurrence of wall slip. However, the blade rotors may cause complicated flow. The geometric height of the rough-structure design will also affect its restrictive effect on wall slip.

The coaxial cylindrical rheometer has better temperature control ability for cement slurry samples and ensures its uniformity. Compared with the parallel-plate rheometer, the cement slurry overflow from the edge is avoided. It is more suitable for the measurement of the rheological properties of most cement pastes.

The storage modulus obtained by SAOS has good consistency with the yield stress change of cement slurry. SAOS is a promising technique for monitoring the development of microstructure in real time. For LAOS with larger strain, the Lissajous curve is the most suitable to interpret the data, and its shape evolution and LAOS rheological parameters can be well connected.

The viscosities measured by the capillary viscometer and the rotational rheometer are similar. However, it is difficult to characterize its rheology through two or even three parameters of the existing rheology model. Falling sphere viscometers are difficult to apply to cement paste because of the complex flow behavior.

The rheometer can quickly and accurately evaluate the rheological properties of cement paste. Therefore, it has been widely used in infrastructure construction and experimental research. However, there is no clear specification for evaluating the applicability of each rheometer and the quantitative relationship between rheological parameters. More research is needed on this aspect.

REFERENCES

Alderman, N. J., Meeten, G. H., and Sherwood, J. D. (1991). "Vane rheometry of bentonite gels." *Journal of Nonnewtonian Fluid Mechanics*, 39(3), 291–310.

Alonso, M. M., Palacios, M., Puertas, F., et al. (2007). "Effect of polycarboxylate admixture structure on cement paste rheology." *Materiales De Construccion*, 57(286), 65–81.

ASTM C 1749-17a. Standard guide for measurement of the rheological properties of hydraulic cementious paste using a rotational rheometer. American Society for Testing and Materials, 2017.

ASTM D 445-18. Standard Test Method for Kinematic Viscosity of Transparent and Opaque Liquids (and Calculation of Dynamic Viscosity).

Barnes, H. A. (1990). "The vane-in-cup as a novel rheometer geometry for shear thinning and thixotropic materials." *Journal of Rheology*, 34(6), 841–866.

Betioli, A. M., Gleize, P. J. P., Silva, D. A., et al. (2009). "Effect of HMEC on the consolidation of cement pastes: Isothermal calorimetry versus oscillatory rheometry." *Cement & Concrete Research*, 39(5), 440–445.

Cardoso, F. A., Fujii, A. L., Pileggi, R. G., and Chaouche, M. (2015). "Parallel-plate rotational rheometry of cement paste: Influence of the squeeze velocity during gap positioning." *Cement & Concrete Research*, 75, 66–74.

Carotenuto, C., and Minale, M. (2013). "On the use of rough geometries in rheometry." *Journal of Non-Newtonian Fluid Mechanics*, 198, 39–47.

Carotenuto, C., Marinello, F., and Minale, M. (2012). "A new experimental technique to study the flow in a porous layer via rheological tests." *AIP Conference Proceedings*, 1453(1), 29–34.

Cepuritis, R., Skare, E. L., Ramenskiy, E., et al. (2019). "Analyzing limitations of the FlowCyl as a one-point viscometer test for cement paste." *Construction and Building Materials*, 218, 333–340.

Chen, S., and Rothstein, J. P. (2004). "Flow of a wormlike micelle solution past a falling sphere." *Journal of Non-Newtonian Fluid Mechanics*, 116(2–3), 205–234.

Cho, K. S., Hyun, K., Ahn, K. H., et al. (2005). "A geometrical interpretation of large amplitude oscillatory shear response." *Journal of Rheology*, 49(3), 747–758.

DIN EN ISO 12058-1. Determination of viscosity using a falling-ball viscometer-Part 1: Inclined-tube method.

Demko, J. M. (1989). "Development of an ASTM standard test method for measuring engine oil viscosity using capillary viscometers at high-temperature and high-shear rates." *ASTM Special Technical Publication*, 1068, 11.

Deng, D., Zhu, R., Peng, J., et al. (2013). "Effect of superplasticizers and limestone powders on shear thickening behavior of cement paste." *Journal of Building Materials*, 16(5), 744–752.

Erdem, T. K., Ahari, R. S., et al. (2015). "Thixotropy and structural breakdown properties of self consolidating concrete containing various supplementary cementitious materials." *Cement & Concrete Composites*, 59, 26–37.

Ferraris, C. F. (1999). "Measurement of the rheological properties of cement paste: A new approach." *Journal of Research of the National Institute of Standards and Technology*, 104(5), 333–342.

Ferraris, C. F., Beaupr, D., et al. Comparison of concrete rheometers: International tests at MB (Cleveland OH, USA) in May, 2003. US Department of Commerce, National Institute of Standards and Technology, 2004.

Ferraris, C. F., Brower, L. E., Banfill, P., et al. Comparison of concrete rheometers: International test at LCPC (Nantes, France) in October, 2000. US Department of Commerce, National Institute of Standards and Technology, 2001.

Ferraris, C. F, and Gaidis, J. M. (1992). "Connection between the rheology of concrete and rheology of cement paste." *ACI Materials Journal*, 89(4), 388–393.

Ferraris, C. F., Geiker, M., Martys, N. S., and Muzzatti, N. (2007). "Parallel-plate rheometer calibration using oil and computer simulation." *Journal of Advanced Concrete Technology*, 5(3), 363–371.

Ferraris, C. F., Martys, N. S., and George, W. L. (2014). "Development of standard reference materials for rheological measurements of cement-based materials." *Cement and Concrete Composites*, 54, 29–33.

Ferraris, C. F., Obla, K. H., and Hill, R. (2001). "The influence of mineral admixtures on the rheology of cement paste and concrete." *Cement & Concrete Research*, 31(2), 245–255.

Feys, D., Cepuritis, R, Jacobsen, S., et al. (2018). "Measuring rheological properties of cement pastes: Most common techniques, procedures and challenges." *RILEM Technical Letters*, 2, 129–135.

Fulmer, E. I., and Williams, J. C. (2002). "A method for the determination of the wall correction for the falling sphere viscometer." *Journal of Physical Chemistry*, 40(1), 143–149.

Giacomin, A. J., Bird, R. B., Johnson, L. M., et al. (2011). Large-amplitude oscillatory shear flow from the corotational Maxwell model. *Journal of Non-Newtonian Fluid Mechanics*, 166, 1081–1099.

Glab, C., Wca, B., Lca, B., et al. (2020). "Rheological properties of fresh concrete and its application on shotcrete - ScienceDirect." *Construction and Building Materials*, 243, 118180.

Guoju, K., Zhang, J., Xie, S., et al. (2020). "Rheological behavior of calcium sulfoaluminate cement paste with supplementary cementitious materials." *Construction and Building Materials*, 243, 118234.

Haimoni, A., and Hannant, D. J. (1988). "Developments in the shear vane test to measure the gel strength of oilwell cement slurry." *Advances in Cement Research*, 1(4), 221–229.

Hanehara, S., and Yamada, K. (1999). "Interaction between cement and chemical admixture from the point of cement hydration, absorption behavior of admixture, and paste rheology." *Cement & Concrete Research*, 29(8), 1159–1165.

Harboe, S, Modigell, M., et al. (2011). "Wall slip of semi-solid A356 in Couette rheometers." *AIP Conference Proceedings*, 1353(1), 1075–1080.

Harboe, S., Modigell, M., and Pola, A. (2012). "Wall slip effect in Couette rheometers." *International Conference on Semisolid Processing of Alloys and Composites S2P 2012*, 192, 353–358.

Hartman Kok, P. J. A., Kazarian, S. G., Lawrence, C. J., et al. (2002). "Near-wall particle depletion in a flowing colloidal suspension." *Journal of Rheology*, 46(2), 481–493.

Hu, C, de Larrard, F., Sedran, T., et al. (1996). "Validation of BTRHEOM, the new rheometer for soft-to-fluid concrete." *Materials & Structures*, 29(10), 620–631.

Huang, T., Yuan, Q., and He, F., et al. (2020). "Understanding the mechanisms behind the time-dependent viscoelasticity of fresh C 3 A-gypsum paste." *Cement and Concrete Research*, 133, 106084.

Hyun, K, Kim, S. H., Ahn, K. H., et al. (2002). "Large amplitude oscillatory shear as a way to classify the complex fluids." *Journal of Non-Newtonian Fluid Mechanics*, 107(1–3), 51–65.

Hyun, K, Wilhelm, M, Klein, C. O., et al. (2011). "A review of nonlinear oscillatory shear tests: Analysis and application of large amplitude oscillatory shear (LAOS)." *Progress in Polymerence*, 36(12), 1697–1753.

Kallus, S., Willenbacher, N., Kirsch, S., et al. (2001). "Characterization of polymer dispersions by Fourier transform rheology." *Rheologica Acta*, 40(6), 552–559.

Kalyon, D. M, and Malik, M. (2012). "Axial laminar flow of viscoplastic fluids in a concentric annulus subject to wall slip." *Rheologica Acta*, 51(9), 805–820.

Keentok, M., Milthorpe, J. F, and O'Donovan, E. (1985). "On the shearing zone around rotating vanes in plastic liquids: Theory and experiment." *Journal of Non-Newtonian Fluid Mechanics*, 17(1), 23–35.

Khayat, K. H. (1998). "Viscosity-enhancing admixtures for cement-based materials — An overview." *Cement and Concrete Composites*, 20(2–3), 171–188.

Koehler, E, and Fowler, D. (2004). "Development of a portable rheometer for fresh portland cement concrete." *Technical Reports, International Center for Aggregates Research, The University of Texas at Austin, Austin, United States*.

Läuger, J., and Stettin, H. (2010). "Differences between stress and strain control in the nonlinear behavior of complex fluids." *Rheologica Acta*, 49(9), 909–930.

Lewis, J. A., Matsuyama, H., Kirby, G., et al. (2000). "Polyelectrolyte effects on the rheological properties of concentrated cement suspensions." *Journal of the American Ceramic Society*, 83(8), 1905–1913.

Liu, Y., Shi, C., Jiao, D., and An, X. (2017). "Rheological properties, models and measurements for fresh cementitious materials-a short review." *Journal of the Chinese Ceramic Society*, 45(05), 708–716.

Liu, Y., Shi, C., Yuan, Q., et al. (2020). "The rotation speed-torque transformation equation of the Robertson-Stiff model in wide gap coaxial cylinders rheometer and its applications for fresh concrete." *Cement and Concrete Composites*, 107, 103511.

Ma, K., Feng, J., Long, G., et al. (2016). "Effects of mineral admixtures on shear thickening of cement paste." *Construction & Building Materials*, 126, 609–616.

Maybury, J., Ho, J. C. M., and Binhowimal, S. A. M. (2017). "Fillers to lessen shear thickening of cement powder paste." *Construction & Building Materials*, 142, 268–279.

Mca, B, Lei, Y, Yan, Z. A., et al. (2020). "Yield stress and thixotropy control of 3D-printed calcium sulfoaluminate cement composites with metakaolin related to structural build-up." *Construction and Building Materials*, 252, 119090.

Mendes, P. R. D. S., Alicke, A. A., and Thompson, R. L. (2014). "Parallel-plate geometry correction for transient rheometric experiments." *Applied Rheology*, 24(5), 52721.

Mooney, M. (1931). "Explicit formulas for slip and fluidity." *Journal of Rheology*, 2(2), 210.

Morinaga, S. (1973). "Pumpability of concrete and pumping pressure in pipelines." *Proceedings of Rilem Seminar, Leeds*, 3, 1–39.

Munro, R. G., Piermarini, G. J., Block, S. (1979). "Wall effects in a diamond-anvil pressure-cell falling-sphere viscometer." *Journal of Applied Physics*, 50(5), 3180–3184.

Nachbaur, L., Mutin, J. C., Nonat, A., et al. (2001). "Dynamic mode rheology of cement and tricalcium silicate pastes from mixing to setting." *Cement & Concrete Research*, 31(2), 183–192.

Nehdi, M, and Rahman, M. A. (2004). "Estimating rheological properties of cement pastes using various rheological models for different test geometry, gap and surface friction." *Cement & Concrete Research*, 34(11), 1993–2007.

Nickerson, C. S., and Kornfield, J. A. (2005). "A "cleat" geometry for suppressing wall slip." *Journal of Rheology*, 49(4), 865–874.

Nicolò, R. S., Davaille, A., Kumagai, I., et al. (2020). "Interaction between a falling sphere and the structure of a non-Newtonian yield-stress fluid." *Journal of Non-Newtonian Fluid Mechanics*, 284, 104355.

Ortega-Avila, J. F., Pérez-González, J., Marín-Santibáñez, B. M., et al. (2016). "Axial annular flow of a viscoplastic microgel with wall slip." *Journal of Rheology*, 60(3), 503–515.

Papo, A., and Caufin, B. (1991). "A study of the hydration process of cement pastes by means of oscillatory rheological techniques." *Cement and Concrete Research*, 21(6), 1111–1117.

Patterson, G. D., and Rabouin, L. H. (1958). "Capillary viscometer for high-temperature measurements of polymer solutions." *Review of Scientific Instruments*, 29(12), 1086–1088.

Pawelczyk, S., Kniepkamp, M., Jesinghausen, S., et al. (2020). "Absolute rheological measurements of model suspensions: Influence and correction of wall slip prevention measures." *Materials*, 13(2), 467.

Proske, T., Rezvani, M., and Graubner, C. A. (2020). "A new test method to characterize the pressure-dependent shear behavior of fresh concrete." *Construction and Building Materials*, 233, 117255.

Qian, Y., Ma, S. W., Kawashima, S., and Schutter, G. D. (2019). "Rheological characterization of the viscoelastic solid-like properties of fresh cement pastes with nanoclay addition." *Theoretical and Applied Fracture Mechanics*, 103, 102262.

Rosa, N. D. L., Poveda, E., Ruiz, G., et al. (2020). "Determination of the plastic viscosity of superplasticized cement pastes through capillary viscometers." *Construction and Building Materials*, 260, 119715.

Rosquoët, F., Alexis, A., Khelidj, A., et al. (2003). "Experimental study of cement grout: Rheological behavior and sedimentation." *Cement and Concrete Research*, 33(5), 713–722.

Roussel, N., Ovarlez, G., Garrault, S., et al. (2012). "The origins of thixotropy of fresh cement pastes." *Cement and Concrete Research*, 42(1), 148–157.

Rushing, T. S., and Hester, R. D. (2003). "Low-shear-rate capillary viscometer for polymer solution intrinsic viscosity determination at varying temperatures." *Review of Scientific Instruments*, 74(1), 176–181.

Saak, A. W., Jennings, H. M., and Shah, S. P. (2001). "The influence of wall slip on yield stress and viscoelastic measurements of cement paste." *Cement & Concrete Research*, 31(2), 205–212.

Schüller, R. B., and Salas-Bringas, C. (2007). "Fluid temperature control in rotational rheometers with plate-plate measuring systems." *Psychologie*, 15, 159–163.

Schultz, M. A., and Struble, L. J. (1993). "Use of oscillatory shear to study flow behavior of fresh cement paste." *Cement and Concrete Research*, 23(2), 273–282.

Shamu, T. J., and Hkansson, U. (2019). "Rheology of cement grouts: On the critical shear rate and no-slip regime in the Couette geometry." *Cement and Concrete Research*, 123, 105769.

Simon, R. (2018). "Large amplitude oscillatory shear: Simple to describe, hard to interpret." *Physics Today*, 71(7), 34–40.

Sun, B. J., Gao, Y. H., and Liu, D. Q. (2007). "Experimental study on rheological property for cement slurry and numerical simulation on its annulus flow." *Journal of Hydrodynamics (Ser.A)*, 22(3), 317–324.

Sun, Z., Voigt, T., and Shah, S. P. (2006). "Rheometric and ultrasonic investigations of viscoelastic properties of fresh Portland cement pastes." *Cement & Concrete Research*, 36(2), 278–287.

Théau, C., and Mohend, C., et al. (2016). "Rheological behavior of cement pastes under large amplitude oscillatory shear." *Cement & Concrete Research*, 89, 332–344.

Thompson, A. M. (1949). "A falling-sphere viscometer for use with opaque liquids." *Journal of Scientific Instruments*, 26(3), 75.

Ukrainczyk, N., Thiedeitz, M., Krnkel, T., et al. (2020). "Modeling SAOS yield stress of cement suspensions: Microstructure-based computational approach." *Materials*, 13(12), 2769.

Wallevik, J. E. (2009). "Rheological properties of cement paste: Thixotropic behavior and structural breakdown." *Cement & Concrete Research*, 39(1), 14–29.

Wang, W., Zhu, H., De Kee, D., et al. (2010). "Numerical investigation of the reduction of wall-slip effects for yield stress fluids in a double concentric cylinder rheometer with slotted rotor." *Journal of Rheology*, 54(6), 1267–1283.

Yoshimura, A. (2000). "Wall slip corrections for Couette and parallel disk viscometers." *Journal of Rheology*, 32(1), 53–67.

Yang, H., Lu, C., and Mei, G. (2018). "Shear-thickening behavior of cement pastes under combined effects of mineral admixture and time." *Journal of Materials in Civil Engineering*, 30(2), 04017282.

Yoshimura, A. S., and Prud'homme, R. K. (1988). "Viscosity measurements in the presence of wall slip in capillary, Couette, and parallel-disk geometries. *SPE Reservoir Engineering*, 3(02), 735–742.

Yuan, Q, Lu, X, Khayat, K. H., et al. (2017). "Small amplitude oscillatory shear technique to evaluate structural build-up of cement paste." *Materials & Structures*, 50(2), 112.

Zhu, H., Martys, N. S., Ferraris, C., et al. (2010). "A numerical study of the flow of Bingham-like fluids in two-dimensional vane and cylinder rheometers using a smoothed particle hydrodynamics (SPH) based method." *Journal of Non-Newtonian Fluid Mechanics*, 165(7–8), 362–375.

Chapter 6

Concrete rheometers

6.1 INTRODUCTION

For a long time in the past, the workability or, more precisely, flow property of fresh conventional concrete has been tested predominately by the slump test (C143, 2008). However, the applications of self-compacting concrete (SCC), high-performance concrete (HPC), high flowable concrete, and other types of concrete introduce a wide range of materials into engineering practice. The compositional complexity of concrete mix makes its flow behavior very sensitive to slight changes in mixture proportions. The slump test becomes a less reliable indicator of the workability of fresh concrete.

So far, approximately 100 tests have been developed to measure concrete flow over the past few decades (Roussel, 2011). These tests fall into the empirical approach and the scientific approach. The empirical tests have been described in Chapter 4. The scientific approach describes the material itself and can understand the intrinsic properties of fresh concrete, which is called rheology. Through rheology, the flow and deformation property (rheological property) of fresh concrete can be defined strictly in terms of physical constants with the fundamental unit, and basic rheological principles can guide the study of the physical and analytical models of the material.

During the 1970s, measuring apparatus with rotating vanes or coaxial cylinders had been used to make the theoretical analysis of the flow behavior of fresh concrete (Banfill, 2006). Based on the two-point test, Tattersall and Banfill (1983) developed a method for measuring the power requirements during the mix. It can be deployed both in the lab and on-site, standing for a significant step forward to characterize the flow properties of fresh concrete. The two-point test measures the values of shear stress under a minimum of two shear rates, and then calculates the rheological parameters of the material.

In the past few decades, a large variety of materials and admixtures, together with many new processing methods, have been introduced into the field of concrete. The rheology of fresh concrete has been comprehensively studied, and its applications are greatly extended. It is possible for us to predict the fresh properties, design and select materials, and achieve the required concrete performance.

According to the US National Institute of Standards and Technology (NIST), all the flow tests can be classified into four groups (Hackley and Ferraris, 2001): confined flow tests, free flow tests, vibration tests, and rotational rheometers. The first three types belong to empirical tests. The use of rotational rheometers is the only method that can determine the exact rheological properties of a test sample in a fundamental rheological unit.

This chapter introduces several prototypes of rotational rheometers for concrete. The basic principles and measurement procedures are introduced, and errors and artifacts are discussed. Finally, the relations of rheological parameters measured by different rheometers are summarized.

6.2 TESTS METHODS AND PRINCIPLES

There are several rheometry methods, including rotational method, capillary method, drainage vessel method, and oscillation method. The most commercially available rheometers for fresh concrete are the rotational rheometers. Rotational rheometry can be classified regarding geometrical design and mode of operation. Four geometries used in rotational rheometers are coaxial, parallel plates, cone-plate, and impeller (See Figure 6.1). In a coaxial rheometer, the fluid is placed in a cylindrical cup, and a coaxial but smaller cylinder is submerged in the fluid. In a parallel-plate rheometer, two disks are positioned parallel to each other. Moreover, the impeller rheometers are based on rotating an impeller, which has various shapes. The test sample is placed between disks. The material is sheared as one disk rotates, while the other remains stationary. Compared to the parallel-plate rheometer, the cone-plate rheometer replaces one plate with a cone. However, since the coarse aggregates at the bottom of the container often block the immersion of the cone, this type of rheometer is rarely used for fresh concrete.

Several concrete rheometers are shown in Figure 6.2. The ICAR rheometer and BML Viscometer are coaxial cylinder types. The BTRHEOM is parallel-plate type, and the Tattersall two-point and IBB rheometers are impeller type. The Tattersall two-point rheometer (MK II) is one of the first instruments to use impeller geometry. It is also the earliest attempt to measure the rheological properties of fresh concrete with the Bingham model. The Tattersall two-point rheometer has two different impeller types, which allow measuring a wide range of concrete mixtures. IBB is, in fact, the automated version of MK III, which is the modified model of the two-point rheometer (Beaupre, 1994). BML Viscometer and ConTec are similar devices. Viscometer 5 is heavy and suitable for laboratory research, while the ICAR and BTRHEOM can be used in the laboratory or on-site.

In the rest of this section, three typical concrete rheometers are introduced. The basic information including their geometries, principles, and other related issues will be described in detail and compared.

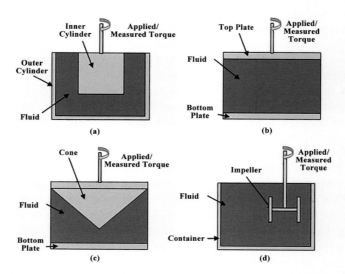

Figure 6.1 Typical rheometer geometry configurations: (a) coaxial cylinder, (b) parallel plate, (c) cone-plate, and (d) impeller.

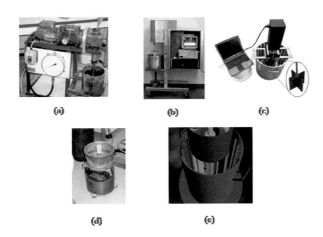

Figure 6.2 Concrete rheometers (Brower and Ferraris, 2001): (a) Tattersall two-point rheometer, (b) IBB rheometer, (c) ICAR rheometer, (d) BTRHEOM rheometer, and (e) BML Viscometer.

6.2.1 Coaxial cylinder rheometer

Two basic operation modes, i.e., stress-controlled and rate-controlled modes, are available to convert the above geometries into a rotational rheometer. The stress-controlled mode measures the resulting shear rate by controlling the stress input, while the rate-controlled model measures the resulting shear stress through controlling shear rate input (Schramm, 1994). Some rheometers can work in both test modes, while most commercial concrete rheometers usually adopt the rate-controlled mode.

Operationally, either the inner or the outer cylinder is rotated with the counterpart (the other cylinder) fixed. For the rheometers with the inner cylinder rotating, i.e., Searle-type rheometers, the torque is measured at the inner cylinder. For the rheometers with the outer cylinder rotating, i.e., Couette type, the inner cylinder is freely suspended from a torsion wire, the resistance of the flow causes wire deflections on the inner cylinder, and then the torque is recorded. Moreover, both the drive on the rotor (inner cylinder) and the torque detector of Searle-type rheometers act on the same rotating axis and may lower the measuring accuracy compared to Couette type. However, as technology advances, the differences between the Searle-type and the Couette-type rheometers are no longer significant.

The inner cylinder of concrete rheometers is generally substituted by a vane or ribs to facilitate the placement of fresh concrete. The ICAR rheometer is a typical commercial Searle rheometer, while the ConTec Viscometer 5 is a typical Couette rheometer.

The measurement procedure involves increasing and decreasing the speed (in preset discrete increments) while measuring the torque at each speed. The problem is to relate speed and torque to shear rate and shear stress to calculate the exact rheological properties in fundamental units. Therefore, a transformation equation should be derived so that the regression analysis can acquire the parameters of a given rheological model.

Compared to the radius of the inner cylinder, the width of the annulus for most of the general-purpose rheometer is relatively narrow. Therefore, calculating the shear rate and shear stress using an average radius will not introduce too many errors. However, these narrow-gaped rheometers do not apply to concrete. The maximum aggregate size should be considered for the concrete rheometer. Studies showed that the minimum gap size should be at least three times the aggregate size so that the inhomogeneity of the material can be eliminated to a large extent during the test (Koehler and Fowler, 2004).

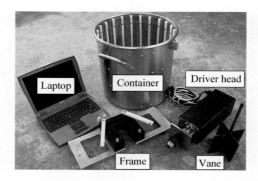

Figure 6.3 Components of ICAR rheometer.

6.2.1.1 Searle rheometer

The ICAR rheometer is a mobile, portable rheological test device for fresh concrete, as shown in Figure 6.3. It was developed at the International Center for Aggregate Research (ICAR), at The University of Texas at Austin. The instrument is appropriate for moderately and highly flowable fresh concrete with a slump value greater than 50–75 mm, especially SCC. It tests the rheological properties of fresh concrete and mortar with a maximum aggregate size of 40 mm, depending on the radius of the out cylinder. The ICR rheometer is designed based on a wide-gap, coaxial cylinder rheometer. It is made of a driver head that contains an electric motor and torque meter, a cross vane that is held by the clamp on the driver acting as an inner cylinder to prevent slip, a frame to affix the driver-vane assembly to the top of the cylinder, a laptop to operate the driver, record the torque during the test, and compute the rheological parameters, and a fresh concrete container which sticks many vertical strips around the inside wall to prevent slipping of the fluid along the container perimeter during the test. The size of the cylinder and the length of the vane shaft are chosen according to the nominal maximum size of the aggregate. The vane height and diameter are both 127 mm.

ICAR rheometer can perform two types of tests, i.e., stress growth test and flow curve test. During the stress growth test, the cross vane is rotated, for example, at a low velocity of 0.025 rev/s. The initial torque increase is measured in terms of time. The maximum torque measured in this test is used to determine the static yield stress.

The flow curve test calculates the dynamic yield stress and plastic viscosity. At the beginning of the trial, the vane is rotated at the maximum velocity to break down the possible thixotropic structure and ensure a consistent shearing history for each test sample. Then the vane speed slows down in a specified number of steps. The vane speed is kept constant for each stage, and both the average speed and the torque are logged. Usually, at least six steps are suggested. However, the number of steps can be chosen by the experimenter. Afterward, the plot of torque versus rotation speed of the vane is drawn. The rheological model parameters are calculated based on the least square regression of its transformation equation.

6.2.1.2 Couette rheometer

One typical Couette rheometer is ConTec Viscometer 5 (see Figure 6.4). It is one of the highly improved models of the BML rheometer. The outer cylinder (container) rotates at a series of given angular velocities during the measurement, while the inner cylinder remains

Figure 6.4 ConTec Viscometer 5.

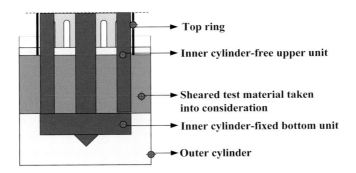

Figure 6.5 Schematic cross section of the ConTec Viscometer 5 (Heirman et al., 2009a).

still. The inner cylinder is composed of two parts, as shown in Figure 6.5. The upper unit is stationary and records the applied torque from the test material, and the lower unit is fixed guard cylinders to eliminate the influence of the end effect (which will be discussed in Section 6.2.1.4) from the measurements. This specially designed geometry ensures that the material in the gap between the upper part of the inner cylinder and the outer cylinder is subjected to the perfect Couette flow.

During the test, the material is pre-sheared at the maximum rotation speed applied for 30 s, followed by a decrease in rotational speed in given steps from the highest testing speed to the lowest speed. The average of the torque and rotational speed measured for each step will be calculated, and one set of data points are generated, provided that the torque reaches a steady state.

6.2.1.3 Principle

One of the biggest challenges in coaxial cylinder rheometry is the Couette inverse problem, which is about how to convert the rheological measurement (T, N) to the fundamental rheological unit. So far, there are two approaches to this problem: one is the numeric method, and the other is an analytical method.

Because most inverse problems have no analytical solution, the numeric method is usually the only approach. Although this method does not need to assume the exact rheological expression for fluid, excellent programming skills and extensive theoretical knowledge about the inverse problem are required, which are challenging.

On the other hand, by assuming the flow distribution of fluid and the geometry of the rheometer, the analytical method attempts to derive an equation between the rotational speed and torque for some simple rheological models. The transformation equations then can fit the test data to get the model parameters. Generally, the fitting result can be highly accurate when using an appropriate model. Therefore, the analytical method is the first choice for the Couette inverse problem.

The actual flow of fresh concrete in a coaxial cylinder rheometer is very complicated, so some presumed condition is necessary to simplify the problem:

a. The material is considered a homogeneous fluid and is invariable with time during the test.
b. Only the material in the space between the two cylinders (annulus) is considered.
c. The material in the annulus is in a stable and laminar flow state, independent of the vertical direction, and the test is executed under equilibrium conditions.
d. The flow is purely circular, and there is no flow in the radial direction.
e. Any end effects at the top or bottom of the cylinder are ignored.
f. The inertial effects are ignored.
g. There is no slippage between a cylinder and the material. The velocity of the adjacent material is equal to the one of the cylinder.

The top view of a coaxial cylinder rheometer is shown in Figure 6.6. For the Searle rheometer, the shear rate of a rotation flow at radius r in cylindrical coordinate is (Macosko, 1994):

$$\dot{\gamma} = r \frac{d\omega}{dr} \tag{6.1}$$

According to the Cauchy stress principle (Malvern, 1969), the shear stress at any distance r in the annulus with the height of h can be expressed as follows:

$$\tau = \frac{T}{2\pi r^2 h} \tag{6.2}$$

Equation (6.2) indicates that the shear stress gradually decreases as the distance from the axis of the rheometer increases. Because fresh concrete is a yield stress material, the material will not flow until the shear stress applied to the material exceeds the yield stress. As the shear stress grows, the adjacent material near the inner cylinder starts to flow. Under this condition, only part of the material close to the inner cylinder (shear zone) undergoes the shear flow. The shear flow stops at a certain imaginary radius $r = R_p$, where the shear stress is equal to the yield stress. The material outside this imaginary radius remains still (plug zone). As the shear stress keeps on growing, the region where the shear flow occurs

Concrete rheometers   129

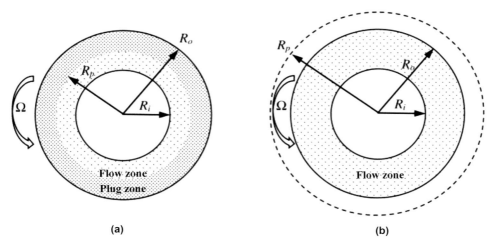

Figure 6.6 Top view of a coaxial cylinder rheometer and flow in the annulus: (a) shearing at low rotating speed and (b) shearing with high rotation speed.

is expanded outward. The boundary between the sheared zone and the unsheared zone is called the plug radius R_p. If the shear stress is large enough, the shear zone will reach or even theoretically become larger than the outer cylinder. Here, the material in the entire annulus is sheared. The material in contact with the surface of the inner cylinder has the same angular velocity as the inner cylinder.

Based on Eq. (6.1) and Eq. (6.2), the constitutive equation of the Bingham model can be written as follows:

$$\frac{T}{2\pi r^2 h} = \tau_0 + \eta \left(r \frac{d\omega}{dr} \right) \tag{6.3}$$

Rearranging the above equation leads to Eq. (6.4):

$$\left(\frac{T}{2\pi r^2 h} - \tau_0 \right) \frac{1}{r} dr = -\eta d\omega \tag{6.4}$$

Suppose all the material in the gap flows, the inner cylinder rotates at a constant angular velocity of Ω, and the out cylinder remains stationary. Equation (6.4) is integrated across the entire flow zone (the whole annulus), from $\omega = \Omega$ at $r = R_i$ to $\omega = 0$ at $r = R_s$:

$$\int_{R_i}^{R_o} \left(\frac{T}{2\pi r^3 h} - \frac{\tau_0}{r} \right) dr = -\int_0^{\Omega} \eta d\omega \tag{6.5}$$

The result of the above integration is:

$$\Omega = \frac{T}{4\pi h \eta} \left(\frac{1}{R_i^2} - \frac{1}{R_o^2} \right) - \frac{\tau_0}{\eta} \ln \left(\frac{R_o}{R_i} \right) \tag{6.6}$$

Equation (6.6) is the transformation equation of the Bingham model, named as Reiner–Riwlin equation. It is customary to replace the angular velocity Ω (in rad/s) with the rotational velocity N (in rps), $\Omega = 2\pi N$.

$$N = \frac{T}{8\pi^2 h \eta}\left(\frac{1}{R_i^2} - \frac{1}{R_o^2}\right) - \frac{\tau_0}{2\pi \eta} \ln\left(\frac{R_o}{R_i}\right) \tag{6.7}$$

or

$$T = \frac{8\pi^2 h}{\left(\frac{1}{R_i^2} - \frac{1}{R_o^2}\right)} \eta N + \frac{4\pi h \ln\left(\frac{R_o}{R_i}\right)}{\left(\frac{1}{R_i^2} - \frac{1}{R_o^2}\right)} \tau_0 \tag{6.8}$$

Although Eqs. (6.7) and (6.8) are derived from the Searle-type rheometer, they can be used for the Couette rheometers. If the material in the entire annulus is in a shear flow state, the relationship between rotational velocity (N) and torque (T) can be defined as $N = AT - B$. Therefore, the Reiner–Riwlin equation can be plotted as a straight line with its slope defined in terms of plastic viscosity, cylinder radii, and cylinder height, and the intercept defined in terms of yield stress, viscosity, and cylinder radii. Given the rheometer data versus rotational velocity, the slop (A) and intercept (B) of the straight-line fit can be determined and transformed into yield stress (τ_0) and plastic viscosity (η) using the following equation:

$$\eta = \frac{1}{8\pi^2 h A}\left(\frac{1}{R_i^2} - \frac{1}{R_o^2}\right)$$

$$\tau_0 = \frac{2\pi \eta B}{\ln\left(\frac{R_o}{R_i}\right)} \tag{6.9}$$

The same approach can apply to the Herschel–Bulkley fluid and the modified Bingham model. Heirman et al. (2006) gave a transformation equation for the Herschel–Bulkley fluid, which is written as follows:

$$\left[\left(\frac{T}{2\pi R_i^2 h K} - \frac{\tau_0}{K}\right)^{1/n}\left[n - \hat{\phi}\left(1 - \frac{T}{2\pi R_i^2 h \tau_0}, 1, \frac{1}{n}\right)\right]\right.$$
$$\left. -\left(\frac{T}{2\pi R_o^2 h K} - \frac{\tau_0}{K}\right)^{1/n}\left[n - \hat{\phi}\left(1 - \frac{T}{2\pi R_o^2 h \tau_0}, 1, \frac{1}{n}\right)\right]\right] = 4\pi N \tag{6.10}$$

However, the derivation was incomplete in literature (Heirman et al., 2006), due to that no good algorithm can be used to obtain the rheological parameters using Eq. (6.10). Heirman gave an approximate solution for the Herschel–Bulkley model:

$$T = \frac{4\pi h \ln\left(\frac{R_o}{R_i}\right)}{\left(\frac{1}{R_i^2} - \frac{1}{R_o^2}\right)} \tau_0 + \frac{2^{2n+1} \pi^{n+1} h K}{n^n \left(\frac{1}{R_i^{2/n}} - \frac{1}{R_o^{2/n}}\right)^n} N^n \tag{6.11}$$

Recently, Liu et al. (2020) derived the transformation equation of the Herschel–Bulkley model in case the material in the gap is entirely sheared, which can be presented as follows:

$$N = \frac{n\left[\left(\dfrac{T}{2\pi h K R_i^2}\right)^{\frac{1}{n}} \cdot {}_2F_1\left(-\dfrac{1}{n},-\dfrac{1}{n};1-\dfrac{1}{n};\dfrac{2\pi h \tau_0 R_i^2}{T}\right) - \left(\dfrac{T}{2\pi h K R_o^2}\right)^{\frac{1}{n}} \cdot {}_2F_1\left(-\dfrac{1}{n},-\dfrac{1}{n};1-\dfrac{1}{n};\dfrac{2\pi h \tau_0 R_o^2}{T}\right)\right]}{4\pi} \quad (6.12)$$

Feys et al. (2013) proposed a transformation equation for the modified Bingham model, which is shown in Eq. (6.13):

$$T = \frac{4\pi h \ln\left(\dfrac{R_o}{R_i}\right)}{\left(\dfrac{1}{R_i^2} - \dfrac{1}{R_o^2}\right)}\tau_0 + \frac{8\pi^2 h}{\left(\dfrac{1}{R_i^2} - \dfrac{1}{R_o^2}\right)}\mu N + \frac{8\pi^3 h}{\left(\dfrac{1}{R_i^2} - \dfrac{1}{R_o^2}\right)}\frac{(R_o + R_i)}{(R_o - R_i)} cN^2 \quad (6.13)$$

Li et al. (2019) suggested that Eq. (6.13) is only an approximate solution and derived the transformation equation in case the entire material in the gap flows under shearing. But generally, these equations are too complex to be applied in practice; thus, the Reiner–Riwlin equation is still the first choice for fresh concrete.

6.2.1.4 Measuring errors and artifacts

Interpreting data from the rheological test is quite complicated. This section discusses three major sources of measurement or interpretive errors, namely, plug flow, particle migration, and end effect.

1. Plug flow

    If only a part of the material is sheared, the shear stress at the boundary R_p equals the yield stress τ_0. The velocities at the boundary R_i and R_p are Ω and 0, respectively. For the Bingham fluid, integrating Eq. (6.4) over the flow domain is:

$$\int_{R_i}^{R_p}\left(\frac{T}{2\pi r^3 h} - \frac{\tau_0}{r}\right) dr = -\int_0^{\Omega} \eta\, d\omega \quad (6.14)$$

The result of the above integration turns out to be:

$$\Omega = \frac{T}{4\pi h \eta}\left(\frac{1}{R_i^2} - \frac{1}{R_p^2}\right) - \frac{\tau_0}{\eta}\ln\left(\frac{R_p}{R_i}\right) \quad (6.15)$$

Replacing the τ in Eq. (6.2) with τ_0, and the angular velocity Ω with the rotational speed N ($N=\Omega/2\pi$, in rps), Eq. (6.15) can be written as follows:

$$N = \frac{T}{8\pi^2 h \eta}\left(\frac{1}{R_i^2} - \frac{2\pi h \tau_0}{T}\right) - \frac{\tau_0}{4\pi \eta}\ln\left(\frac{T}{2\pi h \tau_0 R_i^2}\right) \quad (6.16)$$

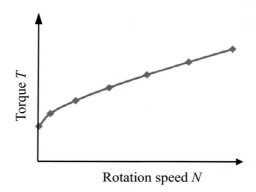

Figure 6.7 The rotation speed versus torque curve for a typical Bingham fluid.

From the above discussion, it can be seen that the N–T function of the Bingham model in coaxial rheometer can be plotted as a stepwise curve (see Figure 6.7), with a straight line with an initial curved part. The range of the nonlinear function depends on the dimension of the gap, yield stress, and height of the cylinder.

The parameters of the Bingham model can be estimated by applying piecewise fitting in different sections of the test data. However, it is not easy for most engineers.

Considering Eq. (6.7) is valid only if the shear flow occurs to all the material in the gap, it means that the plug radius should be at least equal to the radius of the outer cylinder. According to Eq. (6.2), the minimum torque for this will be $T = 2\pi R_o^2 h \tau_0$. Therefore, a straightforward method to obtain the Bingham parameters is by using Eq. (6.7) to all the test data, calculating the minimum torque for the entire shear flow, eliminating the invalid data, and repeating the above procedures until all the data left are qualified for data fitting.

Wallevik (2001) and Heirman et al. (2009b) proved that neglecting the plug flow in the ConTec Viscometers induces slight error for Bingham and Herschel–Bulkley materials, respectively.

2. Particle migration

Although fresh concrete is considered as a homogeneous material, the density difference between coarse particles and mortar matrix exists. During the rheological test, coarse aggregates are pushed away from the zone near the inner cylinder because of the highest momentum (shear rate) (Wallevik, 2003), and particle migration may occur. The severity of the particle migration decreases with the shortened test duration time, low shear rate, relatively narrow gap size, low yield stress, and high plastic viscosity of the material. Particle migration can cause the layer near the inner cylinder to get devoid of coarse aggregate, and then the material becomes heterogeneous. Especially for large gap sizes and high yield stress, particle movement from the sheared zone may increase the range and packing density of the plug zone (Feys and Khayat, 2013).

Wallevik et al. (2015) proposed a simple method to evaluate the influence of particle migration. The procedure involves the calculation of the thickness of the sheared zone via plug flow identification. If the width of the shear zone is smaller than or near the maximum size of the aggregate for the majority of time during the rheological measurement, it may imply a remarkable particle migration. Here, significantly lower rheological properties can be observed, and the rheological measurements would no longer be valid.

3. End effects

The derivation of transformation equations and related discussions presented above were based on the distribution of shear rate and shear stress in the gap of the coaxial cylinder rheometers. However, the inner cylinder should be immersed in the testing material. Because the material is filled above and below the inner cylinder and sheared during the testing, extra shear stress will be added to the total recorded torque. These end effects must be eliminated when calculating the rheological parameters.

One approach is to change the geometry of the rheometer so that shear stress applied to the cylinder ends can be minimized or eliminated. For Couette rheometers like ConTec 5 or BML Viscometer, fixed guard cylinders can be placed below (and above, if necessary). These guard cylinders are adjacent to the inner cylinder but not connected to it, allowing the inner cylinder to rotate at minor angles while measuring the stress exerting on the side. However, this solution is not suitable for a Searle-type rheometer. A double-gap coaxial cylinder rheometer is an option since the cylindrical surface is much larger on the sides than on its ends. Thus, the end effects can be omitted. Yan and James (1997) suggested that the volume below the cylinder can be considered as a parallel-plate rheometer while the volume in the annulus as a coaxial cylinder rheometer. Therefore, the total torque equals the torque from the coaxial cylinder rheometer part, plus the torque from the parallel-plate rheometer part. This method applies to the FHPCM rheometer for fresh concrete.

In order to determine the contribution of end effects to the total torque, Dzuy and Boger (1985) compared three approximate solutions with experimental measurements. They used three different distribution assumptions of shear stress acting on the end of the vane. One is the uniform distribution, the other is the power-law distribution, and the third is the prior unknown distribution. They determined that the third one is the most accurate method. However, this method needs at least two different vanes to make an experimental measurement. Whorlow (1992) suggested measuring the torque at a given rotational speed with the gap of the rheometer filled to a different height. Because the end effects are the same for all the tests, the plot of torque versus immersed vane height should be a straight line. The intercept represents the amount of torque caused by the end effect. This method is also suitable for fully immersed cylinders of different heights. Nevertheless, end effects usually vary for different materials. Individual calibrations are needed for almost every material with at least two vanes, making the test procedure very complicated.

Given this problem, Laskar et al. (2007; 2011) established a new transformation equation, which takes the end effects into account. Figure 6.8a and b illustrates the shear rate distribution profile with a stationary surface in horizontal and vertical directions. The geometrical dimensions of the vane and cylinder are shown in Figure 6.8c. The total torque exerted on the inner cylinder can be considered as the summation of the torques of all regions.

3.1 Volume $CDBA$ (V_1)

Supposing the angular velocity of the vane is ω, the shear (strain) rate along the radial direction is

$$\dot{\gamma} = \omega R_i / R_i = \omega \qquad (6.17)$$

The torque contribution is

$$T_1 = (\tau_0 + \eta\omega) 2\pi R_i^2 H \qquad (6.18)$$

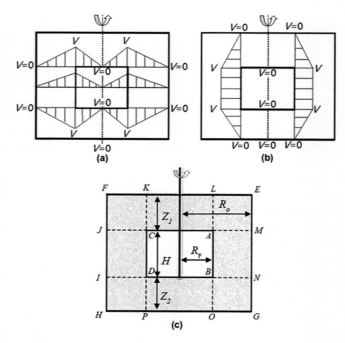

Figure 6.8 (a–c) Velocity profile along with the horizontal and vertical directions (Laskar and Bhattacharjee, 2011).

3.2 Volume $DPOB$ (V_2)

Assuming there is a circular element dr along BD at radius r from the rotational axis of the vane, the linear velocity at this radius is equal to $r\omega$, and the shear rate is $\dot{\gamma} = \omega r / Z_2$, the torque on this elemental disc can be written as follows:

$$dT = (\tau_0 + \mu\dot{\gamma})2\pi r^2 dr \tag{6.19}$$

The torque T_2 is:

$$T_2 = \int_0^{R_i}\left(\tau_0 + \eta\frac{\omega r}{Z_2}\right)\cdot 2\pi r^2 dr = \frac{2\pi R_i^3}{3}\tau_0 + \frac{\pi R_i^4 \omega}{2Z_2}\eta \tag{6.20}$$

3.3 Volume $KCAL$ (V_3)

Similar to 3.2, the torque T_3 can be expressed as follows:

$$T_3 = \int_0^{R_i}\left(\tau_0 + \eta\frac{\omega r}{Z_1}\right)\cdot 2\pi r^2 dr = \frac{2\pi R_i^3}{3}\tau_0 + \frac{\pi R_i^4 \omega}{2Z_1}\eta \tag{6.21}$$

3.4 Volume of hollow-cylinder $IHPD$-$BOGN$ (V_5)

As shown in Figure 6.9, for an elemental layer of thickness dz at a height z from the bottom of the cylindrical surface $DPOB$, the velocity along the radial direction on the surface of $DPOB$ is written as: $v_r = \frac{z}{Z_2}\omega R_i = \frac{vz}{Z_2}$. Define the effective

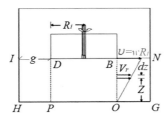

Figure 6.9 Velocity profile in the gap between vane and outer cylinder (Laskar and Bhattacharjee, 2011).

gap of annulus g as a constant of R_o-R_i; the shear stress at the height of z from the bottom is given as $\tau_r = \tau_o + \mu \dfrac{vz}{Z_2 g}$. The force on the elemental area will be $dF = \left(\tau_0 + \mu \dfrac{vz}{Z_2 g}\right) \cdot 2\pi R_i dz$, and the total force is given as follows:

$$F = \int_0^{Z_2} dF = 2\pi R_i \left(\tau_o + \frac{\eta v}{2g}\right) Z_2 \tag{6.22}$$

And the torque of the part is:

$$T_4 = R_i F = 2\pi R_i^2 Z_2 \left(\tau_0 + \frac{\omega R_i \eta}{2g}\right) \tag{6.23}$$

3.5 Volume of hollow-cylinder *FJCK-LAME* (V_4)

The torque T_4 can be deduced in a similar approach as in 3.4 above, and the result is shown in Eq. (6.24):

$$T_5 = R_1 \cdot \int_0^{Z_1} dF = 2\pi R_i^2 Z_1 \left(\tau_0 + \frac{\omega R_i \eta}{2g}\right) \tag{6.24}$$

3.6 Volume in the gap of *JIDC-ABNM* (V_6)

The velocity and shear rate on the surface of CDBA are $v = \omega R_i$ and v/g, respectively. The torque of this part is:

$$T_6 = \left(\tau_0 + \eta \frac{v}{g}\right) \cdot 2\pi R_i^2 H \tag{6.25}$$

Assuming that the vane is at the center of the testing material, $Z_1 = Z_2 = Z$, and $T_2 = T_3$, $T_4 = T_5$. According to the literature (Barnes and Carnali, 1990, Dzuy and Boger, 1985, Sherwood and Meeten, 1991), material in volume V_1 does not shear during the test. Therefore, $T_1 = 0$. The summation of the above torques of different parts results in Eq. (6.26):

$$T = 2T_2 + 2T_4 + T_6 \tag{6.26}$$

Figure 6.10 Serrated shapes of the inner cylinder of the ConTec Viscometer 5.

By replacing T_2, T_4, and T_6 in Eq. (6.26) with Eq. (6.20), Eq. (6.24), and Eq. (6.25), the following relationship can be derived:

$$T = 4\pi R_i^2 \left(\frac{H}{2} + Z + \frac{R_i}{3} \right) \tau_0 + \frac{\pi^2 R_i^3}{15} \left(\frac{R_1}{2Z} + \frac{H+Z}{R_o - R_i} \right) \eta N \tag{6.27}$$

The equation above shows that the total torque T is a linear function of rotational speed N. Therefore, the Bingham parameters can be calculated by fitting the experimental data with Eq. (6.27).

4. Hydrodynamic pressure

In order to avoid slippage between the concrete and the rheometer, the internal wall of the rheometer needs to be designed with serrated shapes (Figure 6.10), allowing the coarse aggregate to be part of the internal boundary. The coarse aggregates will sit in the space between the serration, and no slippage boundary can be observed. However, well-fined flow occurs between the blades for the four blades-vane rheometers, introducing extra errors in calculating the shear rate. Wallevik (2014) indicated that hydrodynamic pressure would apply to the blades' wall boundary for a four blades-vane rheometer and contribute to torque, just like the viscous stresses. Therefore, the total torque is the summation of the torque caused by viscous stress and the torque caused by hydrodynamic pressure. Numerical simulation suggested that hydrodynamic pressure accounts for 80% of the total torque recorded by the rheometer. The conclusion is valid for the Newtonian, Bingham, and Herschel–Bulkley fluids. This emphasizes the necessity of thoroughly studying the influence of hydrodynamic pressure on different types of rheometers.

6.2.2 Parallel-plate rheometer

A parallel-plate rheometer is generally used to measure the rheological properties of polymeric materials. The distance between the two plates is usually less than 1 mm. However, it is not applicable to concrete because of particles, sand, and coarse aggregates. The distance between the two plates should be 5–10 times the maximum particle size of concrete mixture, i.e., 50–100 mm. One solution is to fill the concrete in a cylindrical container with a blade rotation. A seal is needed to keep the material between the blades to ensure the material is sheared without a leakage. Because the container introduces extra shear resistance to the material, the analytical solution to decide the shear rate and shear stress cannot be applied.

One typical parallel-plate rheometer for concrete is BTRHEOM, developed at LCPC (Laboratoire Central des Ponts et Chaussées). The rheometer is specially designed to measure moderately to highly flowable concrete (slump at least 100 mm, up to SCC). By assuming complete concrete slippage at the wall of the container, its influence can be ignored. The maximum size of aggregate is up to 25 mm. It holds about 7 L specimens of concrete and can be used in the lab and on a construction site.

6.2.2.1 Geometry

The prototype of the BTRHEOM rheometer is illustrated in Figure 6.11. It consists of a 120 mm radius hollow cylinder with two parallel blades installed at the top and bottom of the cylinder. The vertical distance between the blades is 100 mm. The bottom blade is fixed as the top blade rotates. Both the blades are designed with openings, so that slippage can be avoided. A motor is mounted below the cylinder and connected to the top plate vertically with a shaft. The radius of the shaft is 20 mm. A vibrator can apply to the sample material in the container to either consolidate concrete or measure the vibration's effect on the rheological parameters. However, the rheological measurement cannot be carried out during the vibration. As the top blade rotates at a series of different rotational velocities, torque caused by the resistance of the concrete being sheared is measured through the top blade.

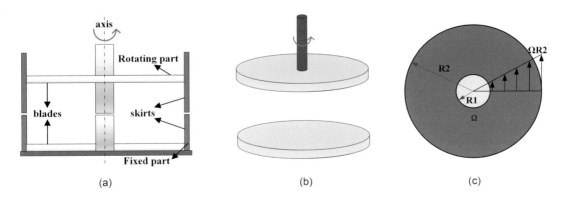

Figure 6.11 (a–c) Prototype of the BTRHEOM rheometer (Hu et al., 1996).

An accompanying software program, ADRHEO, operates the rheometer (rotation speed and vibration), collects the measurements (torque and rotation speed), and calculates the rheological parameters from the raw data. The maximum measurable torque value is about 14 N·m. The range of rotation speed is from 0.63 rad/s (0.1 rev/s) up to 6.3 rad/s (1 rev/s), though 0.63 rad/s (0.1 rev/s) and 5.02 rad/s (0.8 rev/s) are chosen as the lower and upper limits for testing, respectively. A text output file will be generated after the test, including the torque values at each rotation speed and the calculated rheological parameters, either the Bingham or the Herschel–Bulkley parameters (depending on the different versions of ADRHEO software).

6.2.2.2 Principle

Before initial testing, the BTRHEOM rheometer needs to be calibrated for rotational speed, torque, and vibration frequency according to the operation manual. A seal is used to ensure that no concrete flows into the region between the base container and upper rotating cylinder before each testing. Then a rotational calibration test is executed for further refinement. For each set of seals, a rheology test is performed with water to determine the frictional resistance of the seal. The ADRHEO software uses the results to eliminate the friction effects of the seals in the following rheological test of concrete or mortar samples.

The ADRHEO software controls the entire testing process. After the container is filled with the testing material, an optional operation is to vibrate it for 15 s to consolidate the concrete (vibration can also be applied during the measurement). The frequency of this pre-vibration ranges from 35 to 55 Hz. It should be noted that pre-vibration is not for concrete with low yield stress, such as SCC. Then measurement starts. The test is composed of one or two sequential down ramps (up ramps are also possible but rarely used, except for the thixotropy test). Each down ramp includes five to ten measurement points, i.e., torques at decreasing rotation speed are collected. For each measurement data point, the rotation speed remains constant for about 20 s so that a torque measurement can be stabilized and recorded.

Under the no-slip boundary condition, the rotational velocity at a radius r is Ωr on the upper blade and zero on the lower blade in cylindrical coordinates. The shear rate at radius r can be written as follows:

$$\dot{\gamma}(r) = \frac{\Omega r}{H} \tag{6.28}$$

Equation (6.28) indicates that the shear rate alters along the radial direction, and simple shear can be achieved at any radius r in a parallel-plate rheometer. As a result, the local shear rate in between the two blades is known no matter what the material constitutive law is.

It can also be shown that the shear stress $\tau(R)$ at the edge of the geometry ($r = R$) can be presented as follows:

$$\tau(R) = \frac{T}{2\pi R^2}\left[3 + \frac{\Omega}{T}\frac{\partial T}{\partial \Omega}\right] \tag{6.29}$$

Let $r = R$, and the share rate at the edge of the geometry is $\dot{\gamma}(R) = \frac{\Omega R}{H}$. Because of the derivative term of the torque regarding the rotational velocity in Eq. (6.29), numerous accurate $T(\Omega)$ data need to be collected to calculate $\tau = (\dot{\gamma})$, which is relatively difficult to use.

Another approach is to replace the shear stress and shear rate in a rheological model by Eqs. (6.28) and (6.29), and establish a transformation equation between the torque T and the rotational speed N. The rheological properties can be estimated via data regression from a single $T(\Omega)$ measurement.

According to the experiment, the relationship between the torque T and the rotation speed N can be expressed as follows:

$$T = T_0 + AN^n \tag{6.30}$$

where T_0 is the flow resistance, A is the viscosity factor, and n is the flow index factor. This expression is in a similar function form to the Herschel–Bulkley model.

Suppose fresh concrete is a Herschel–Bulkley material; its constitutive equation is $\tau = \tau_0 + K\dot{\gamma}^n$. If the power exponent index equals 1, the Herschel–Bulkley model will be turned into the Bingham model ($\tau = \tau_0 + \eta\dot{\gamma}$).

Integrating the contribution of all the surface elements to the torque leads to the following equations (De Larrard et al., 1998):

$$\begin{cases} T_0 = \dfrac{2\pi}{3}\left(R_2^3 - R_1^3\right)\tau_0 \\ A = \dfrac{(2\pi)^{n+1}}{(b+3)h^n}\left(R_2^{n+3} - R_1^{n+3}\right)K \end{cases} \tag{6.31}$$

The above equations can be inverted to get the material parameters by nonlinear regression of the T–N curve, which results in:

$$\begin{cases} \tau_0 = \dfrac{3}{2\pi\left(R_2^3 - R_1^3\right)}T_0 \\ K = 0.9\dfrac{(n+3)}{(2\pi)^{n+1}}\dfrac{h^n}{\left(R_2^{n+3} - R_1^{n+3}\right)}A \end{cases} \tag{6.32}$$

For the Bingham fluid, the yield stress would be τ_0, and the equivalent plastic viscosity η would be:

$$\eta = \dfrac{3K}{n+2}\dot{\gamma}_{max}^{n-1} \tag{6.33}$$

where $\dot{\gamma}_{max} = \dfrac{\Omega_{max}R_2}{h}$ is the maximum shear rate recorded in the measurement.

One thing that should be mentioned here is that the BTRHEOM cannot shear the specimen adequately at the angular velocity of 0.015 rad/s (0.1 rev/s) or lower. The data point will be excluded as an outlier for data analysis.

6.2.2.3 Measuring errors and artifacts

One advantage of using the rheometer is that the entire material on a horizontal plane is sheared uniformly for a given loading condition. As a result, plug flow, which often appears in coaxial cylinder rheometers, does not occur in the BTRHEOM rheometer.

The shear-induced sedimentation of particles is a possible measurement artifact. Note that the vertical gap between the two parallel plates is about 100 mm, and then the gravity effect caused by the difference between the cement–sand mix and aggregate cannot be ignored. During the rheological test, particles, especially coarse aggregates, tend to settle down from the zone near the upper plate. Other possible artifacts include the slip risk and wall effect. Due to the limited volume of sheared specimens, these artifacts are minimal. Therefore, no considerable systematic error will be introduced after careful calibration. Considering all the friction-related effects, a 10% correction applies to factor K, as it is usually done for plastic viscosity (Hu et al., 1996).

Another interesting point is about the calculation of Bingham parameters. Although Eqs. (6.32) and (6.33) can be used to obtain the yield stress and plastic viscosity, these two rheological parameters can be calculated directly from Eq. (6.30) provided $n=1$. Unfortunately, the calculation results are different using these two approaches. Therefore, further research is needed to decide which method is better.

6.2.3 Other rheometers

6.2.3.1 CEMAGREF-IMG rheometer

The CEMAGREF-IMG rheometer is a large coaxial cylinder rheometer, and it can load about 500 L of concrete for testing (see Figure 6.12). During the test, the inner cylinder rotates, while the outer cylinder remains stationary. The surface of the inner cylinder is made of a metallic grid to minimize the slippage of concrete, and vertical blades are welded on the internal wall of the outer cylinder. A rubber gasket is attached to the bottom of the inner cylinder to prevent material leakage from the slit between the bottom of the container and the cylinder.

However, because of its large volume, some plug flow is expected when testing concrete materials. Therefore, Eq. (6.7) can calculate the rheological parameters of the Bingham material. Alternatively, Eq. (6.16) should be used if plug flow occurs.

6.2.3.2 Viskomat XL

Schleibinger developed Viskmat XL based on many years of experience with rheometer for mortar and fresh concrete. The Viskmat XL is a vane rheometer, which can rotate in both clockwise and counterclockwise directions (See Figure 6.13). Unlike the ICAR rheometer, the Viskomat XL has frame vanes. Usually, it is run under a shear rate-controlled mode, and the rotational speed can be programmed in linear steps, either increasing or decreasing. Also, a logarithmic or oscillating mode is possible as an option. Alternatively, it can be run in a shear stress-controlled mode. Therefore, the torque can be predefined over time, so the rotational speed can automatically be controlled to achieve the preset torque (shear stress). Viskomat XL equips a double-wall container, where cooling liquid can circulate between the walls. Via this, the temperature of the specimen can be controlled. It also has a high time resolution, and the sample rate can be set from 0.005 s to 10 min. Overall, it is a versatile apparatus for the rheological measurement of mortar and fresh concrete.

6.2.3.3 The IBB rheometer

The IBB rheometer is based on the existing device (MKIII) developed by Tattersall (see Figure 6.2b). It is completely automated and uses a computer program to drive an impeller to rotate in fresh concrete. The software can analyze the testing results and calculate the rheological parameters of the Bingham material.

Concrete rheometers 141

Figure 6.12 The CEMAGREF-IMG rheometer (Banfill et al., 2001).

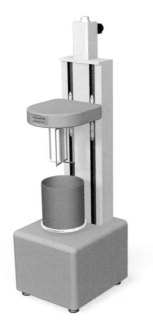

Figure 6.13 The Viskmat XL.

Figure 6.14 Rheometer developed by Yuan and Shi.

6.2.3.4 Rheometer developed in China

Yuan and Shi have developed a prototype rheological instrument, which is basically a vane rheometer (Yuan and Shi, 2018). It is commercialized by a Chinese company a few years ago (see Figure 6.14). In comparison with other rheometers, this one is more automated. After loading concrete in the container, the shaft will drop automatically to a specified height and start rotation in rampway to build a torque–rotation speed curve. The fresh concrete specimen is first sheared in the highest speed to eliminate thixotropy, and then ramp downwards. The Bingham model and other rheological models can be used to transform torque–rotation speed data into yield stress and viscosity by software embedded in the rheometer. A cylindrical rotator is also equipped to measure the properties of the lubrication layer. And the software has a friendly interface for each function, which can facilitate the operation of the rheometer.

6.2.3.5 The modifications of the BTRHEOM rheometer

As for the parallel-plate rheometer, the UIUC concrete rheometer was built at the University of Illinois (Beaupré et al., 2004). The significant improvements are to help install and clean the device during the test. Besides, other researchers modified the BTRHEOM (Struble et al., 2001, Szecsy, 1997), but these modified rheometers have not been commercialized. The major improvements include suspending the top plate from a shaft held above the concrete and eliminating the center axis inside the concrete container.

6.2.3.6 Other instruments

Laskar and Bhattacharjee (2011) of India developed a cross vane rheometer; however, more researchers focused on designing the spindle with a different geometry. For example, Gerland et al. (2019) modified Viskomat NT with a ball probe. They used a simulation-based approach to determine the yield stress and plastic viscosity of UHPC and SCC from the rheometer measurement. Soualhi et al. (2017) from France also designed a portable rheometer with new vane geometry. Other spindle types include helical, H-shape, and double spiral.

Although rotational rheometers can give an accurate description of the rheological behavior of fresh concrete, they are not suitable for online and continuous applications. On the other hand, the rheological measurements of fresh concrete during mixing and transportation have aroused wide concern. One example is the attempt to correlate the output of the concrete-mixing truck to values obtained by rotational rheometers. The output could be a watt meter or hydraulic pressure, which is known as the "slump meter". However, the experimental errors might be higher than usual, making it more difficult to characterize the rheological values of the concrete-mixing truck. Wallevik et al. (2020) used a series of computer simulations to analyze the relationship between the power required to rotate the drum of a concrete-mixing truck and the rheological properties of fresh concrete. They found that the power could be calculated using different Bingham parameters (i.e., yield stress τ_0 and plastic viscosity μ). With those simulations, the rotation speed–power of the drum curves were plotted, and the intersection value G and the slope H were calculated, which can be used to represent the truck's rheological properties. By power used per unit mass, a possible relationship between the truck's rheological values G and H and the Bingham parameters τ_0 and μ could be established. Although these two equations are not completely accurate, it proves that using the vehicle as a rheometer is technically feasible.

6.3 MEASURING PROCEDURES

6.3.1 Preparation of specimen

For rheological testing, the mixing process of concrete materials is similar to those for other properties. The concrete specimen should be homogeneous so that stable and reproducible rheological measurements can be achieved during the test. In addition, adequate raw materials should be prepared to ensure the test accuracy; usually, the capacity of the mixture should be at least two times the volume of the container for rheological testing.

6.3.2 The testing procedures of ICAR

The operation interface of ICAR rheometer software is shown in Figure 6.15. All the operations can be managed through a single screen. In this interface, you can specify the name of the storage file and its storage path, set the geometry parameters, and execute the stress growth test and flow curve test. If something goes wrong, the software can terminate the trial.

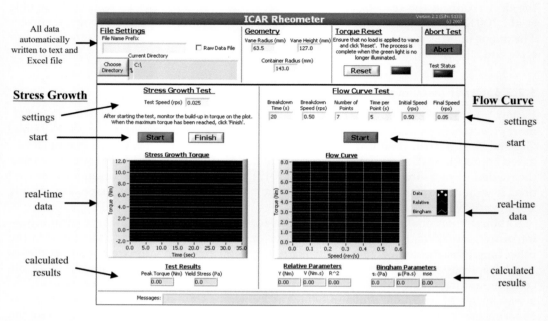

Figure 6.15 The ICAR rheometer software interface.

The testing procedures for stress growth test include the following steps:

1. Assemble all the components of the ICAR rheometer correctly according to the manual;
2. Place the testing mixture into the container. Ensure the mix and the vertical ribs on the internal wall of the container have the same height. Insert the cross vane into the mix;
3. Ensure the default geometrical parameters set in the operation interface are the actual sizes of the vane and the container;
4. Input the rotation speed value. The vane speed for the stress growth test is typically between 0.01 and 0.05 rps, with 0.025 rps by default;
5. Click the "Reset" button to ensure the initial torque of the vane shaft is zero;
6. Start the test;
7. Once the peak of the torque–time curve appears and the torque gradually decreases, click the "Finish" button to complete the test.

The testing procedures for the flow curve test are as follows:

1. After the stress growth test is completed, the flow curve test of the mixture can be carried out;
2. In order to destroy the flocculation structure of the mixture and provide a consistent shearing history, it is necessary to let the vane rotate at maximum speed. Usually, the breakdown time and the breakdown speed are set to 20 s and 0.50 rps, respectively, before starting the flow curve test;
3. Set the initial rotational speed for the flow curve test as 0.50 rps and the end speed as 0.05 rps, which is divided into seven test points evenly, each of which lasts for 5 s;
4. Click the "Start" button to start the flow curve test;
5. If the mixture is so viscous or the yield stress is so high that the vane cannot rotate at the maximum speed of 0.50 rps, immediately click the "Abort" button to stop the test;

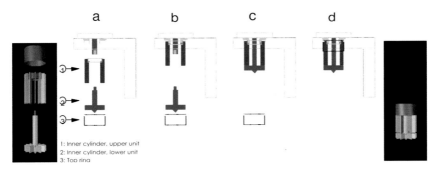

Figure 6.16 The assembly of the inner cylinder of Viscometer 5 (Banfill et al., 2001).

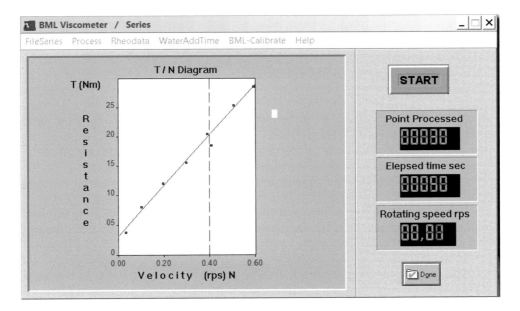

Figure 6.17 The main operation interface of ConTec Viscometer 5.

6. Once the test is finished, the flow curve will be plotted and illustrated in the interface windows, and the linear fitting and Bingham parameters will be calculated. Then, testing data will be written into an output file for further analysis.

6.3.3 The testing procedures of ConTec Viscometer 5

Before testing, the inner cylinder should be mounted to the axis of the testing unit, as illustrated in Figure 6.16. Each test takes about three minutes, and the whole test is controlled by a special software named FRESHWIN (see Figure 6.17). During the trial, the specimen is sheared for about 1 min. Then, the container can be emptied for the next test.

All the parameters needed for testing can be inputted into the software as a basic setup. As the test is finished, the output of the test result will be plotted (see Figure 6.17), and the rheological parameters in the fundamental unit will be calculated.

Similar to the flow curve test of the ICAR rheometer, the flow curve test uses seven points to calculate the rheological parameter. The rotational speed equally decreases between the highest and lowest speeds. For each point, the testing period is 5 s, including the first 1.5 s transient interval and 3.5 sampling interval. Then, rotational speed increases to the maximum speed for 2 s afterward and continues to rotate for 5 s at 2/3 maximum speed so that the aggregate segregation factor can be evaluated (segregation test). The testing procedures include the following steps:

1. Fill the concrete mixture into the container, and lower the inner cylinder until it inserts into the container to a fixed depth. Ensure the mix has the same height as the stripes welded on the internal wall of the container.
2. Execute the FRESHWIN software, select "Process" from the menu, and select "Parameters" from the drop-down menu. Choose the existed test name in the "Name" input box, or click the "Add" button to set up a new test.
3. Ensure to input the correct values in the "Cylinder dimensions" group box. By default, the height of the inner cylinder is 0.2 m, the radius of the inner cylinder is 0.1 m, and the radius of the outer cylinder is 0.145 m.
4. Select the correct equation in the "Equation" group box. "Reiner-Rivlin equation" can be applied in most cases.
5. Input proper parameters in the "Run time parameters" and "Beater Control" group boxes.
6. Click the "OK" button, close the process parameter setting, and return to the initial interface of the FRESHWIN software. Click the "Start" button to execute the test. The inner cylinder will descend and insert into the mix. After the pre-shearing, the flow curve test will be conducted and followed by the segregation test.
7. Select "Save" from the "File" menu, and store the data on the disk after finishing the test.
8. The inner cylinder will rise automatically, and the container will be taken off, emptied, and ready for the next test. If time permits, the inner cylinder should be disassembled and cleaned after every two tests.

For each test, the rheological parameters will be calculated automatically. Besides, the diagram of rotation velocity vs. torque will be drawn, and linear regression will be used to establish an equation between rotation velocity and torque. A particular point on the line as the rotation velocity is 2/3 of the maximum velocity (see Figure 6.18), will be calculated. Based on this, the segregation factor will be calculated using the following equation:

$$\text{Seg} = \frac{H - H'}{H} \cdot 100\% \tag{6.34}$$

The aggregate segregation test examines if the aggregate is segregated from the concrete mix. If the segregation factor Seg is <5%, it means that no apparent segregation in the trial, concrete mix is very stable and consistent. The concrete mix is unstable for Seg greater than 10%, and aggregate segregation significantly influences the test result. Further tests will be needed in such cases.

The Viscometer 5 can perform the thixotropy test as well. The test type "Thixotropy" from the drop-down list in the name box will perform the test. The rest setting is similar to the flow curve test. The standard loading procedure for thixotropy testing is shown in Figure 6.19. The material should be previously at rest for a pre-determined period to detect the effect of thixotropy.

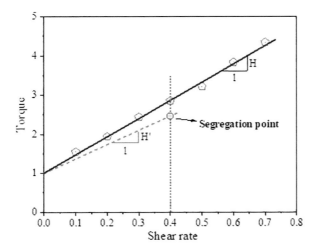

Figure 6.18 The torque–rotational speed diagram (Wallevik, 2006).

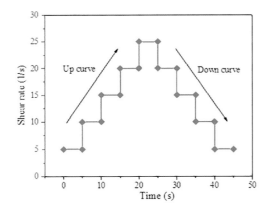

Figure 6.19 The rate of shear history to evaluate the thixotropy by Viscometer 5 (Wallevik, 2006).

6.3.4 The testing procedures of the BTRHEOM rheometer

Before preliminary testing, the BTRHEOM rheometer needs to be calibrated for rotational speed, torque, and vibration frequency according to the operation manual. Two seals are used to ensure that no concrete flows into the region between the base container and upper rotating cylinder before each testing. Then, a rotational calibration test is executed for further refinement. For each set of seals, a rheology test is performed with water to determine the frictional resistance of the seal. The ADRHEO software uses the results to eliminate the friction effects of the seals in the following rheological test of concrete or mortar samples.

The ADRHEO software controls the entire testing process. After the container is filled with the testing material, an optional operation is to vibrate it for 15 s to improve the compactability of the concrete. The vibration can also be used during the measurement for the same purpose. The frequency of this pre-vibration ranges from 35 to 55 Hz. It should be noted that pre-vibration is not applicable for concrete with low yield stress, such as SCC.

Then, measurement starts. The test comprises one or two sequential down ramps (up ramps are also possible but rarely used, except for the thixotropy test). Each down ramp includes five to ten measurement points, i.e., torques at decreasing rotation speed are collected. For each measurement data point, the rotation speed remains constant for about 20s so that a torque measurement can be stabilized and recorded.

6.4 DATA COLLECTION AND PROCESSING

Fresh concrete is generally considered a homogeneous material during the test with a rheometer. However, fresh concrete is, in fact, a heterogeneous material that consists of various particles and water. The torque and rotational speed readings often show significant and frequent fluctuations. As a result, it is pretty common that the rheological parameters differ in repeated tests. Usually, it is acceptable if the error is less than 20%. Otherwise, further trials will be compulsory.

6.4.1 Static yield stress test

The time–torque curve can be plotted based on the data recorded in the stress growth test. As presented in Figure 6.20, the highest point of the curve is the yield torque. The ascend section before the highest point is linearly elastic. After reaching the yield torque, the internal flocculation structure is destroyed, and the specimen starts to flow. As a result, the torque value decreases. The shear stress corresponding to the yield torque is the static yield stress, which can be calculated using the following equation:

$$\tau_{S0} = \frac{2T_m}{\pi D^3 \left(\dfrac{H}{D} + \dfrac{1}{3}\right)} \tag{6.35}$$

where τ_{S0} is the static yield stress, T_m is the maximum (yield) torque, D and H are the diameter and height of the inner cylinder, respectively.

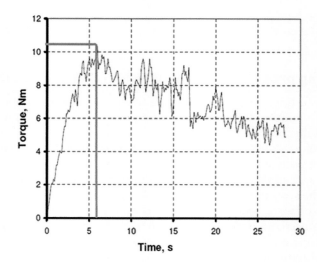

Figure 6.20 The typical toque–time curve of concrete mixture.

Concrete rheometers 149

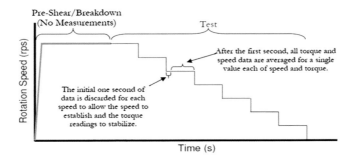

Figure 6.21 The time-rotation speed schematic diagram of ICAR rheometer.

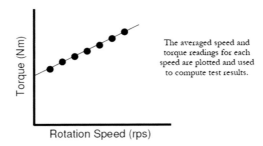

Figure 6.22 The rotation speed–torque diagram.

6.4.2 The flow curve test

The flow curve test measures the shear stress at different shear rates. Figure 6.21 illustrates the ideal rotation speed–time curve of the ICAR rheometer for the testing. Numerous torque and speed sample points at each rotational speed are collected during the testing. For each speed, the torque and averaged speed readings are plotted (see Figure 6.22). These data can establish the relationship between torque and rotational speed, and compute the yield stress and plastic viscosity.

According to Eq. (6.8), a linear fit between torque and rotation speed can be made using the seven torque–rotation speed data pairs. It is:

$$T = G + H \cdot N \tag{6.36}$$

where T is the torque in N·m, N is the rotation speed in rps, G is the intercept related to the yield stress, and H is the slope associated with the plastic viscosity. Therefore, the yield stress and plastic viscosity are as follows:

$$\begin{cases} \tau_0 = \dfrac{\left(\dfrac{1}{R_i^2} - \dfrac{1}{R_o^2}\right)}{4\pi h \ln\left(\dfrac{R_o}{R_i}\right)} G \\ \eta = \dfrac{\left(\dfrac{1}{R_i^2} - \dfrac{1}{R_o^2}\right)}{8\pi^2 h} H \end{cases} \tag{6.37}$$

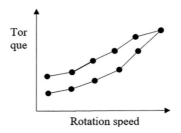

Figure 6.23 The hysteresis loop of the thixotropy.

It should be pointed out that if the correlation coefficient of the above linear regression is over 0.9, the yield stress and plastic viscosity are convincible. The concrete mix can be considered as the Bingham fluid. Otherwise, some outliers may be removed, or other transformation equations may be applied.

6.4.3 Thixotropy test

The thixotropy of a concrete mix is usually assessed by the hysteresis loop area enclosed by the up and down ramps of the rotational speed and torque curves. The rotational speed–torque hysteresis loop is presented in Figure 6.23. The larger the hysteresis loop area, the greater the thixotropic effect.

6.5 RELATION OF RHEOLOGICAL PARAMETERS MEASURED BY DIFFERENT RHEOMETERS

Two international comparison tests of concrete rheometers were held in 2000 and 2003, respectively (Banfill et al., 2001, Beaupré et al., 2004). It was found that each rheometer gave distinct values of the Bingham parameters in terms of yield stress and plastic viscosity. To a large extent, these differences were attributed to the wall slip and friction. There were linear correlation functions between-rheometer for both the yield stress and plastic viscosity (see Tables 6.1–6.4). Tables 6.1 and 6.2 were based on the comparison test in October 2000. Tables 6.3 and 6.4 were based on the comparison test in May 2003.

They also found a linear relationship between the slump test and the yield stress for all the rheometers. In both comparison tests, all the mixtures can be sorted statistically in the same sequence by all the rheometers for the yield stress and plastic viscosity. All the rheometers can evaluate the rheological behavior of concrete based on the Bingham parameters.

Hočevar et al. (2013) compared the testing results of concrete mixtures using both ConTec Viscometer 5 and ICAR rheometer. They found that the yield stress values and the plastic viscosity calculated by using ConTec Viscometer 5 differ significantly from those of the ICAR rheometer. There are linear relationships between the yield stress calculated by the ConTec Viscometer 5 and ICAR rheometer, and the same is true for the plastic viscosity. However, the yield stress values given by the ICAR rheometer are 42% higher than those by ConTec Viscometer 5 on average. In comparison, the plastic viscosity values provided by the ICAR are 43% lower compared to the Viscometer 5. They also noticed that the test results of the ICAR rheometer are less repeatable than those of the ConTec Viscometer 5. Both rheometers showed higher repeatability for measuring the yield stress than for the plastic viscosity.

Concrete rheometers 151

Table 6.1 Between-rheometer linear correlation functions for the yield stress (Banfill et al., 2001)

A; B	BML (Pa)	BTRHEOM (Pa)	CEMAGREF-IMG (Pa)	IBB (N.m)	Two-point (Pa)	Slump (mm)
BML (Pa)	–	1.85; 300.9	1.98; 179.89	0.008; 0.334	1.010; 7.007	−0.18; 248.2
BTRHEOM (Pa)	0.50; −122.0	–	0.974; −93.6	0.004; −0.91	0.54; −153.9	−0.09; 273.0
CEMAGREF-IMG (Pa)	0.45; −40.7	0.91; 204.7	–	0.0049; −0.824	0.56; −99.79	−0.09; 261.1
IBB (N.m)	79.7; 126.2	155.3; 504.3	163.4; 316.9	–	95.4; 75.4	−15.8; 231.6
Two-point (Pa)	0.87; 46.3	1.72; 338.3	1.75; 194.6	0.008; 0.114	–	−0.17; 244.9
Slump (mm)	−5.59; 1387.3	−10.77; 2942.2	−11.1; 2898.1	−0.063; 14.7	−5.99; 1466.8	–

Note: In Tables 6.1–6.4, the rheometers in the column head are Y and those in the row head are X. In each cell, the coefficients of the equation $Y = AX + B$ are shown. The rheometers are listed in alphabetical order.

Table 6.2 Between-rheometer linear correlation functions for the plastic viscosity (Banfill et al., 2001)

A; B	BML (Pa.s)	BTRHEOM (Pa.s)	CEMAGREF-IMG (Pa.s)	IBB (N.m.s)	Two-point (Pa.s)	Mod slump (Pa.s)
BML (Pa.s)		1.202; 6.20	2.06; −31.19	0.089; 5.3	0.37; 13.84	−0.08; 51.6
BTRHEOM (Pa.s)	0.59; 11.16		1.01; −9.91	0.056; 6.07	0.36; 7.20	−0.11; 75.24
CEMAGREF-IMG (Pa.s)	0.47; 15.7	0.81; 15.50		0.081; 5.36	0.35; 12.68	−0.039; 51.7
IBB (N.m.s)	10.37; −50.9	13.2; −62.60	11.9; −62.9		4.43; −10.97	−0.009; 10.5
Two-point (Pa.s)	0.926; 20.06	1.87; 9.88	1.59; −6.19	0.0961; 6.64		−0.034; 36.8
Mod slump (Pa.s)	−2.03; 263.1	−0.79; 233.3	−1.79; 269.9	−16.4; 331.1	−0.69; 210.9	

Table 6.3 Between-rheometer linear correlation functions for the yield stress (Beaupré et al., 2004)

A; B	BML (Pa)	BTRHEOM (Pa)	Two-point (Pa)	IBB (N.m)	IBB portable (N.m)	Slump (mm)
BML (Pa)		1.66; 229.9	1.34; 76.6	0.0090; 0.062	0.0113; 0.221	−0.258; 288.8
BTRHEOM (Pa)	0.48; −64.0		0.61; 1.17	0.0040; −0.363	0.0053; −0.493	−0.134; 310.6
Two-point (Pa)	0.688; −36.16	1.225; 156.8		0.0067; −0.342	0.0081; −0.267	−0.194; 302.2
IBB (N.m)	102.4; 9.51	178.2; 249.2	123.6; 114		1.12; 0.383	−24.25; 280.6
IBB portable (N.m)	85.5; −11.7	154.4; 191.9	113.5; 61.84	0.844; −0.196		−22.56; 291.8
Slump (mm)	−3.58; 1050	−6.14; 2018	−4.75; 1468	−0.032; 9.52	−0.04; 11.87	

Table 6.4 Between-rheometer linear correlation functions for the yield stress (Beaupré et al., 2004)

A; B	BML (Pa.s)	BTRHEOM (Pa.s)	Two-point (Pa.s)	IBB (N.m.s)	IBB portable (N.m.s)
BML (Pa.s)		1.84; 44.3	2.15; −12.4	0.698; −10.85	0.683; −7.10
BTRHEOM (Pa.s)	0.272; 6.39		0.387; 21.9	0.151; −2.47	0.138; 0.1692
Two-point (Pa.s)	0.227; 22.3	0.233; 98.47		0.246; −0.52	0.210; 3.14
IBB (N.m.s)	0.927; 23.1	1.32; 92.76	3.52; 10.3		0.815; 3.88
IBB portable (N.m.s)	1.17; 17.6	1.65; 85.9	4.26; -6.65	1.11; −3.03	

6.6 SUMMARY

This chapter reviewed several types of rheometers for fresh concrete and mortar. So far, all the rheometers can be divided into three groups:

1. Coaxial rheometers: ICAR, ConTec Viscometer, Viskomat XL, BML, and CEMAGREF-IMG.
2. Parallel rheometers: BTRHEOM and UIUC.
3. Mixer with an impeller: IBB and other two-point apparatuses.

All the concrete rheometers are rotational and can measure torque values at different rotational speeds. Since torque–rotational speed equations are derived from rheological models, the data recorded by a rheometer can be used to calculate the rheological parameters. Fresh concrete samples for rheological testing don't require special treatment, but testing procedures should be strictly followed to avoid errors and artifacts. It should be noted that although correlations between the test results from any pair of rheometers can be observed, these results vary significantly. Therefore, more research is needed to find out what causes these differences and design better rheometers.

REFERENCES

Banfill, P., et al. (2001). Comparison of concrete rheometers: International test at LCPC (Nantes, France) in October, 2000. National Institute of Standards and Technology Interagency Report (NISTIR) 6819.

Banfill, P. (2006). "Rheology of fresh cement and concrete." Rheology Reviews 2006, The British Society of Rheology.

Barnes, H., and Carnali, J. E. (1990). "The vane-in-cup as a novel rheometer geometry for shear thinning and thixotropic materials." *Journal of Rheology*, 34(6), 841–866.

Beaupre, D. (1994). "Rheology of high performance shotcrete", Ph.D. Thesis, University of British Columbia.

Beaupré, D., et al. (2004). Comparison of concrete rheometers: International tests at MBT (Cleveland OH, USA) in May 2003. National Institute of Standards and Technology Interagency Report (NISTIR) 7154.

Brower, L., and Ferraris, C. F. (2001). Comparison of concrete rheometers: International test at LCPC (Nantes, France) in October, 2000. Gaithersburg, MD, USA: US Department of Commerce, National Institute of Standards and Technology.

C143, A. (2008). *Standard test method for slump of hydraulic-cement concrete*. ASTM International, West Conshohocken (PA): American Society for Testing and Materials.

De Larrard, F., et al. (1998). "Fresh concrete: A Herschel-Bulkley material." *Materials and Structures*, 31(7), 494–498.

Dzuy, N., and Boger, D. V. (1985). "Direct yield stress measurement with the vane method." *Journal of Rheology*, 29(3), 335–347.

Feys, D., et al. (2013). "Extension of the Reiner–Riwlin equation to determine modified Bingham parameters measured in coaxial cylinders rheometers." *Materials and Structures*, 46(1–2), 289–311.

Feys, D., and Khayat, K. H. (2013). "Comparison and limitations of concrete rheometers." *The Proceedings of the 7th International RILEM Symposium on Self-Compacting Concrete*, Paris, France.

Gerland, F., et al. (2019). "A simulation-based approach to evaluate objective material parameters from concrete rheometer measurements." *Applied Rheology*, 29(1), 130–140.

Hackley, V., and Ferraris, C. F. (2001). The use of nomenclature in dispersion science and technology. NIST Recommended Practice Guide, SP 960-3.

Heirman, G., et al. (2006). "Contribution to the solution of the Couette inverse problem for Herschel-Bulkley fluids by means of the integration method." *The 2nd International Symposium on Advances in Concrete through Science and Engineering*.

Heirman, G., et al. (2009a). Integration approach of the Couette inverse problem of powder type self-compacting concrete in a wide-gap concentric cylinder rheometer: Part II. Influence of mineral additions and chemical admixtures on the shear thickening flow behaviour. *Cement and Concrete Research*, 39(3), 171–181.

Heirman, G., et al. (2009b). "Influence of plug flow when testing shear thickening powder type self-compacting concrete in a wide-gap concentric cylinder rheometer." *The 3rd International RILEM Symposium on Rheology of Cement Suspensions such as Fresh Concrete*. Edited by O.H. Wallevik, S. Kubens and S. Oesterheld. Reykjavik, Iceland.

Hočevar, A., et al. (2013). "Rheological parameters of fresh concrete–comparison of rheometers." *Građevinar*, 65(02), 99–109.

Hu, C., et al. (1996). "Validation of BTRHEOM, the new rheometer for soft-to-fluid concrete." *Materials and Structures*, 29(10), 620–631.

Koehler, E., and Fowler, D. W. (2004). Development of a portable rheometer for fresh Portland cement concrete. International Center for Aggregates Research Report ICAR 105-3F.

Laskar, A., et al. (2007). "Design of a new rheometer for concrete." *Journal of ASTM International*, 5(1), 1–13.

Laskar, A., and Bhattacharjee, R. (2011). "Torque–speed relationship in a concrete rheometer with vane geometry." *Construction and Building Materials*, 25(8), 3443–3449.

Li, M., et al. (2019). "Integration approach to solve the Couette inverse problem based on nonlinear rheological models in a coaxial cylinder rheometer." *Journal of Rheology*, 63(1), 55–62.

Liu, Y., et al. (2020). "An amendment of rotation speed-torque transformation equation for the Herschel-Bulkley model in wide-gap coaxial cylinders rheometer." *Construction and Building Materials*, 237, 117530.

Macosko, C. (1994). *Rheology: Principles, measurements, and applications*. Wiley-VCH, Weinheim, Germany.

Malvern, L. (1969). *Introduction to the mechanics of a continuous medium*. Prentice-Hall International Incorporated, New Jersey, United States.

Roussel, N. (2011). *Understanding the rheology of concrete*. Woodhead Publishing, Sawston, United Kingdom.

Schramm, G. (1994). *A practical approach to rheology and rheometry*. Haake Karlsruhe, Baden-Württemberg, Germany.

Sherwood, J., and Meeten, G. H. (1991). "The use of the vane to measure the shear modulus of linear elastic solids." *Journal of Non-Newtonian Fluid Mechanics*, 41(1–2), 101–118.

Soualhi, H., et al. (2017). "Design of portable rheometer with new vane geometry to estimate concrete rheological parameters." *Journal of Civil Engineering and Management*, 23(3), 347–355.

Struble, L., et al. (2001). "Concrete rheometer." *Advances in Cement Research*, 13(2), 53–63.

Szecsy, R. (1997). "Concrete rheology", Ph.D. Thesis, the University of Illinois at Urbana-Champaign.

Tattersall, G., and Banfill, P. F. G. (1983). *The rheology of fresh concrete*. Pitman Books Limited, London, United Kingdom.

Wallevik, J. (2001). Rheological measurements on a high yield value Bingham fluid with low agitation. The Annual transactions of the Nordic Rheology Society.

Wallevik, J. (2003). "Rheology of particle suspensions: Fresh concrete, mortar and cement paste with various types of lignosulfonates", Ph.D. Thesis, The Norwegian University of Science and Technology.

Wallevik, J. (2014). "Effect of the hydrodynamic pressure on shaft torque for a 4-blades vane rheometer." *International Journal of Heat and Fluid Flow*, 50, 95–102.

Wallevik, J., et al. (2020). "Concrete mixing truck as a rheometer." *Cement and Concrete Research*, 127, 105930.

Wallevik, O. (2006). The ConTec BML Viscometer, Viscometer 4&5 operating manual.

Wallevik, O., et al. (2015). "Avoiding inaccurate interpretations of rheological measurements for cement-based materials." *Cement and Concrete Research*, 78, 100–109. doi:10.1016/j.cemconres.2015.05.003.

Whorlow, R. (1992). *Rheological techniques*, 2nd ed. Ellis Horwood Limited, Chichester, United Kingdom.

Yan, J., and James, A. E. (1997). "The yield surface of viscoelastic and plastic fluids in a vane viscometer." *Journal of Non-Newtonian Fluid Mechanics*, 70(3), 237–253.

Yuan, Q. and Shi, C. et al. (2018), A testing method for static and dynamic rheologcial parameters of concrete, Chinese Patent. CN201810860119.4,

Chapter 7

Mixture design of concrete based on rheology

7.1 INTRODUCTION

The workability, mechanical properties, durability, and ecological efficiency should be taken into account during the concrete mixture design process. As a result, the optimizing mix proportion of concrete is more of an art than a science. For traditional optimization of mixture design, trial-and-error or single factor variable methods are usually used to satisfy several performance requirements. With the developments and applications of high-performance concrete and self-compacting concrete (SCC), a large number of mixture design methods have been proposed, such as statistical mixture design method (Simon et al., 1997), compressive packing method (de Larrard and Sedran, 1994, de Larrard and Sedran, 2002), densified mixture design algorithm (Chang, 2004, Chang et al., 2001, Hwang and Hung, 2005), and methods based on the minimum paste theory (Ji et al., 2013) or artificial neural network (Ji et al., 2006). Every existed design method has its own advantages and drawbacks (Ashish and Verma, 2019, Shi et al., 2015a). Besides, in these concrete composition optimization methods, all the available raw materials should be tested intensively to achieve an optimal mixture proportion (Jiao et al., 2018b).

Unfortunately, contradictions and conflicts always exist when altering the concrete proportion to satisfy the several performances at the same time. For example, highly flowable concrete could be obtained by increasing the water-to-cementitious ratio, and replacing rough crushed stones with round-shaped smooth natural pebbles. While for the high-strength concrete, it is necessary to decrease the water-to-cementitious ratio and use the roughly crushed stone aggregates. Furthermore, the accurate assessment of fresh concrete properties is an indispensable part of concrete engineering and science due to the fact that the fresh concrete exerts a significant effect on the quality of forming and casting processes and also the hardened concrete (Tattersall and Banfill, 1983, Tattersall, 1991). The traditional evaluation methods of fresh concrete properties such as slump, V-funnel, or L-box are empirical, operator-sensitive, and lacking scientific basis. For the concrete mixtures with distinct rheological parameters, concrete might exhibit the same behavior when measured using the traditional testing methods (Banfill, 2006). Consequently, the fundamental physical parameters such as yield stress and plastic viscosity could be used to characterize the fresh properties of concrete, which could give a more accurate description of the flow and deformation of concrete (Jiao et al., 2017).

According to the curves of shear stress against shear rate, the rheological properties of fresh concrete can be evaluated by several rheology models, such as the Bingham model, Herschel–Bulkley model (De Larrard et al., 1998), and modified Bingham model (Feys et al., 2008, Feys, et al., 2009). The Bingham model is the most popular concrete model, which can obtain the yield stress and plastic viscosity of concrete by linearly fitting the points of shear stress and shear rate. The yield stress is the minimum shear stress when concrete starts to flow, and its magnitude depends on the solid volume fraction, packing fractions, size and surface roughness of solid particles, interactions between particles and superplasticizer, etc. (Jiao, et al., 2017, Jiao, et al., 2019, Yammine, et al., 2008). The plastic viscosity is the ratio between shear stress and shear rate under the steady state, and it is greatly affected by the colloidal particle interaction forces, Brownian forces, hydrodynamic forces, and viscous forces between particles (Barnes, et al., 1989, Kovler and Roussel, 2011). The yield stress and plastic viscosity are effective parameters to characterize the flowability and dynamic and static stability, evaluate structure, and predict the pumpability and formwork filling of fresh concrete (De Schutter and Feys, 2016, Jiao, et al., 2020, Thiedeitz, et al., 2020).

Overall, the rheological properties of fresh concrete play important roles in engineering applications, the revolution of mechanical strength, and even the durability of concrete. Thus, many researchers have tried to design high-performance concretes from a rheological point of view. This chapter first presents a brief overview of the existing mixture design methods of concrete from the viewpoint of rheology. The principles of the vectorized-rheograph approach, paste rheology criteria, concrete rheology approach, excess paste thickness-rheology method, and simplex centroid design method are illustrated in detail. Afterward, representative examples for some typical mentioned mixture design methods are given.

7.2 PRINCIPLES OF MIXTURE DESIGN METHODS BASED ON RHEOLOGY

7.2.1 Vectorized-rheograph approach

The vectorized-rheograph approach is proposed by Wallevik (2003) and Wallevik and Wallevik (2011) by means of rheograph and workability box. Rheograph describes the plots of the relationship between yield stress on the y-axis and plastic viscosity on the x-axis with the changes in constituents, materials properties, time, etc. It is a convenient and essential tool to understand the changes in the rheological properties of cement-based materials. The term "rheograph" is first proposed by Bombled in 1970 (Bombled, 1970), and a very systematic and comprehensive application of rheograph was produced by Wallevik in 1983 (Wallevik, 1983). Rheograph can be used to describe the influence of concrete constituents on rheological properties. A typical rheograph is presented in Figure 7.1. Each constituent plays its unique role in changing the rheological parameters of fresh concrete (Jiao et al., 2017). For example, increasing water content at a constant cement content reduces the yield stress and plastic viscosity of concrete, while adding superplasticizer significantly decreases the yield stress but has less influence on the viscosity (Hu and Wang, 2011, Jiao et al., 2018a, Kostrzanowska-Siedlarz and Gołaszewski, 2016). It should be mentioned that rheograph is generally applied in mortar and concrete, and it is not applicable to cement paste. This is due to the more pronounced thixotropic structural build-up and structural breakdown of cement paste than mortar and concrete (Jiao et al., 2021, Wallevik and Wallevik, 2011). In addition, the influence of aggregates on the rheological properties cannot be completely described by rheograph, due to the simultaneous change of other parameters. However, some general trends can still be observed from rheograph. For example, from Figure 7.2, we can see that replacing crushed aggregates with rounded ones reduces the yield stress slightly, while the plastic viscosity shows a significant decrease. Increasing sand fraction at the fixed total content of aggregates will reduce the plastic viscosity but increase the yield stress.

The workability box is a specific area in a rheograph, illustrating the certain domain of yield stress and plastic viscosity for a kind of concrete used in specialized construction applications (Wallevik and Wallevik, 2011). It should be noted that there is no specific shape for

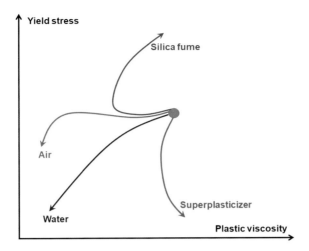

Figure 7.1 Rheograph illustrating the effect of adding different constituents to a reference mixture. (Adapted from Wallevik and Wallevik, 2011.)

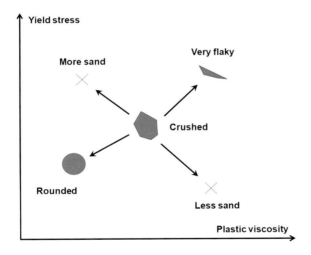

Figure 7.2 Rheograph describing the effect of aggregate shape and sand fraction. (Adapted from Wallevik and Wallevik, 2011.)

a workability box. It can be composed of a two-dimensional polygon, or regions without an exact and clear boundary. Therefore, the "box" here is a loose description of certain domains and boundaries. Several workability boxes can be included in a single rheograph. Typical relative workability boxes (without exact values of yield stress and plastic viscosity) of concrete for different construction applications are presented in Figure 7.3. The workability box is an effective tool to help technicians to design specific concrete successfully. For example, the plastic viscosity of conventional vibrated concrete (CVC) with a slump lower than 170 mm, for the on-site cases in Iceland, Norway, and Denmark (Wallevik and Wallevik, 2011), is around the region of 20–40 Pa.s, and the yield stress is generally higher

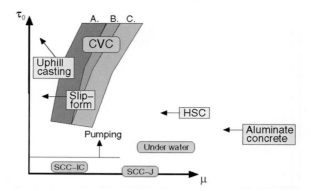

Figure 7.3 Typical relative workability boxes of concrete for different construction applications. No exact values of yield stress and plastic viscosity are provided. (Adapted from Wallevik and Wallevik, 2011.)

than 300 Pa. If the rheological parameters of a prepared CVC are located outside the box, a risk of structure failure could possibly appear at job site due to improper consistency. In this case, the constituents of the concrete should be adjusted based on rheographs to put the rheological parameters back into an advisable range, i.e., vectorized-rheograph approach.

The vectorized-rheograph approach is based on the fact that the reduced or increased yield stress (or plastic viscosity) can be achieved by vectorially changing the components. A representative description of this approach by an example of SCC is illustrated in the following sentences. In order to satisfy the requirements of flowing easily and maintaining stability, the low plastic viscosity of SCCs should combine with sufficiently high yield stress, while the yield stress of viscous SCCs should be close to zero. That is, both the yield stress and the plastic viscosity should be properly controlled in view of flowability and stability. However, traditional workability testing methods such as slump flow spread and V-funnel time cannot indicate how to change the yield stress and plastic viscosity simultaneously. Through the tool of rheology, the aforementioned rheological properties can be controlled. If a reference concrete with high plastic viscosity is defined and lower plastic viscosity is desired, there are several methods to alter the plastic viscosity of the concrete, as shown in Figure 7.4. Adding more water directly will be beneficial to reduce the plastic viscosity and yield stress, but has a negative effect on the stability. Another possible solution is adding the combination of a stabilizer (to increase the yield stress) and water (to reduce the plastic viscosity and yield stress). These two vector additions of constituents are defined as the vectorized-rheograph approach (Wallevik and Wallevik, 2011). Other vector steps can be combined to achieve this goal. For example, the plastic viscosity can be significantly reduced by the addition of superplasticizer (see Figure 7.1) and by replacing part of crushed aggregates with rounded ones (see Figure 7.2). Overall, changing the constituents in a combination manner in rheograph can be used to adjust the rheological properties of fresh concrete conveniently and feasibly.

7.2.2 Paste rheology criteria

The paste rheology criteria were first proposed by Saak and colleagues to design the mixture proportions of SCC (Saak, et al., 2001). Fresh concrete can be regarded as aggregate particles suspending in a flowing cement paste medium. The workability of SCC depends on the rheological properties of the suspending medium, as well as the volume fraction and physical properties of aggregates. According to the critical state that a single spherical

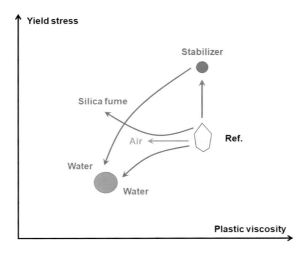

Figure 7.4 Several ways to reduce the plastic viscosity of concrete. (Adapted from Wallevik and Wallevik, 2011.)

particle can be suspended avoiding both the dynamic and static segregations, a series of theoretical relationships between rheological parameters and the density difference between aggregates and paste medium could be established to determine the minimum yield stress and plastic viscosity of the cementitious paste. The upper bound of yield stress and plastic viscosity were experimentally calibrated based on the critical state of poor deformability of concrete. Consequently, a suitable range of rheological properties meeting the requirements of excellent segregation resistance and acceptable workability could be obtained by altering the rheology of the suspending paste.

The concept of the paste rheology model was afterward extended (Bui et al., 2002), by focusing on the blocking and minimum paste volume to obtain acceptable segregation resistance, good workability, and optimum superplasticizer dosage. After considering the volume fraction and characteristics of aggregates such as particle size distribution, average particle diameter, and particle shape, the average aggregate space d_{ss} (m), which is twice the excess paste layer thickness, can be calculated as follows:

$$d_{ss} = d_{av}\left[\sqrt[3]{1 + \frac{V_{paste} - V_{void}}{V_{concr} - V_{paste}}} - 1\right] \tag{7.1}$$

where V_{paste} is the minimum volume of paste (m³), V_{void} is the void content of aggregates (m³), V_{concr} is the volume of concrete (m³), and d_{av} is the average diameter of the aggregate particles (m), which is defined as follows:

$$d_{av} = \frac{\sum_i d_i m_i}{\sum_i m_i} \tag{7.2}$$

where d_i and m_i are the diameter (m) and the weight (kg) of aggregate in fraction i, respectively. Based on the aggregate spacing and average aggregate diameter, the yield stress and

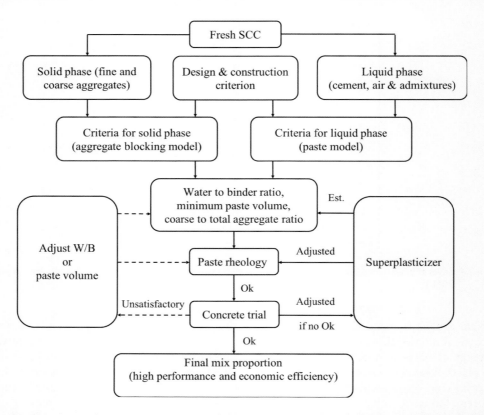

Figure 7.5 Flow chart of mixture design based on paste rheology criteria. (Adapted from Shi et al., 2015a.)

plastic viscosity of the cement paste are optimized to design SCC. The flow chart of the mixture design method is described in Figure 7.5. Theoretically, for a fixed average aggregate diameter, a minimum volume of cement paste is required to fill the voids between aggregates and lubricate the aggregate particles to ensure the flowability and segregation resistance of concrete. Concrete mixtures with higher average aggregate spacing should have cement paste with higher yield stress and higher plastic viscosity to meet the required flowability and favorable segregation stability. At a given aggregate grading, the measured rheological properties can be adjusted to a satisfactory zone by varying the aggregate content and paste content, as depicted in Figure 7.6. The paste rheology model can reduce laboratory work and materials waste significantly, providing a theoretical basis for quality control of concrete (Ashish and Verma, 2019, Shi et al., 2015a).

It should be worth noting that the paste rheology model can also be used to design steel fiber-reinforced self-compacting concrete (Ferrara et al., 2007). The incorporated fibers can be regarded as an "equivalent spherical particle" with an equivalent diameter $d_{\text{eq-fiber}}$, as shown in Eq. (7.3), which can be obtained from the specific surface area equivalence under the assumption that the surface area of fibers corresponds to the value of an equal mass of spheres with the same specific weight of aggregates:

$$d_{\text{eq-fibers}} = \frac{3L_f}{1+2\dfrac{L_f}{d_f}} \dfrac{\gamma_{\text{fiber}}}{\gamma_{\text{aggregate}}} \tag{7.3}$$

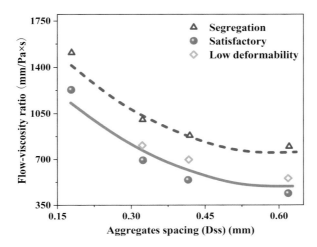

Figure 7.6 Model lines for measured flow viscosity ratio and aggregate spacing, at a fixed average diameter of 4.673 mm. (Adapted from Bui et al., 2002.)

where L_f and d_f are the length and diameter of the fiber (m), respectively, and γ_{fiber} and $\gamma_{aggregate}$ are the specific weight of fibers and aggregates, respectively. In this case, the average equivalent diameter of solid particles in fiber-reinforced skeleton can be modified to:

$$d_{av} = \frac{\sum_i d_i m_i + d_{eq\text{-}fiber} m_{fiber}}{\sum_i m_i + m_{fiber}} \tag{7.4}$$

where m_{fiber} is the mass of fibers (kg). After determination of the paste content and the void ratio of the graded solids including aggregates and fibers, the average solid space d_{ss} can be calculated by Eq. (7.1). The modified paste rheology model is an efficient tool to design steel fiber-reinforced SCC with required fresh properties by considering various contents and types of steel fibers (Ferrara et al., 2007).

Fresh SCC can also be regarded as a two-phase suspension containing mortar and coarse aggregate, and mortar can be considered as fine aggregate particles suspending in the cement paste phase, as presented in Figure 7.7. The workability of SCC depends on the content and physical nature of gravel, as well as the rheological characteristics of mortar, and this combining effect can be evaluated by the mortar film thickness. For a given mortar film thickness, the workability of SCC is mainly determined by the rheological properties of mortar. Similarly, mortar is composed of paste and sand. For a given sand content, the rheological properties of mortar are dependent on the rheological characteristics of the paste medium.

Based on this theoretical viewpoint, Wu and An (2014) developed a simple and efficient mixture design method for SCC using the paste rheology criteria. The flow chart of this method is shown in Figure 7.8. The theoretical foundation of this method is the fact that the rheological properties of mortar affect the bearing force and movement conditions of aggregate particles, and thus the workability and segregation resistance of SCC. To ensure the high flowability of SCC, the gravel particles suspended in the mortar should have the ability to move. Under flowing conditions, the segregation and settlement of gravel particles

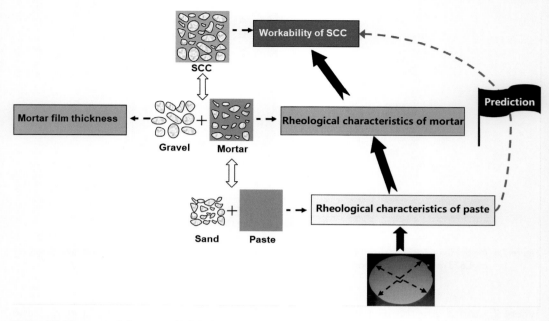

Figure 7.7 Theoretical diagram of SCC. (Adapted from Wu and An, 2014.)

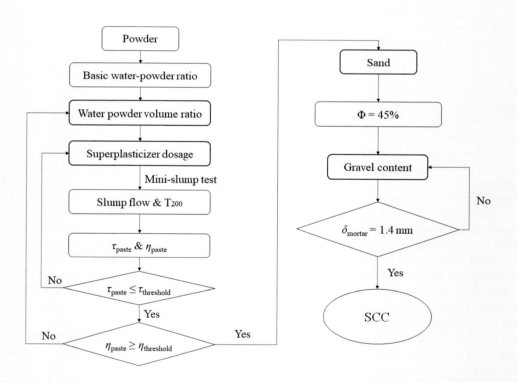

Figure 7.8 Flow chart of mixture design method proposed by Wu and An. (Adapted from Wu and An, 2014.)

should be avoided. Considering the flowability and segregation resistance of SCC, there is an upper limit for the mortar yield stress and a lower limit for the mortar viscosity (Wu and An, 2014), which can be described as follows:

$$\tau_{mortar} \leq \frac{\sqrt{2}\Delta\rho \cdot gr^2}{3\delta_{mortar}} \tag{7.5}$$

$$\eta_{mortar} \geq \frac{2r^2 g \cdot \Delta\rho \cdot T_f}{9H} \tag{7.6}$$

where τ_{mortar} and η_{mortar} are the yield stress (Pa) and viscosity (Pa.s) of the mortar, respectively; $\Delta\rho$ is the density difference between coarse aggregates and suspending mortar (kg/m^3); g is the gravitational acceleration (m/s^2); r is the average radius of coarse aggregates (m); δ_{mortar} is the excess mortar film thickness (m); T_f is the flowing time (s); and H is the height of the slump cone (m). According to the correlations of rheological characteristics between cement paste and mortar (Chidiac and Mahmoodzadeh, 2009, Toutou and Roussel, 2006), the rheological properties of cement paste should meet the following criteria to ensure the flowability and segregation:

$$\tau_{paste} \leq \tau_{threshold} = \frac{\sqrt{2}\Delta\rho \cdot gr^2}{3\delta_{mortar}} \bigg/ \left(1 - \varphi/\varphi_{max}\right)^{-n} \tag{7.7}$$

$$\eta_{paste} \geq \eta_{threshold} = \frac{2r^2 g \cdot \Delta\rho \cdot T_f}{9H} \bigg/ \left(1 - \varphi/\varphi_{max}\right)^{-[\eta]\cdot\varphi_{max}} \tag{7.8}$$

where τ_{paste} and η_{paste} are the yield stress (Pa) and plastic viscosity (Pa.s) of the paste, respectively; n is a coefficient with a value of 4.2 applied in Toutou and Roussel (2006); $[\eta]$ is the intrinsic viscosity of aggregates with the value of 2.5 for rigid spheres; and φ and φ_{max} are the aggregate volume fraction and maximum particle volume fraction (%), respectively. By establishing the abovementioned threshold equations, the self-compacting area of cement paste and mortar could be easily obtained following the design procedure in Figure 7.8 without using mathematical software (Nie and An, 2016). It should be noted that in their study, the sand content is fixed at 45% of the mortar volume and the excess mortar thickness is controlled at 1.4 mm obtained from sieving experiments, considering the reliability and economy of SCC in engineering practice. Under constant sand content and excess mortar thickness, the workability of SCC is mainly governed by the rheological properties of the paste.

Furthermore, this paste rheological threshold formula is recently modified by considering the replacement of cement with supplementary cementitious materials (such as fly ash and limestone powder), and acceptable experimental accuracy of the modified method is obtained (Li et al., 2020, Li et al., 2021b). Moreover, the threshold theory of mortar rheological parameters can also be used to design self-compacting lightweight aggregate concrete (Li et al., 2017, Li et al., 2021a). This design method not only provides an effective method to predict the workability of SCC from the rheological properties of paste but also reduces the amount of laboratory work and testing time significantly. However, because the physical properties of aggregates have a great effect on the rheological properties and workability of concrete, the characteristics of aggregates should be taken into account in establishing the threshold equations.

7.2.3 Concrete rheology method

The plastic viscosity of an SCC mix can be estimated by the known plastic viscosity of the paste medium and the contributions of each solid phase, i.e., micromechanical constitutive model (Ghanbari and Karihaloo, 2009). On this basis, a rigorous mixture design method for preparing normal and high-strength SCC with and without steel fibers has been proposed (Deeb and Karihaloo, 2013, Karihaloo and Ghanbari, 2012). However, no practical guidelines on how to choose a suitable mix are provided. In this case, Abo Dhaheer et al. (2015a) modified the mixture-proportioning method and further developed mixture design criteria for SCC based on the desired plastic viscosity and compressive strength. Note that the yield stress of SCC is generally in the order of 10 Pa, and thus the plastic viscosity, which is the most important parameter influencing the workability and stability, is considered. This proposed design method includes the following steps:

1. Choose a target plastic viscosity for the SCC mix in the range of 3–15 Pa.s;
2. According to the compressive strength requirement expressed by Abrams-type relation (e.g., $f_{cu}(28d) = \dfrac{195}{12.65^{w/cm}}$), calculate the water-to-cementitious ratio (w/cm);
3. Referring to the guidelines of EFNARC (Concrete, 2005), determine the water content from the range of 150–210 kg/m³, and then calculate the mass of cementitious materials (cm) in kg/m³;
4. Assume a trial superplasticizer (SP) dosage;
5. Estimate the plastic viscosity of the cement paste from the w/cm and SP/cm, which can be referred to Table 7.1;
6. Calculate the mass of solid ingredients, i.e., filler, fine aggregate, and coarse aggregate based on their volume fractions;
7. Check the total volume of the produced mix. Scale the ingredient masses until achieving a total volume of 1 m³;
8. Calculate the plastic viscosity of the SCC mixture using Eq. (7.9), and compare it with the selected one in step 1. Adopt the mix proportions if the difference is within ±5%. If not, adjust the volume fractions of solid ingredients in step 6, and repeat steps 7 and 8.

$$\eta_{mix} = \eta_{paste} \cdot \left(1 - \varphi_{filler}/\varphi_{max}\right)^{-1.9} \cdot \left(1 - \varphi_{fine\ agg}/\varphi_{max}\right)^{-1.9} \cdot \left(1 - \varphi_{coarse\ agg}/\varphi_{max}\right)^{-1.9} \quad (7.9)$$

where η_{mix} and η_{paste} are the plastic viscosity of the SCC mix and the corresponding paste, respectively. φ_{filler}, $\varphi_{fine\ agg.}$, $\varphi_{coarse\ agg}$, and φ_{max} are the volume fraction of filler, fine aggregate, coarse aggregate, and maximum particle volume fraction, respectively. The flow chart of

Table 7.1 Estimated plastic viscosity of cement paste (i.e., cementitious materials + SP + water + air) (Abo Dhaheer, et al., 2015a)

w/cm	η_{paste} (mPa.s)	$\eta_{paste + airvoids}$ (mPa.s)
0.63	104	110
0.57	176	180
0.53	224	230
0.47	286	290
0.40	330	340
0.35	365	370

Mixture design of concrete based on rheology 165

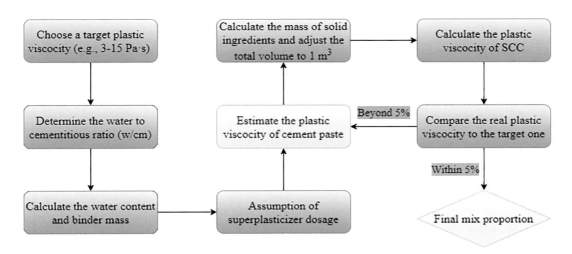

Figure 7.9 Procedure of the mixture design method proposed by Abo Dhaheer et al. (Adapted from Abo Dhaheer et al., 2015a.)

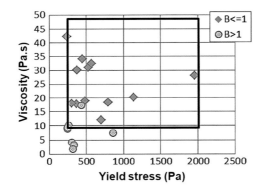

Figure 7.10 Relationship between rheological parameters and bleeding resistance of concrete. B is the ratio between the surface bleeding volume and the total water volume in the sample, and B less than 1 means good bleeding resistance. (Adapted from Xie et al., 2013.)

the abovementioned mixture design method is shown in Figure 7.9. Experimental results for various volumetric ratios of paste to solid particles prepared by this design chart show that all the mixtures satisfied the self-compacting criteria and achieved the desired target plastic viscosity and compressive strength (Abo Dhaheer, et al., 2015b).

In the existing literature, there was another mixture design method based on the rheological properties of concrete, which can be found in Xie et al. (2013). Specifically, the yield stress and plastic viscosity of concrete can be used as a bridge to connect the workability (such as flowability, bleeding resistance, and segregation resistance) and the components of concrete. By establishing the relationship between workability aspects and rheological parameters, with an example between rheological parameters and bleeding resistance in Figure 7.10, and the relationship between rheological properties and the concrete proportions, with an example in Figure 7.11, the appreciate ranges of the components for the excellent workability of concrete can be obtained. For example, for the exemplified C20 concrete

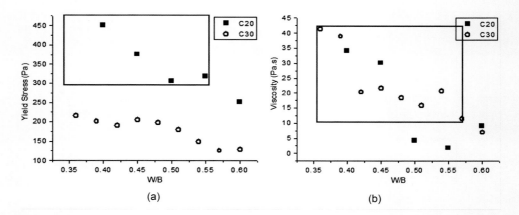

Figure 7.11 Relationship between water-to-binder (*w/b*) ratio and rheological parameters: (a) yield stress and (b) plastic viscosity. (Adapted from Xie et al., 2013.)

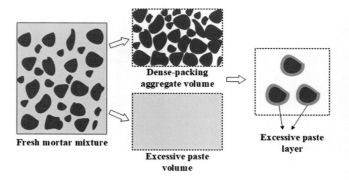

Figure 7.12 Schematic diagram of the fresh concrete model (Reinhardt and Wüstholz, 2006).

with good bleeding resistance, the suitable yield stress ranges from 280 to 2000 Pa, and the viscosity is in the range of 11–43 Pa.s. Under this proper region of yield stress and viscosity, it can be seen from Figure 7.11 that the appropriate range of *w/b* is 0.4–0.55.

7.2.4 Excess paste theory

Cement paste in fresh concrete can be divided into two parts. One is for the filling of the voids between aggregate particles, and the other is for coating aggregate surfaces with lubrication effect, which is defined as an excess paste (Hu, 2005, Jiao, et al., 2017, Reinhardt and Wüstholz, 2006), as shown in Figure 7.12. The excess paste layer is generally used to lubricate the gravel aggregate particles and provide the flowability of the concrete mixture (Jiao, et al., 2017). Similarly, from the viewpoint of paste, it can be regarded as cementitious particles suspending in water solution, and the water coating the solid particles is called excess water. The water film thickness, excess paste thickness, and mortar film thickness are the key factors controlling the flowability, rheological properties, and cohesiveness and adhesiveness of fresh paste, mortar, and concrete (Kwan and Li, 2012, Li and Kwan, 2011).

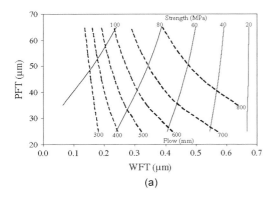 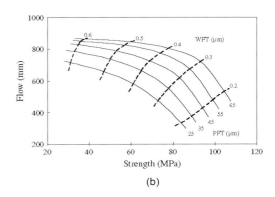

Figure 7.13 Concrete mix proportion design charts developed by Li and Kwan: (a) estimating strength and flowability based on WFT and PFT and (b) estimating WFT and PFT according to flowability and strength. (Adapted from Li and Kwan, 2013.)

Li and Kwan (2013) stated that the influence of concrete mix parameters (such as w/c and paste volume) and particle size distribution of aggregates can be converted into the factors of water film thickness (WFT) and paste film thickness (PFT). On the basis of this theoretical foundation, they proposed two design charts to optimize concrete mix proportions, as shown in Figure 7.13. Through conducting a series of experiments for various WFT and PFT, the contours of flowability and strength can be plotted, as shown in Figure 7.13a. From this chart, the flowability and strength of concrete can be estimated by locating the coordinates of WFT and PFT. Figure 7.13a can also be transformed into Figure 7.13b with the contours of WFT and PFT at various flowability and strength. The required WFT and PFT of concrete can be determined by reading the designed flowability and strength. These two design charts are helpful to prepare concrete mixes for high-performance concrete and understand the effect of WFT and FFT. For example, it can be seen from Figure 7.13b that the minimum PFT for a high-strength concrete with 100 MPa is 55 μm. By contrast, the minimum PFT for a high-flowability concrete with a flowability of 650 mm and a strength of 40 MPa can be determined as 25 μm. Based on the estimated PFT, the total paste volume of the concrete can be determined by the volume of paste required to fill the voids between aggregates and the paste volume coating the surface of aggregates with the determined PFT.

7.2.5 Simplex centroid design method

The simplex centroid design method is commonly used in industrial product formulations, e.g., food processing, chemical formulations, textile fibers, and pharmaceutical drugs (Scheffé, 1958). Given that concrete can be viewed as a concentrated suspension consisting of cementitious paste, fine aggregate, and coarse aggregate, the simplex centroid design can also be used to evaluate the effects of components on the properties of concrete. For example, the quantitative mathematical relationship between compressive strength and the compositions of mortar or concrete with ternary cementitious compositions could be obtained by using the simplex centroid design method (Douglas and Pouskouleli, 1991, Sun and Yan, 2003, Wang and Chen, 1997). The heat of hydration, porosity, calcium hydroxide content, and alkali reactivity of concrete with a ternary cementitious system could also be evaluated (Shi et al., 2015b, Shi et al., 2016). Obviously, the simplex centroid design method is a very powerful tool for the study of concrete.

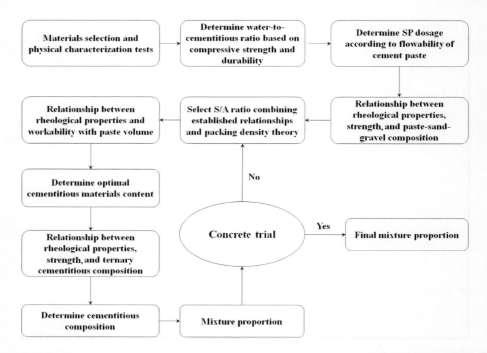

Figure 7.14 Flow chart of the mixture design method based on simplex centroid design. (Adapted from Jiao et al., 2018b.)

Through using the simplex centroid design method, Jiao et al. (2018b; 2018a) recently proposed an advanced mixture design approach according to the quantitative relationships between mixture proportion parameters and rheological properties. The flow chart of this concrete mixture design method is shown in Figure 7.14. The water-to-binder ratio (w/b) is determined according to the requirements of mechanical properties and durability. The dosage of superplasticizer, if needed, is determined by the saturation dosage from the fluidity test of cement paste. The optimal paste volume is obtained by establishing the relationship between rheological properties and workability with various paste volumes. With regard to the aggregate composition, it is determined by dividing fresh concrete into three components, i.e., paste, fine aggregate, and coarse aggregate. By establishing the quantitative relationships between performance parameters (such as flowability, yield stress, plastic viscosity, and compressive strength) and the paste-fine aggregate-coarse aggregate ternary components, together with the closest test, the optimal sand-to-aggregate ratio meeting the requirements of flowability, rheological properties and mechanical strength can be determined. The binder composition if containing three components (e.g., cement, fly ash, and slag) is also determined similarly. A representative relationship between flowability and three cementitious components is expressed in Eq. (7.10):

$$S = 1.75x_1 + 0.694x_2 + 1.389x_3 + 0.0222x_1x_2 + 0.00278x_1x_3 + 0.35x_2x_3 - 0.00053x_1x_2x_3 \quad (7.10)$$

where S is the flowability (slump) in mm, and x_1, x_2 and x_3 are the mass proportion of cement, fly ash, and slag, respectively. Based on the relation in Eq. (7.10), the flowability contours of the ternary components can be obtained, which will be presented later.

The optimal binder composition can be obtained from the overlapped area of different areas of various properties. The simplex centroid design method is an effective way to optimize the mixture design of concrete to strike a good balance between flowability, rheological properties, stability, mechanical properties, and economy. This proposed methodology with reduced time and labor has been extended to design the high-performance concrete and even 3D printing concrete with multiple performance requirements (Liu, et al., 2019, Shi, et al., 2018b).

7.3 TYPICAL EXAMPLES OF MIXTURE DESIGN

In the current section, typical examples of mixture design procedures of paste rheology criteria (designing for SCC and fiber-reinforced SCC), concrete rheology method, and simplex centroid design method are explained in detail.

7.3.1 Paste rheology criteria proposed by Wu and An

A pure limestone powder SCC is given as an example of the mixture design method of paste rheology criteria. The physical properties of the limestone powder are not given. Quartz sand with continuous gradation ranging from 0.075 to 4.75 mm is used. The maximum particle size of crushed limestone gravel is 19 mm. For more information, the reader is referred to Wu and An (2014).

The water to limestone powder volume ratio is first set as 0.8. At a superplasticizer dosage of 0.6%, the slump flow and $T200$ from the mini-slump test are 270 mm and 1.2 s, respectively. The corresponding yield stress and plastic viscosity of the paste are 0.93 Pa and 0.14 Pa.s, respectively. Based on the flowability and segregation criteria in Eqs. (7.7) and (7.8), the calculated threshold values of yield stress and plastic viscosity are 1.08 Pa and 0.26 Pa.s, respectively, with satisfactory yield stress but insufficient viscosity. In this case, the water to powder ratio should be adjusted based on Figure 7.8, and here, the water to limestone powder volume ratio is reduced by 10% to increase the viscosity of the paste, i.e., $V_w/V_p = 0.72$. By adjusting the superplasticizer dosage, the slump flow is 277 mm and the $T200$ is 2.5 s when the superplasticizer dosage is 0.63%. By following the same step, the yield stress and plastic viscosity are calculated as 0.84 Pa and 0.30 Pa.s, respectively, with the corresponding threshold values of 1.02 Pa and 0.25 Pa.s, respectively. That is, both the yield stress and the plastic viscosity of the paste meet the rheological criteria. By setting the sand volume and excess mortar thickness, the mixture proportion of the SCC can be determined. The specific procedure is shown in Figure 7.15. After preparing a trial mix according to the aforementioned steps, the slump flow and the V-funnel time are 690 mm and 10.91 s, respectively. After stopping flow from the slump test, there is no segregation of gravel particles observed even at the periphery of the mixture.

7.3.2 Paste rheology model proposed by Ferrara et al.

As mentioned earlier, the paste rheology criteria can also be used to design steel fiber-reinforced SCC. A specific example of this design method is demonstrated in this section. A Type I Ordinary Portland cement with a Blaine specific surface area of 352 m²/kg and a specific gravity of 3.15 is used. A class C fly ash with a specific gravity of 2.74 and a polycarboxylate-based superplasticizer (SP) are used. The maximum particle size of the coarse aggregate is 9.5 mm. A kind of steel fiber (Dramix65/35) with a length of 35 mm and an aspect ratio of 65 is utilized. More information on the raw materials can be found

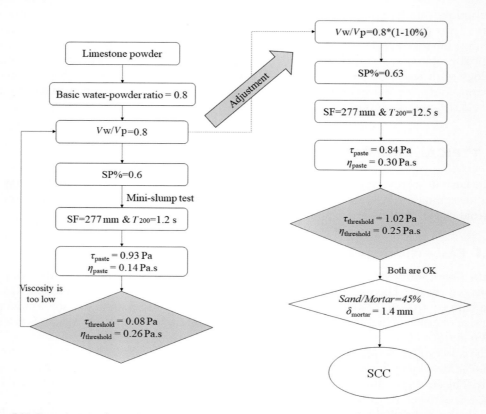

Figure 7.15 Preparation of pure limestone powder SCC. (Adapted from Wu and An, 2014.)

in Ferrara et al. (2007). The design procedure is summarized in Figure 7.16, and the detailed steps will be illustrated in the following pages.

1. **Establish the model of cement paste:** After selecting the raw materials for cement paste, the mini-slump flow diameter and the viscosity of the cement paste with various water-to-binder (*w/b*) ratios and SP dosages are first measured. Afterward, the relationship between the ratio of paste mini-slump flow diameter to viscosity and both the water-to-binder ratio and the SP dosages are plotted. Typical examples of their relationships are shown in Figure 7.17.
2. **Optimization of the solid skeleton:** For the plain concrete, fine aggregate and coarse aggregate are sieved and then optimized based on the following equation with q equal to 0.5:

$$P(d) = \frac{d^q - d_{min}^q}{d_{max}^q - d_{min}^q} \tag{7.11}$$

For the steel fiber-reinforced concrete, the fibers are regarded as equivalent spherical particles. The equivalent diameter can be determined according to Eq. (7.3). Similar to the case of plain concrete, the grading curve of the solid skeleton with fibers is theoretically optimized by Eq. (7.11). Typical examples of optimization grading of the

Mixture design of concrete based on rheology 171

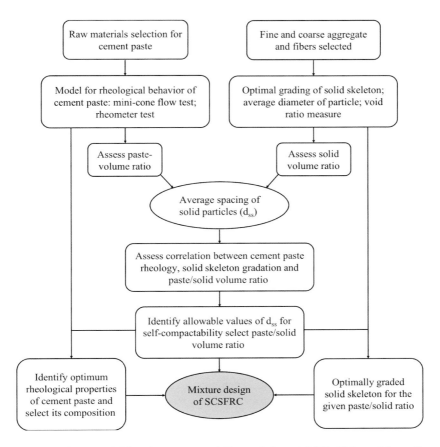

Figure 7.16 Paste rheology model for designing steel fiber-reinforced SCC. (Adapted from Ferrara et al., 2007.)

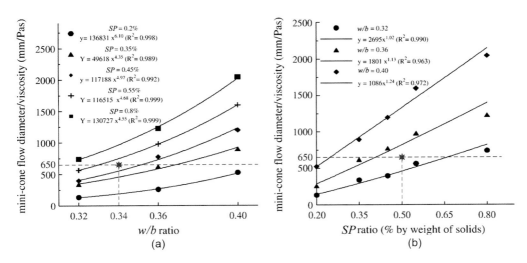

Figure 7.17 Typical relationships between paste mini-slump flow diameter/viscosity ratio and (a) w/b ratio and (b) SP dosage. (Adapted from Ferrara et al., 2007.)

172 Rheology of Fresh Cement-Based Materials

solid skeleton are presented in Figure 7.18. Based on the optimized grading curve, the average aggregate diameter and the void ratio can be determined.

3. **Calculate the average aggregate spacing d_{ss}:** According to Eq. (7.1), the average aggregate spacing can be obtained.
4. **Establish the correlation between paste rheology, solid grading, and paste/solid volume ratio:** The key feature of the rheology paste model is developing a suitable correlation between paste rheology, average aggregate spacing, flowability, and rheological stability characteristics of concrete. Under various average aggregate spacing, prepare a series of plain concrete and fiber-reinforced concrete. By measuring the slump flow diameter, T_{50}, visual segregation index, and passing ability, the rheological criteria based on paste slump flow diameter/viscosity ratio and average aggregate spacing are established. A typical example of the correlation is shown in Figure 7.19. The model

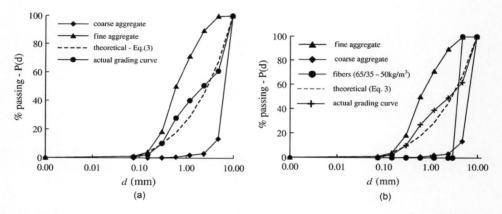

Figure 7.18 Typical examples of solid skeleton optimization (a) plain concrete and (b) fiber-reinforced concrete. (Adapted from Ferrara et al., 2007.)

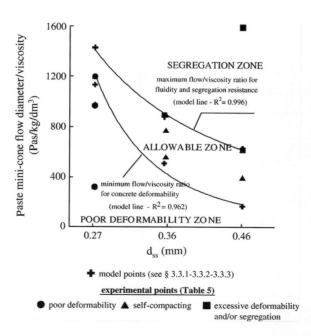

Figure 7.19 Typical correlation between paste flow diameter/viscosity ratio and aggregate spacing. (Adapted from Ferrara et al., 2007.)

Table 7.2 Mixture design of steel fiber-reinforced SCC and its flow properties (Ferrara, et al., 2007)

| Mix | V_p | d_{ss} | Solid skeleton (% by mass) | | | Flow (mm) | T_{50} (s) | VSI |
			Fibers	CA	FA	Exp.	Exp.	
1	0.366	0.360	2.913	43.62	53.46	650	4;4;3	1.5,1.5,1

lines divide the plane into three areas, i.e., poor deformability, self-compactability, and segregation regions. That is, sufficient deformability and segregation resistance can be guaranteed at the appropriate slump flow diameter–viscosity ratio of the paste.

5. **Design a new mix proportion**: If a steel fiber-reinforced SCC with good self-compacting properties, for example, is desired, the paste flow diameter/viscosity ratio and the average aggregate spacing can be set as 650 and 0.360 mm, respectively. The paste volume fraction is similar to that during establishing the model procedure. According to Figure 7.17, the water-to-binder ratio is equal to 0.34 and the SP dosage is equal to 0.5% for preparing the cement paste. The mixture design and the corresponding flow properties of the designed steel fiber-reinforced SCC are shown in Table 7.2. The results indicate that it is reliable to design steel fiber-reinforced SCC by using the proposed mix design approach.

7.3.3 Concrete rheology method of Abo Dhaheer et al.

An SCC with 28 d target compressive strength of 70 MPa is used to exemplify the mixture proportion method proposed by Abo Dhaheer et al. (2015a). The cement is a local type II cement with a specific gravity of 2.95. Slag with a specific gravity of 2.40 is used. River sand with a specific gravity of 2.65 and crushed gravel with a maximum particle size of 20 mm and a specific density of 2.80 are selected. A limestone powder with a maximum particle size of 0.125 mm and a specific density of 2.40 is used as fine filler. A polycarboxylate ether superplasticizer with a specific gravity of 1.07 is used. The detailed procedure is shown in the following:

1. The desired plastic viscosity for the SCC mix is set as 9 Pa.s;
2. Based on Abrams-type relation, the water to the cementitious ratio (*w/cm*) is calculated as 0.40;
3. Assume the water content is 184 kg/m³, and then the mass of cementitious materials (*cm*) is calculated as 460 kg/m³;
4. Assume the superplasticizer (*SP*) dosage is 0.65% (i.e., 3 kg/m³);
5. Based on Table 7.1, the plastic viscosity of the cement paste with a *w/cm* of 0.40 is estimated as 0.34 Pa.s;
6. Calculate the volume fraction and mass of solid ingredients: Eq. (7.9) is first rewritten as follows:

$$\eta_{mix} = \eta_{paste} \cdot \left(1 - \varphi_{LP}/0.524\right)^{-1.9} \cdot \left(1 - \varphi_{FA}/0.63\right)^{-1.9} \cdot \left(1 - \varphi_{CA}/0.74\right)^{-1.9} \quad (7.12)$$

Assuming

$$u = \left(\frac{\eta_{mix}}{\eta_{paste}} \cdot 0.524^{-1.9} \cdot 0.63^{-1.9} \cdot 0.74^{-1.9}\right)^{\frac{1}{-1.9}} \quad (7.13)$$

Then, Eq. (7.12) can be changed to:

$$u = (0.524 - \varphi_{LP}) \cdot (0.63 - \varphi_{FA}) \cdot (0.74 - \varphi_{CA}) \tag{7.14}$$

At the plastic viscosity for the SCC mix η_{mix} of 9 Pa.s and the viscosity for the paste η_{paste} of 0.34 Pa.s, Eq. (7.14) can be modified to:

$$u = (0.524 - \varphi_{LP}) \cdot (0.63 - \varphi_{FA}) \cdot (0.74 - \varphi_{CA}) = 0.044 \tag{7.15}$$

Then,

$$\begin{aligned} \varphi_{LP} &= 0.524 - 0.353 t_1 \\ \varphi_{FA} &= 0.63 - 0.353 t_2 \\ \varphi_{CA} &= 0.74 - 0.353 t_3 \end{aligned} \tag{7.16}$$

where $t_1 \cdot t_2 \cdot t_3 = 1$. For the first instance, choose $t_1 = 1$, $t_2 = 1$, and $t_3 = 1$, and then the volume fractions of the solid phases are

$$\begin{aligned} \varphi_{LP} &= 0.524 - 0.353 t_1 = 0.171 \\ \varphi_{FA} &= 0.63 - 0.353 t_2 = 0.277 \\ \varphi_{CA} &= 0.74 - 0.353 t_3 = 0.387 \end{aligned} \tag{7.17}$$

The amount of limestone filler, fine aggregate, and coarse aggregate can be calculated based on their densities if assuming the volume fraction of trapped air bubbles is 0.02:

$$\begin{cases} \varphi_{LP} = \dfrac{\dfrac{LP}{\rho_{LP}}}{\left(\dfrac{c}{\rho_c} + \dfrac{s}{\rho_s} + \dfrac{w}{\rho_w} + \dfrac{SP}{\rho_{SP}} + 0.02\right) + \dfrac{LP}{\rho_{LP}}} \rightarrow LP = 184 \text{ kg/m}^3 \\[2em] \varphi_{FA} = \dfrac{\dfrac{FA}{\rho_{FA}}}{\left(\dfrac{c}{\rho_c} + \dfrac{s}{\rho_s} + \dfrac{w}{\rho_w} + \dfrac{SP}{\rho_{SP}} + \dfrac{LP}{\rho_{LP}} + 0.02\right) + \dfrac{FA}{\rho_{FA}}} \rightarrow FA = 456 \text{ kg/m}^3 \\[2em] \varphi_{CA} = \dfrac{\dfrac{CA}{\rho_{CA}}}{\left(\dfrac{c}{\rho_c} + \dfrac{s}{\rho_s} + \dfrac{w}{\rho_w} + \dfrac{SP}{\rho_{SP}} + \dfrac{LP}{\rho_{LP}} + \dfrac{FA}{\rho_{FA}} + 0.02\right) + \dfrac{CA}{\rho_{CA}}} \rightarrow CA = 1098 \text{ kg/m}^3 \end{cases} \tag{7.18}$$

7. Check the total volume of the produced mix:

$$\text{Total volume} = \frac{c}{\rho_c} + \frac{s}{\rho_s} + \frac{w}{\rho_w} + \frac{SP}{\rho_{SP}} + \frac{LP}{\rho_{LP}} + \frac{FA}{\rho_{FA}} + \frac{CA}{\rho_{CA}} + 0.02 = 1.013 \text{ m}^3 \qquad (7.19)$$

As the total volume of the mix is not equal to 1 m^3, the amounts of each material are adjusted as follows:

$$\begin{cases} c = 345/1.013 = 340.5 \text{ kg/m}^3 \\ s = 115/1.013 = 113.5 \text{ kg/m}^3 \\ w = 184/1.013 = 181.6 \text{ kg/m}^3 \\ SP = 3/1.013 = 2.96 \text{ kg/m}^3 \\ LP = 184/1.013 = 182 \text{ kg/m}^3 \\ FA = 456/1.013 = 450 \text{ kg/m}^3 \\ CA = 1098/1.013 = 1084 \text{ kg/m}^3 \end{cases} \qquad (7.20)$$

8. Calculate the updated volume fraction of each material due to the change in amounts:

$$\begin{cases} \varphi_{LP} = \dfrac{\dfrac{LP}{\rho_{LP}}}{\left(\dfrac{c}{\rho_c} + \dfrac{s}{\rho_s} + \dfrac{w}{\rho_w} + \dfrac{SP}{\rho_{SP}} + 0.02\right) + \dfrac{LP}{\rho_{LP}}} \rightarrow LP = 0.172 \\[2em] \varphi_{FA} = \dfrac{\dfrac{FA}{\rho_{FA}}}{\left(\dfrac{c}{\rho_c} + \dfrac{s}{\rho_s} + \dfrac{w}{\rho_w} + \dfrac{SP}{\rho_{SP}} + \dfrac{LP}{\rho_{LP}} + 0.02\right) + \dfrac{FA}{\rho_{FA}}} \rightarrow FA = 0.277 \\[2em] \varphi_{CA} = \dfrac{\dfrac{CA}{\rho_{CA}}}{\left(\dfrac{c}{\rho_c} + \dfrac{s}{\rho_s} + \dfrac{w}{\rho_w} + \dfrac{SP}{\rho_{SP}} + \dfrac{LP}{\rho_{LP}} + \dfrac{FA}{\rho_{FA}} + 0.02\right) + \dfrac{CA}{\rho_{CA}}} \rightarrow CA = 0.387 \end{cases} \qquad (7.21)$$

Table 7.3 Mixture proportions of SCC mix designed by Abo Dhaheer et al. (2015a)

	Ingredient (kg/m³)							η_{mix} (Pa.s)	Difference
	C	S	W	SP	LP	FA	CA		
Density	2950	2400	1000	1070	2400	2650	2800	-	-
Before	345	115	184	3.0	184	456	1098	-	-
1st adjust	340.5	113.5	181.6	2.96	182	450	1084	8.88	−1.3%
2nd adjust	355.5	118.5	189.7	3.10	144	729	788	8.77	−2.6%

Then, the estimated plastic viscosity of the mix can be calculated by Eq. (7.9) or Eq. (7.12):

$$\eta_{mix} = \eta_{paste} \cdot \left(1 - \varphi_{LP}/0.524\right)^{-1.9} \cdot \left(1 - \varphi_{FA}/0.63\right)^{-1.9} \cdot \left(1 - \varphi_{CA}/0.74\right)^{-1.9}$$

$$= 0.34 \cdot \left(1 - 0.172/0.524\right)^{-1.9} \cdot \left(1 - 0.277/0.63\right)^{-1.9} \cdot \left(1 - 0.387/0.74\right)^{-1.9} \quad (7.22)$$

$$= 8.88 \text{ Pa.s}$$

The mix proportions before and after the first adjustment are shown in Table 7.3. The difference in the estimated plastic viscosity from the target one is 1.3%, which is within the acceptable range. However, the amount of coarse aggregate exceeds the limit guideline (Concrete, 2005). Therefore, the mix proportions need to be adjusted by choosing different arbitrary values of t_1, t_2, and t_3. For example, let $t_1 = 1.1$, $t_2 = 0.7$, and $t_3 = 1.3$, and repeat steps 6–8. The determined mass content of each ingredient and the calculated plastic viscosity of the mix is 8.77 Pa.s. The difference between the calculated plastic viscosity and the target one is 2.6%, which is within the range of ±5%. The mix proportions after the second adjustment are acceptable.

7.3.4 Simplex centroid design method proposed by Jiao et al.

The simplex centroid design method has been practically applied in many engineering cases, such as the Dongting Lake Bridge project (Shi et al., 2018b), Fengtai District Guogongzhuang Depot Project (1518–632), the Farmers' Houses and Supporting Facilities Project at Xiaohongmen Village, and the Construction Project of Beijing Subway Line 8 (Shi et al., 2018a). A specific example of the mixture design of a concrete with a compressive strength of 30 MPa will be given in the section. It should be mentioned that only the determination process of cementitious composition by using the simplex centroid design method is illustrated in detail.

P·I 42.5 cement with a specific gravity of 3.15 and an average particle size of 16.49 μm is used. The average particle size of fly ash and slag is 25.82 and 11.03 μm, respectively. Crushed limestone particles with a specific gravity, bulk density, and maximum particle size of 2.77, 1630 kg/m³, and 20 mm, respectively, are used as coarse aggregate. Fine aggregate is river sand with a fineness modulus of 2.82. For more details, the reader is referred to Jiao et al. (2018b). If the compressive grade of the concrete is 30 MPa, and a ternary cementitious system with cement, fly ash, and slag is used, the water-to-binder ratio, total binder content, and sand-to-aggregate ratio are determined as 0.5, 440 kg/m³, and 0.38, respectively, according to the flow chart in Figure 7.14. Seven concrete mixtures with various levels of cementitious components can be produced according to the simplex centroid design method. By measuring the flowability, rheological properties, and mechanical strength and drawing the contour figures of each performance parameter, for example, the yield stress and plastic viscosity contours are shown in Figure 7.20, the appropriate range area of the cementitious materials composition can be obtained according to the requirements for different properties of the concrete.

For the desired C30 concrete in this example, the compressive strength at 3 and 28 days should be higher than 15 and 38 MPa, respectively. The slump value should be higher than 170 mm, and the yield stress and plastic viscosity should be as low as possible under excellent stability and homogeneity. From the contours of each performance, several critical lines of each property could be acquired to meet these required performance values, as shown in Figure 7.21. The overlapped area of the five different areas in Figure 7.21 is regarded as

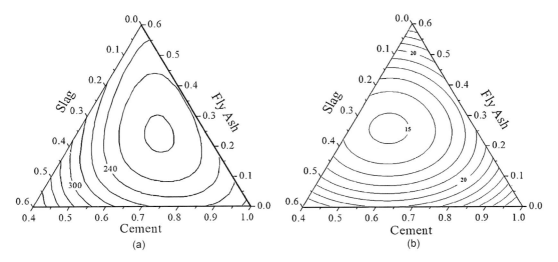

Figure 7.20 Contours of concrete with cement–fly ash–slag ternary system: (a) yield stress and (b) plastic viscosity. (Adapted from Jiao et al., 2018b.)

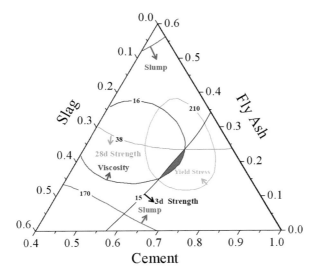

Figure 7.21 Determination of cementitious composition. (Adapted from Jiao et al., 2018b.)

the optimal combination of cementitious materials composition, i.e., the optimum contents of fly ash and slag are 15%–20% and 15%–25%, respectively. It is worth mentioning that some other properties such as drying shrinkage and carbonation depth could also be added to Figure 7.21 to modify the appreciation range to satisfy all the required performances. The results showed that it is an effective way to optimize the mixture design of concrete based on the rheological properties using the simplex centroid design method (Jiao et al., 2018b).

7.4 SUMMARY

Rheological properties of fresh concrete play important roles in flowability, workability, stability, mechanical strength, and even durability of the concrete. The engineering processes of concrete, such as transporting, pumping, and formwork casting, are also influenced by the rheological properties of fresh concrete. In this context, many researchers have tried to design high-performance concretes from the viewpoint of rheological parameters. In this chapter, a brief overview of the existing mixture design methods of concrete based on the rheology has been presented. The principles of the vectorized-rheograph approach, paste rheology criteria, concrete rheology approach, excess paste thickness-rheology method, and simplex centroid design method have been illustrated. Several representative examples for the mix design methods of paste rheology criteria, concrete rheology model, and simplex centroid design method have been given.

REFERENCES

Abo Dhaheer, M. S., et al. (2015a). "Proportioning of self-compacting concrete mixes based on target plastic viscosity and compressive strength: Part II - experimental validation." *Journal of Sustainable Cement-Based Materials*, 5(4), 217–232.

Abo Dhaheer, M. S., et al. (2015b). "Proportioning of self–compacting concrete mixes based on target plastic viscosity and compressive strength: Part I - mix design procedure." *Journal of Sustainable Cement-Based Materials*, 5(4), 199–216.

Ashish, D. K., and Verma, S. K. (2019). "An overview on mixture design of self-compacting concrete." *Structural Concrete*, 20(1), 371–395.

Banfill, P. (2006). "The rheology of fresh cement and concrete-rheology review." *British Society of Rheology*, 61, 130.

Barnes, H. A., et al. (1989). *An introduction to rheology*. Elsevier, Amsterdam, the Netherlands.

Bombled, J. (1970). "A rheograph for studying the rheology of stiff pastes: application to cement setting." *Revue des Materiaux de Construction*, 673, 256–277.

Bui, V. K., et al. (2002). "Rheological model for self-consolidating concrete." *Materials Journal*, 99(6), 549–559.

Chang, P. K., et al. (2001). "A design consideration for durability of high-performance concrete." *Cement and Concrete Composites*, 23, 375–380.

Chang, P. K. (2004). "An approach to optimizing mix design for properties of high-performance concrete." *Cement and Concrete Research*, 34(4), 623–629.

Chidiac, S. E., and Mahmoodzadeh, F. (2009). "Plastic viscosity of fresh concrete – A critical review of predictions methods." *Cement and Concrete Composites*, 31(8), 535–544.

Concrete, S.-C. (2005). "The European guidelines for self-compacting concrete." *BIBM, et al.*, 22, 563.

De Larrard, F., et al. (1998). "Fresh concrete: A Herschel-Bulkley material." *Materials and Structures*, 31, 494–498.

de Larrard, F., and Sedran, T. (1994). "Optimization of ultra-high-performance concrete by the use of a packing model." *Cement and Concrete Research*, 24(6), 997–1009.

de Larrard, F., and Sedran, T. (2002). "Mixture-proportioning of high-performance concrete." *Cement and Concrete Research*, 32, 1699–1704.

De Schutter, G., and Feys, D. (2016). "Pumping of fresh concrete: Insights and challenges." *RILEM Technical Letters*, 1, 76–80.

Deeb, R., and Karihaloo, B. L. (2013). "Mix proportioning of self-compacting normal and high-strength concretes." *Magazine of Concrete Research*, 65(9), 546–556.

Douglas, E., and Pouskouleli, G. (1991). "Prediction of compressive strength of mortars made with portland cement-blast-furnace slag-fly ash blends." *Cement and Concrete Research*, 21(4), 523–534.

Ferrara, L., et al. (2007). "A method for mix-design of fiber-reinforced self-compacting concrete." *Cement and Concrete Research*, 37(6), 957–971.

Feys, D., et al. (2008). "Fresh self compacting concrete, a shear thickening material." *Cement and Concrete Research*, 38(7), 920–929.

Feys, D., et al. (2009). "Why is fresh self-compacting concrete shear thickening?" *Cement and Concrete Research*, 39(6), 510–523.

Ghanbari, A., and Karihaloo, B. L. (2009). "Prediction of the plastic viscosity of self-compacting steel fibre reinforced concrete." *Cement and Concrete Research*, 39(12), 1209–1216.

Hu, J. (2005). "A study of effects of aggregate on concrete rheology", Doctoral Thesis, Iowa State University.

Hu, J., and Wang, K. (2011). "Effect of coarse aggregate characteristics on concrete rheology." *Construction and Building Materials*, 25(3), 1196–1204.

Hwang, C.-L., and Hung, M.-F. (2005). "Durability design and performance of self-consolidating lightweight concrete." *Construction and Building Materials*, 19(8), 619–626.

Ji, T., et al. (2006). "A concrete mix proportion design algorithm based on artificial neural networks." *Cement and Concrete Research*, 36(7), 1399–1408.

Ji, T., et al. (2013). "A mix proportion design method of manufactured sand concrete based on minimum paste theory." *Construction and Building Materials*, 44, 422–426.

Jiao, D., et al. (2021). "Thixotropic structural build-up of cement-based materials: A state-of-the-art review." *Cement and Concrete Composites*, 122, 104152.

Jiao, D., et al. (2017). "Effect of constituents on rheological properties of fresh concrete-A review." *Cement and Concrete Composites*, 83, 146–159.

Jiao, D., et al. (2018a). "Influences of shear-mixing rate and fly ash on rheological behavior of cement pastes under continuous mixing." *Construction and Building Materials*, 188, 170–177.

Jiao, D., et al. (2018b). "Mixture design of concrete using simplex centroid design method." *Cement and Concrete Composites*, 89, 76–88.

Jiao, D., et al. (2019). "Effects of rotational shearing on rheological behavior of fresh mortar with short glass fiber." *Construction and Building Materials*, 203, 314–321.

Jiao, D., et al. (2020). "Rheological properties of cement paste with nano-Fe_3O_4 under magnetic field: Flow curve and nanoparticle agglomeration." *Materials*, 13(22), 5164.

Karihaloo, B. L., and Ghanbari, A. (2012). "Mix proportioning of self-compacting high- and ultra-high-performance concretes with and without steel fibres." *Magazine of Concrete Research*, 64(12), 1089–1100.

Kostrzanowska-Siedlarz, A., and Gołaszewski, J. (2016). "Rheological properties of high performance self-compacting concrete: Effects of composition and time." *Construction and Building Materials*, 115, 705–715.

Kovler, K., and Roussel, N. (2011). "Properties of fresh and hardened concrete." *Cement and Concrete Research*, 41(7), 775–792.

Kwan, A. K. H., and Li, L. G. (2012). "Combined effects of water film thickness and paste film thickness on rheology of mortar." *Materials and Structures*, 45(9), 1359–1374.

Li, J., et al. (2017). "A mix-design method for lightweight aggregate self-compacting concrete based on packing and mortar film thickness theories." *Construction and Building Materials*, 157, 621–634.

Li, J., et al. (2021a). "Mixture design method of self-compacting lightweight aggregate concrete based on rheological property and strength of mortar." *Journal of Building Engineering*, 43, 102660.

Li, L. G., and Kwan, A. K. H. (2011). "Mortar design based on water film thickness." *Construction and Building Materials*, 25(5), 2381–2390.

Li, L. G., and Kwan, A. K. H. (2013). "Concrete mix design based on water film thickness and paste film thickness." *Cement and Concrete Composites*, 39, 33–42.

Li, P., et al. (2021b). "Improvement of mix design method based on paste rheological threshold theory for self-compacting concrete using different mineral additions in ternary blends of powders." *Construction and Building Materials*, 276, 122194.

Li, P., et al. (2020). "An enhanced mix design method of self-compacting concrete with fly ash content based on paste rheological threshold theory and material packing characteristics." *Construction and Building Materials*, 234, 117380.

Liu, Z., et al. (2019). "Mixture Design Approach to optimize the rheological properties of the material used in 3D cementitious material printing." *Construction and Building Materials*, 198, 245–255.

Nie, D., and An, X. (2016). "Optimization of SCC mix at paste level by using numerical method based on a paste rheological threshold theory." *Construction and Building Materials*, 102, 428–434.

Reinhardt, H. W., and Wüstholz, T. (2006). "About the influence of the content and composition of the aggregates on the rheological behaviour of self-compacting concrete." *Materials and Structures*, 39(7), 683–693.

Saak, A. W., et al. (2001). "New methodology for designing self-compacting concrete." *Materials Journal*, 98(6), 429–439.

Scheffé, H. (1958). "Experiments with mixtures." *Journal of the Royal Statistical Society: Series B (Methodological)*, 20(2), 344–360.

Shi, C., et al. (2015a). "A review on mixture design methods for self-compacting concrete." *Construction and Building Materials*, 84, 387–398.

Shi, C., et al. (2015b). "The hydration and microstructure of ultra high-strength concrete with cement–silica fume–slag binder." *Cement and Concrete Composites*, 61, 44–52.

Shi, C., et al. (2018a). "Mixture design of concrete based on rheology." *4th International Symposium on Design, Performance and Use of Self-Consolidating Concrete (SCC'2018-China)*, Changsha, China, 3–15.

Shi, C., et al. (2018b). "Design of high performance concrete with multiple performance requirements for #2 Dongting Lake Bridge." *Construction and Building Materials*, 165, 825–832.

Shi, Z., et al. (2016). "Factorial design method for designing ternary composite cements to mitigate ASR expansion." *Journal of Materials in Civil Engineering*, 28(9), 04016064.

Simon, M. J., et al. (1997). "Concrete mixture optimization using statistical mixture design methods." *Proceedings of the PCI/FHWA International Symposium on High Performance Concrete*, New Orleans, Louisiana, 230–244.

Sun, W., and Yan, H. (2003). "Studies on the quantitative relationships between compositions of blended cementitious materials and compressive strength of concrete." *Dongnan Daxue Xuebao/Journal of Southeast University(Natural Science Edition)*, 33(4), 450–453.

Tattersall, G. H. (1991). *Workability and quality control of concrete*. CRC Press, Boca Raton, FL.

Tattersall, G. H., and Banfill, P. F. (1983). *The rheology of fresh concrete*. Pitman Books Limited, London, England.

Thiedeitz, M., et al. (2020). "L-box form filling of thixotropic cementitious paste and mortar." *Materials (Basel)*, 13(7), 1760.

Toutou, Z., and Roussel, N. (2006). "Multi scale experimental study of concrete rheology: From water scale to gravel scale." *Materials and Structures*, 39, 189–199.

Wallevik, O. H. (1983). "Description of fresh concrete properties by use of two-point workability test instrument." *NTH, Trondheim*.

Wallevik, O. H. (2003). "Rheology—a scientific approach to develop self-compacting concrete." *Proceedings of the 3rd International Symposium on Self-Compacting Concrete*, Reykjavik, 23–31.

Wallevik, O. H., and Wallevik, J. E. (2011). "Rheology as a tool in concrete science: The use of rheographs and workability boxes." *Cement and Concrete Research*, 41(12), 1279–1288.

Wang, D., and Chen, Z. (1997). "On predicting compressive strengths of mortars with ternary blends of cement, GGBFS and fly ash." *Cement and Concrete Research*, 27(4), 487–493.

Wu, Q., and An, X. (2014). "Development of a mix design method for SCC based on the rheological characteristics of paste." *Construction and Building Materials*, 53, 642–651.

Xie, H., et al. (2013). "Workability and proportion design of pumping concrete based on rheological parameters." *Construction and Building Materials*, 44, 267–275.

Yammine, J., et al. (2008). "From ordinary rhelogy concrete to self compacting concrete: A transition between frictional and hydrodynamic interactions." *Cement and Concrete Research*, 38(7), 890–896.

Chapter 8

Rheology and self-compacting concrete

8.1 INTRODUCTION TO SCC

8.1.1 Brief history of SCC

The concept of self-consolidating or self-compacting concrete, often abbreviated as SCC, was first proposed by Professor Hajime Okamura of the University of Tokyo in 1989 at the Second East-Asia and Pacific Conference on Structural Engineering and Construction. At that time, many buildings/structures made of reinforced concrete in Japan, which was built just after the Second World War, suffered from the durability problem. Investigations of the causes indicated that the durability problem of reinforced concrete is mainly because of insufficient compaction, due to a lack of skilled and qualified workers. Therefore, concrete with high flowability and without vibration was proposed as a solution to combat the very bad quality of concrete structures in Japan (Hayakawa et al., 1993, Kuroiwa et al., 1993).

Admittedly, the development of chemical admixtures, mainly superplasticizers and viscosity-enhancing agents, makes SCC applicable to industry. The advent of superplasticizer makes highly flowable concrete with less water possible. Viscosity-modifying agent significantly improves the segregation resistance of highly flowable concrete. It is first used to produce nondispersive concrete for placing underwater.

Shortly after its invention, SCC made its entry into other parts of the world such as Europe, North America, and China. Soon, academia and industry showed great enthusiasm for SCC. It became a hot research topic worldwide. Gradually, SCC has been applied in the construction industry increasingly. In some special cases, SCC has its obvious advantages over conventional vibrated concrete, such as heavily reinforced sections and locations where vibration cannot reach. Although it is generally accepted that SCC is a type of concrete with no necessity of vibration, different countries or areas have various definitions for SCC:

- Japan: SCC can be compacted into every corner of a formwork, purely by means of its own weight and without the need for vibrating compaction.
- Europe: SCC is an innovative concrete that does not require vibration for placing and compaction. It is able to flow under its own weight, completely filling formwork and achieving full compaction, even in the presence of congested reinforcement. The hardened concrete is dense and homogeneous, and it has the same engineering properties and durability as traditional vibrated concrete.
- US: SCC is highly flowable, non-segregating concrete that can spread into a place, fill the formwork, and encapsulate the reinforcement without any mechanical consolidation.
- China: SCC is highly flowable, homogenous, and stable concrete that can spread into a place and fill up the formwork under its own weight without any external vibration.

All these definitions have one thing in common: SCC can flow on its own weight, and no extra vibration is needed. This means that concrete should flow like a fluid, and this is completely different from conventional vibrated concrete. On the other hand, homogenous and stable concrete is required. It seems contradictory and hard to make. However, striking a balance between fluidity and stability is the major challenge for SCC. In fact, the main differences between SCC and conventional vibrated concrete are fresh properties. SCC with the same hardened properties as conventional vibrated concrete can be designed and produced. Since fresh properties are mainly related to rheology, the rheology science of concrete is thus becoming more and more important due to the popularity of SCC.

During the past decades, SCC researchers have been always very active, and extensive investigations have been conducted worldwide on SCC. A large quantity of literature has been published as journal papers, reports, and conference proceedings. Technical committees devoted to SCC have been formed. More than 10 series of international conferences have been held in Japan, Belgium, Canada, France, Germany, US, and China, as follows:

- 1997—RILEM Technical Committee (TC 174-SCC) on SCC is formed.
- 1998—International Conference on SCC is held in Kochi, Japan.
- 1999—1st International RILEM Symposium on SCC is held in Stockholm, Sweden.
- 2001—ASTM International Subcommittee C 09.47 on Self-Consolidating Concrete is formed.
- 2001—2nd International RILEM Symposium on SCC, Tokyo, Japan.
- 2002—1st North American Conference on SCC, Chicago, Illinois, USA.
- 2003—ACI Technical Committee 237—Self-Consolidating Concrete is formed.
- 2003—3rd International RILEM Symposium on SCC, Reykjavik, Iceland.
- 2005—4th International RILEM Symposium and 2nd North American Conference on SCC, Chicago, USA.
- 2007—5th International RILEM Symposium on SCC, Ghent, Belgium.
- 2008—3rd North American Conference on SCC, Chicago, USA.
- 2010—6th International RILEM Symposium and 4th North American Conference on SCC, Montreal, Quebec, Canada.
- 2013—7th International RILEM Symposium and 5th North American Conference on SCC, Paris, France.
- 2016—8th International RILEM Symposium on Self-Compacting Concrete and 6th North American Conference on Design and Use of Self-Consolidating Concrete, Washington, USA.
- 2019—9th International RILEM Symposium on Self-Compacting Concrete, Dresden, Germany.
- 2022—10th International RILEM Symposium on Self-Compacting Concrete, Changsha, China.

8.1.2 Raw materials of SCC

The raw materials for producing SCC are the same as conventional vibrated concrete. It is composed of four groups of materials, as follows:

8.1.2.1 Powder

High-powder content is often required for SCC. Apart from cement powder, many other powders are used in SCC, either reactive or inert, such as limestone powder, fly ash, slag, silica fume, metakaolin, and even finely ground sand. For economic reasons, most of the other powers are cheaper than cement. Thus, the use of other powders as much as possible can reduce the cost of SCC, and in the meantime, helps reach the greenness of concrete. For technical reasons, the use of other powders has the following advantages: (1) improve the mechanical strength, (2) reduce the heat of hydration and therefore reduce the risk of cracking from thermal expansion, (3) improve the durability of concrete and reduce the risk of damage from the alkali–silica reaction associated with the alkali content of the cement, and (4) improve the stability and rheological behavior of SCC.

8.1.2.2 Chemical admixtures

All the chemical admixtures used for conventional vibrated concrete can be used for SCC to control the specific characteristics of the mixture, such as water reducer, superplasticizer, air-entraining agent, and retarding agent. Combinations of various admixtures are used to modify multiple properties simultaneously. It is worth mentioning that superplasticizer and viscosity-modifying agents are especially important for SCC.

Melamine formaldehyde- and naphthalene formaldehyde-based water reducers were initially used for concrete, which disperses cement particles by electrostatic repulsion. However, these two types of superplasticizers have their own limitations. Due to the low water-reducing rate, higher dosages are necessary for highly flowable concrete. This may result in a delayed rate of hardening and early strength development, which are undesirable in many applications. Furthermore, these two types of superplasticizers may suffer from rapid slump loss, which is not desired for SCC. The new-generation superplasticizers—polycarboxylic ether-based admixtures (PCEs)—are characterized by a high water-reducing rate, high compatibility with cement, and good comprehensive performances. Steric hindrance is believed to be the main dispersion mechanism of PCEs. PCE-based admixtures are almost a necessary chemical admixture for SCC.

Viscosity-modifying admixtures (VMAs) are new types of chemical admixtures introduced for SCC. They were initially used for concrete placed underwater, which can increase the cohesiveness and viscosity of underwater concrete mixtures and eliminate or significantly decrease the separation of mixture constituents underwater. VMAs are used to improve the viscosity and thus the stability of SCC. Also, the robustness of SCC can be improved by the addition of VMAs. There are different varieties of VMAs: some thicken the cement paste, while others specifically thicken the water. Khayat (1998) summarized the mechanism of action into three classes: (1) Adsorption. The long-chain polymer molecules adsorb and fix part of the mixing water, and thereby swelling. This increases the viscosity of the mixing water and thus the cement-based materials. (2) Association. Molecules in adjacent polymer chains can develop attractive forces, causing a gel formation and an increase in viscosity. (3) Intertwining. At low shear rates and high polymer concentration, the polymer chains can intertwine and entangle, thus increasing viscosity.

PCE-based admixtures and VMAs are often used together to increase the flowability and enhance the stability of an SCC mixture. However, special attention should be paid to the interaction between different chemical admixtures (Yuan et al., 2017a).

8.1.2.3 Aggregates

Although SCC contains high-powder content, aggregates comprise between 60% and 80% of the total volume of a concrete mixture. Paste plays a role as a lubrication layer between aggregates, and thus high paste content ensures the flowability of SCC. However, a low volume of aggregate may cause increased shrinkage and cracking risk, which leads to durability problems. Thus, the volume fraction of aggregate is an important parameter for SCC.

Lightweight aggregate, recycled aggregate, and normal aggregate have all been successfully applied in SCC. The maximum size of coarse aggregates used to produce SCC mixtures ranges from 10 to 40 mm, and the maximum size aggregate of 16 or 20 mm is most often used. The choice of the maximum size of aggregate depends on the required passing ability, local availability, and practice.

The packing density of fine and coarse aggregates determines the void space between solid particles that need to be filled with paste. The particle size distribution influences the packing density of the aggregate. The particle shape influences the packing density of aggregate. More rounded aggregates have a relatively higher packing density than angular ones do. In addition, the particle shape influences the mobility of the aggregate as the mixture flows. More rounded particles facilitate the flow of concrete. For fine aggregate, the fineness modulus is often used as an indication of the coarseness of the fine aggregate, and as a key parameter for the quality control of fine aggregate.

8.1.2.4 Water

Clean water is required for SCC.

8.1.3 Mix proportion of SCC

Generally, SCC can be classified into three types: powder type, viscosity agent type, and combination type. The combination type is more often used in practice, due to its robustness (Feys et al., 2008b). The early development of SCC in Japan established the essential criteria for achieving three key properties (Okamura and Ozawa, 1994):

- a low water/cement (or water/powder) ratio with a high dosage of superplasticizer to achieve high flowable capacity without instability or bleeding.
- a paste content sufficient to overfill all the voids in the aggregate skeleton to the extent that each particle is surrounded by an adequate lubricating paste layer, thus reducing the frequency of contact and collision between the aggregate particles during flow.
- a sufficiently low coarse aggregate content to avoid particle bridging and hence blocking of flow when the concrete passes through confined spaces.

During the past decade, many SCC mixture proportioning procedures have been developed or proposed worldwide. It is worth mentioning that most of these methods only focus on the fresh properties of concrete which ensure the compactability of concrete, and hardened properties of concrete are less focused and mainly adopt the conventional methods borrowed from ordinary concrete. These methods are developed based on different principles. However, it can be roughly divided into three groups. The first group is based on laboratory experiments and empirical parameters. The second group is based on the statistical method. The third approach is based on maximum packing and appropriate paste volume, which can ensure the flowability of concrete and as less as possible required paste.

8.1.3.1 Laboratory experiments and empirical parameters

The simple empirical step-by-step method originated from the extensive early work on SCC in Japan (Ozawa et al., 1994). In this method, formulations often seem to rest on the empiricism of the approach, and a series of experiments may end up with, for example, a method that sets a priori a gravel/sand mass ratio of 1, a mixture proportion of 350–500 kg/m^3 cement, etc. This approach is hampered by a lack of reliability and becomes inefficient if one of the parameters varies slightly, and thus it is a very empirical method.

The first step corresponds to synthesizing formulation principles from various published research works and adapting these principles to the manufacturing conditions, where the characteristics of the raw materials must be considered. The second step is verification and validation using the experimental tests to make sure the SCC meets all the requirements. In the meantime, it is indispensable for proving the robustness of the mixture. Take the method developed in Japan as an example. A relatively simple procedure was proposed based on the following principles: limited gravel concentration (to ensure low gravel to paste ratio) and low water/binder ratio. The basic steps in proportioning concrete mixtures are described as follows:

- setting the gravel volume to 50% of the solid volume in the mixture;
- setting the sand proportion to 40% of the mortar volume;
- water/cement volume ratio between 0.9 and 1.0 is determined depending on the cement type; the corresponding mass ratio is very low, between 0.29 and 0.32;
- adjust superplasticizer and final water contents to ensure the self-compacting ability;
- the optimum amount of superplasticizer can always be estimated from studies on mortar.

8.1.3.2 Statistical method

Statistical tools are always useful for stochastic parameters, and concrete mixture parameters are stochastic parameters. This method optimizes mixture proportions with the help of statistical analysis methods for the five fundamental formulation parameters:

- cement content;
- *w/b* ratio;
- superplasticizer addition level;
- viscosity-enhancing agent addition level;
- volume of aggregates.

The team led by Khayat first developed this approach (Khayat et al., 1999). This method needs comprehensive experimental data. Laboratory tests were intended to express the rheological and mechanical properties of fresh and hardened concrete, respectively, such as yield stress, viscosity, and compressive strength at 7 and 28 days of age. The experimental data were used to draw a series of trend lines. Afterward, the trend lines were linked with statistical models. The mixture proportions were optimized for given rheology and strength, which enabled a significant reduction in the number of laboratory tests.

8.1.3.3 Maximum packing density

The mix design method developed by Sedran and de Larrard at LCPC, France (Larrard et al., 1998, Sedran and Delarrard, 1999) tried to produce an optimal mixture as efficiently and simply as possible with the lowest paste volume and the minimum laboratory testing. Only a few preliminary tests on concrete mixes are required. Its main feature is the use of a mathematical model called the compressible packing model, in which the maximum packing density of the particulate materials was theoretically calculated. A software package was also developed for this method. This software can optimize the overall particle size distribution of all powder materials and particle materials based on their size distribution. The paste was assumed to fill up the voids between the aggregates. The space existing between the aggregates is inferred by an additional quantity of paste, called excess paste. The proportion of the excess paste is calculated by fixing the ratio between the volume occupied by the aggregates and the volume occupied by the aggregates covered in paste. It is found that this model is very useful for predicting the effect of aggregate type (natural gravel or crushed rock) on SCC properties and for studying the robustness of concrete to variation in the mixed parameters.

8.1.3.4 Other methods

It is worth mentioning that the mixture design methods detailed above do not allow direct integration of SCC fluidity criteria into the procedure. Many other methods were also developed for the mixture design of SCC around the world. Wallevik and Nielsson (1998) considered the rheological properties to be the dominant requirement of SCC and included rheological testing as an essential and integral part of the mix development process. Domone (2006) developed a mix proportion method based on the experience and understanding of the behavior of SCC gained from other methods of mix design, and only a normally equipped concrete laboratory is required for this test. For more mixture design methods for preparing SCC, the reader is referred to Shi et al. (2015a,b) and Ashish and Verma (2019).

8.1.4 Application of SCC

The application of SCC has many advantages as follows:

- Ease of placement with limited access;
- Provides higher in-place quality and aesthetics;
- Ease of placement and consolidation through dense reinforcement;
- Faster speed of construction and time savings;
- Labor savings;
- Ease of placement and consolidation in a complex structure or shape;
- Improves worker safety and noise reduction.

SCC has been successfully used in precast, cast-in-place, structural, architectural, vertical, horizontal, large, and small projects. During the past several decades, many guidelines or specifications on SCC have been issued across the world, as follows:

- AIJ Recommended Practice for High Fluidity Concrete for Building Construction, Architectural Institute of Japan, 1997;
- Recommendation for Construction of Self-Compacting Concrete, Japan Society of Civil Engineers, 1998;
- The European Guidelines for Self-Compacting Concrete: Specification, Production and Use, Self-Compacting Concrete European Project Group, 2005;
- ACI 237R-07, Self-Consolidating Concrete, American Concrete Institute, 2007;
- JGJ/T 283-2012, technical specification for application of self-compacting concrete, Ministry of Housing and Urban-rural Development of China. 2012;
- Q/CR 596-2017, Specification of self-compacting concrete for China Railway Track System III (CRTS III) ballastless slab track, 2017.

Even though SCC has established its position in the prefabrication industry (around half of the volume produced), its in situ use is struggling to make an impact on construction sites. Despite its numerous advantages, it finds limited application in engineering applications around the world. Several factors are ascribed to explain this slow expansion of SCC use. First, making SCC is somewhat difficult, since the components must be of good quality but are hard to control. The properties of fresh traditional vibrated concrete are affected relatively little by normal variations in the components in terms of size distribution, water content, etc. On the contrary, SCC is much more sensitive to variations in raw materials and environmental parameters. Second, the production tool is not always precise enough for making concretes which are strongly affected by errors in the mixture proportions. Third, the formworks must be well prepared and properly waterproofed, and must, above all, be able to withstand pressures that are a priori higher than those involved in handling vibrated concretes. Finally, the cost for the raw materials of SCC is higher than that of normal concrete, due to the high content of powder and chemical admixtures.

However, as a new type of material, SCC has been used in some special scenarios on construction sites. For instance, SCC for CRTS III ballastless slab track has been widely used in China. Thousands of kilometers of high-speed railway (HSR) lines have been implemented with CRTS III slab ballastless track, and millions of cubic meters of SCC for CRTS III slab track have been consumed in China. The structure section and layout of CRTS III slab ballastless track are shown in Figure 8.1. The tracking form consists of four layers, which are, from top to bottom, prefabricated prestressed-slab, SCC layer, isolated geotextile layer, and base plate. The SCC layer is cast-in-place and is required to have a strong bonding with the above prefabricated slab. Consequently, the two layers function as a composite plate. The loadings of a high-speed train are transferred to the roadbed by this composite plate. That is to say, the interface bonding between above prefabricated slab and cast-in-place bottom SCC should be strong and durable enough to secure the serviceability of the slab track. This is extremely important for this unique slab track.

SCC layer is a thin plate with the dimension of 90 mm × 2500 mm × 5600 mm. The main functions of the SCC layer of CRTS III slab ballastless track are as follows: (1) the thickness of the SCC layer, which is cast-in-place, can be adjusted to assure above the prefabricated slab's position is preset. By doing so, the high smoothness of the track can be attained. (2) SCC layer works with the slab as a composite plate transferring the trainloads. The tracking form implements a concept of "decreasing stiffness from top to bottom".

Normal SCC is often cast into an open and rigid formwork. The upper surface of SCC is smoothed by the process of finishing, so slight bleeding and rising of the bubble are acceptable. Compared with normal SCC, the SCC applied in CRTS III slab ballastless track presents some prominent and unique characteristics:

- SCC is grouted into a flat, narrow, and sealed space with the dimension of 90 mm × 2500 mm × 5600 mm.
- There is a flexible geotextile layer under the SCC layer. SCC flows on the geotextile layer during the casting of SCC. It increases the flow resistance. In addition, geotextile may absorb the water from SCC and affect the flow of SCC.

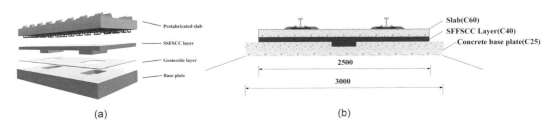

Figure 8.1 The structure section and layout of CRTS III slab ballastless track (Yuan et al., 2016).

- An HSR line is often hundreds of kilometers, and it is usually one-time completion. Construction sites along the HSR line may extend hundreds of kilometers, and various raw materials will be used, and they may present large variations in properties and compositions. It's quite challenging for the quality control of SCC because of the variation in local raw materials.

Due to the abovementioned characteristics, SCC for CRTS III slab ballastless track is different from normal SCC and puts forward stricter requirements. The first characteristic is the most important and unique one. Hence, we name this type of SCC as sealed-space-filling self-compacting concrete (Yuan et al., 2016).

8.2 RHEOLOGY OF SCC

8.2.1 Factors affecting rheology of SCC

As stated above, there are various raw materials used in SCC, and these materials have a different effect on the rheological properties of SCC. This section will focus on the effect of powder materials, chemical admixtures, and recycled aggregate on the rheology of SCC.

8.2.1.1 Fly ash

Fly ash (FA) is an industrial by-product derived from electricity-generating plants. It is generated by burning coal. Its chemical and mineralogical composition mainly depends on the relevant properties of the raw material used as well as on the type of furnace and the way it is collected. FA particles consist of clear glassy spheres and a spongy aggregate. Scanning electron microscopy shows the typical spherical shape of FA particles, some of which are hollow (Figure 8.2). The addition of FA decreases the need for viscosity-modifying chemical admixtures (Heikal et al., 2013). Jalal et al. (2013) concluded that increasing the percentage of class F FA improves the workability/rheology of fresh high-strength SCC, due to the balling effect of spherical particles in FA.

8.2.1.2 Rice husk ash

Rice husk ash (RHA) is an agricultural waste generated from the milling process of paddy. It is abundant in many rice-cultivating countries, e.g., China, India, and Vietnam. A 200 kg rice husk is generated from each ton of paddy rice, which makes about 40 kg of ash on combustion. According to the "Rice Market Monitor" report, the rice husk production in 2011 was approximately 145 million tons. Normally, rice husk from paddy rice mills is burnt in open piles on the fields or disposed directly into the environment, thus causing severe environmental pollution, especially when it is disintegrated in wet conditions. The use of RHA in concrete is a step toward sustainable production.

Le et al. (2015) investigated the effect of mean pore size (MPS) of RHA on the rheological properties of self-compacting high-performance concrete (SCHPC). It was found that the incorporation of RHA increased plastic viscosity and yield stress, and this effect is significant in the case of coarser RHA. The addition of RHA decreased the flowability of SCC mortar (Safiuddin et al., 2010). The incorporation of RHA increased the plastic viscosity, robustness, and segregation resistance, but slightly decreased filling and passability in fresh SCC (Le et al., 2016).

Figure 8.2 Spherical particle of FA.

8.2.1.3 Silica fume

Silica fume (SF) is considered to be one of the most widely used supplementary cementitious materials (SCMs) that greatly reduce the permeability and significantly increase the strength of the mixture. However, its utilization in SCC is limited because of the difficulty in obtaining the desired SCC workability (Hassan et al., 2010).

Due to its extremely fine particles, SF is sometimes used as a viscosity-enhancing agent, just like polymer-type VMA which can be used to increase the resistance to segregation. The chains of VMA polymer connect with each other by van der Waals interaction and block the movement of free water. Benaicha et al. (2015) studied the influence of SF and VMA on the plastic viscosity of SCC. The result indicated that SF and VMA could replace each other based on the availability of the material. SCC mixtures with 10% SF (wt. of cement) or 0.1% VMA (wt. of binder) showed approximately the same rheological properties (yield stress and plastic viscosity). Lu et al. (2015) incorporated SF with different proportions (0%–16%) in the SCC mixture and measured the rheological properties at different times after water addition (TW) (0–120 min). The yield stress increased with increasing TW and SF contents, while by increasing SF content, the plastic viscosity first decreased and then increased, giving the lowest value at 4% SF content. According to Hassan et al. (2010), 8 wt.% replacement of cement with SF showed no effect on the viscosity of SCC, but increased the yield stress. The addition of 2% of SF decreased the plastic viscosity and yield stress, but its further increase deteriorated the flowability of SCC (Ling et al., 2018). This is due to the active nature of SF which promotes the hydration process and the high fineness of SF which increases the water demand. According to Aleksandra and Gołaszewski (2016), both yield stress and plastic viscosity of high-performance SCC decreased by reducing the SF volume, the addition of SP, and increasing the *w/b* ratio.

8.2.1.4 Metakaolin

Metakaolin (MK) is an eco-friendly material that can be prepared from kaolinic clays without carbon dioxide production. The main sources of MK are either kaolin or paper sludge after suitable treatment. Kaolin has a particle size ranging from 0.2 to 15 μm with a BET-specific area of 10,000–29,000 m²/kg. MK reacts with the hydrating cement and forms a modified paste microstructure. In addition to its positive environmental impact, MK improves the workability, durability, and mechanical properties of concrete (Rashad, 2013).

At 2% addition, metakaolin decreased the fundamental rheological properties, but showed an increased effect with further addition (Gang et al., 2018). This can be attributed to the high fineness of MK which increases the water demand. At 2% addition, small particles fill the interfacial spaces of the mixture and release the free water, thus improving the rheology of the SCC mix. Hassan et al. (2010) investigated the rheology of the MK with the replacement of 0–25 wt.% of cement. Both yield stress and plastic viscosity increased by increasing MK content and showed maximum value at the replacement level of 25%. Moreover, MK addition increased the demand for HRWR in SCC mixtures. With the increase of time from 10 to 60 min, the yield stress and plastic viscosity of SCC mixes containing MK as a partial replacement of cement increased, but the increase of plastic viscosity was lower than yield stress (Vejmelková et al., 2010).

8.2.1.5 Blast furnace slag

Slag is a by-product of the iron-manufacturing industry. About 90% of mass content is comprised of CaO, SiO_2, and Al_2O_3. The specific gravity of slag is approximately 2.90, and its bulk density varies from 1200 to 1300 kg/m³. Variation in the specific surface area is noted, as 375–425 m²/kg measured by the Blaine method, 350–450 m²/kg in the United States, and 450–550 m²/kg in the United Kingdom (Jiao et al. 2017). In China, a specific surface area higher than 450 m²/kg was reported (Pal et al., 2003). Due to the reduction of CO_2 emission and improvement in workability, slag has been widely used in cement paste, mortar, and concrete. SCC mixture containing blast furnace slag as partial replacement of cement exhibited Newtonian fluid properties with zero yield stress and significant plastic viscosity (Vejmelková et al., 2010). Sethy et al. (2016) incorporated high-volume (30%–90%) industrial slag in SCC mixes. For all replacement levels, the value of yield stress was nearly zero, while plastic viscosity decreased by increasing slag content. Moreover, the addition of slag significantly reduced the HRWR dosage. Replacement of cement with up to 20% of slag improves the workability of SCC with an optimum content of 15% (Boukendakdji et al., 2012). Generally, the effect of slag on the rheology of concrete mainly depends on its fineness. The finer the particle, the more viscous the SCC.

8.2.1.6 Fibers

Alberti et al. (2019) studied the influence of steel fibers (35 and 50 mm long) and polyolefin fibers (60 mm long) on the rheology of SCC at a volumetric fraction from 0.33% to 1.1%. They stated that a linear increase in yield stress and plastic viscosity was observed by increasing the volumetric fraction of fibers. In addition, the number of fibers and geometry also influenced the fresh-state properties of SCC, while fiber stiffness showed minor impact on fresh properties. A reduction in workability by using hooked-end steel fibers (0–1% vol. matrix) was reported (Silva et al., 2017).

8.2.1.7 Air-entraining agent

In a cold environment, air-entraining agents (AEAs) are used to protect cement-based materials from freeze-thaw cycles. The effect of AEA on the rheology of SCC has been studied by many researchers as summarized below.

Huang et al. (2018) incorporated the rosin resin-type AEA and studied the rheology of SCC. The yield stress increases and plastic viscosity decreases by increasing AEA content, as shown in Figure 8.3. On the one hand, AEA increases the volume of paste and increases the lubrication effect, hence reducing the yield stress value. On the other hand, calcium ions are adsorbed on the cement particle during cement hydration, imparting a positive charge. The negative-charge AEA is prone to be attracted by the cement particles with a positive charge. Consequently, bubble bridges can form between cement particles, which increases the yield stress. However, the latter effect depends on the type of AEAs (Du and Folliard, 2004). The air bubble increases the volume and deformation capacity of the paste but decreases the actual content of cementitious material in the unit volume. Consequently, the cohesive strength of the mixture decreases, leading to a decrease in plastic viscosity (Chia and Zhang, 2004).

8.2.1.8 Superplasticizer

Superplasticizer (SP) is the key constituent of SCC to achieve the desired properties. The influence of SP on the rheology properties of SCC is summarized in the following.

Polycarboxylate (PC)-based SP produces a better dispersion effect than naphthalene sulphonate-based SP for increasing the workability of SCC (Boukendakdji et al., 2012). PC is synthesized from petroleum products. The chemical structure of PC-based SP includes the main backbone length, side-chain length, charge density differentiation, degree of backbone polymerization, and composition of functional groups. The chemical structure of SP influences the rheological properties of the cement-based system. For instance, Andersen et al. (1988) investigated the effect of the molecular weight of SP on concrete. They concluded that SP with the largest molecular weight showed the largest negative zeta potential, and hence the greatest ability to prevent flocculation of cement particles. Mardani-Aghabaglou et al. (2013) studied the effect of polycarboxylate ether-based SP having the same polymer structure and the same main chain, but different molecular weight and side-chain density

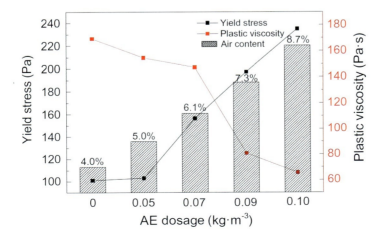

Figure 8.3 Effect of AE dosage on the rheological parameters of SCC (Huang et al., 2018).

of carboxylic acid groups on the rheology of SCC. The apparent yield stress seemed to be affected by the dosage of SP, while other characteristics of SP had no significant effect on apparent yield stress. By increasing the side-chain density, the plastic viscosity decreased. Moreover, the slump retention of SCC seemed to be considerably affected by the side-chain density of the polymer.

Huang et al. (2018) studied the influence of polycarboxylate SP on the rheology of SCC and concluded that both yield stress and plastic viscosity decreased with the addition of SP, as shown in Figure 8.4. The decrease in rheological properties by adding polycarboxylate-based SP was also observed by Ma and Wang (2013). Within the solution, the SP forms a directionally aligned adsorption layer on the surface of cement grains, and thus disperses the fine particles. Consequently, the free water is released due to electrostatic and steric hindrance effects, and yield stress decreases by increasing SP dosage (Ferrari et al., 2010). For conventional concrete, the addition of SP slightly increases the plastic viscosity. The reason is that increasing SP dosage not only decreases the plastic viscosity of paste but also weakens its lubrication effect. This leads to an increase in the friction resistance between aggregate particles, which increases the plastic viscosity of concrete. However, as high sand ratio and powder content are present in the SCC mixture, the coarse aggregate particles are easy to be lubricated due to the presence of enough paste. As a result, the reduction of the lubrication of paste would have less influence on the friction resistance between coarse aggregate particles. Therefore, with the addition of SP in SCC, the plastic viscosity increased (Huang et al., 2018).

Benaicha et al. (2019) investigated the influence of SP on the rheology of SCC mixtures made with cement and limestone. By increasing SP dosage, a decrease in plastic viscosity and yield stress was observed. The authors believed that this effect is due to the release of free water between cement particles and the increase of the water film coating the mixture particles.

Temperature plays an important role in the performance of polycarboxylate-based polymers in SCC. Due to daily and annual variations in temperature, it is necessary to investigate the influence of temperature variation on the robustness of SCC with different types and dosages of SP. To obtain the robustness of the SCC mixture, it is necessary to optimize the charge densities of polycarboxylate-based polymers and the water-to-powder ratio with respect to ambient temperature. Schmidt et al. (2014) studied the effect of temperature on the rheology of SCC with different compositions and polycarboxylate-based polymers of varying anionic charge densities. High-powder SCC mixtures showed robust performance

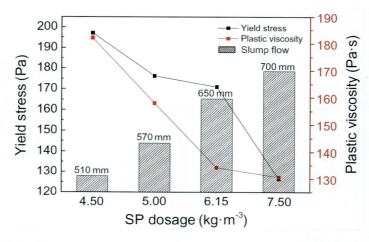

Figure 8.4 Effect of SP dosage on the rheological parameters of SCC (Huang et al., 2018).

at low temperatures, while low-powder SCC mixtures were observed to be favorable at high temperatures. At low temperatures, high-charge-density polycarboxylate-based polymers seemed to be robust, while at high temperatures, it significantly reduced the flow retention. At low temperature, low-charge-density polycarboxylate-based polymers could not generate self-compacting properties but retained the flow performance over a sufficiently long time. According to the authors' opinion, the sound knowledge of the effect of charge density of polycarboxylate-based polymers on the properties of SCC with varying temperatures is imperative for achieving robust and high-performance SCC.

Different SCMs have different SP demands. For instance, Ahari et al. (2015c) found that for the same slump flow, the SP demand increased with the addition of MK, SF, and FAC, while decreased in the case of FAF and blast furnace slag (BFS). This is attributed to the surface texture, fineness, and geometry of different SCMs.

8.2.1.9 Recycled concrete aggregates

A large quantity of construction and demolition waste (CDW) is produced due to demolition and repair works. Therefore, the use of CDW debris as an aggregate for concrete production is beneficial in the economy and environmental perspective. CDW, after treatment from certified recycling plants, is an appropriate product for the production of certain types of structural concrete. Its use is regulated by many standards, such as Italian and Spanish regulations, and according to its components, it may be classified into three main types of materials, i.e., crushed concrete, mixed demolition debris, and crushed masonry. Due to the large amounts of energy consumed by the extraction of natural aggregates, research has been carried out on the potential of using recycled aggregates in the construction industry (Revilla-Cuesta et al., 2020). Recycled aggregates (natural aggregates + adhered cement mortar) have lower density and higher absorption capacity (especially due to adhering mortar) as compared to natural aggregates. As a result, the properties of fresh concrete are affected. To control the casting process effectively, it is necessary to determine the mixing water before the production of concrete. The mixing water is composed of effective water and water absorbed by the aggregates. Thus, it is necessary to add a certain amount of water to saturate the recycled aggregates, before or during mixing, to get the desired workability. Another characteristic of recycled concrete aggregates (RCA) that can affect the rheology of self-compacting recycled concrete (SCRC) is the higher internal friction of recycled aggregates due to their high surface texture (Gonz et al., 2018). Natural aggregates may be treated as angular aggregates (little wear on the particle surface), while recycled aggregates may be treated as sub-angular aggregates (evidence of some wear). In addition, recycled aggregates have a more porous and rougher surface texture than natural aggregates. The fines content in recycled aggregates is higher as compared with the natural aggregates.

González-Taboada et al. (2017a) investigated the rheology of SCRC with recycled coarse aggregate (fraction size (FS) ranged between 4 and 11 mm, and fineness modulus (FM) was 6.47), with a replacement of 20%, 50%, and 100%. The absorption capacity of RCA was compensated and investigated by three methods, i.e., dry aggregate and extra water, pre-soaked aggregate, and aggregate with 3% natural moisture and extra water. The results indicate that the "dry aggregate and extra water" is the most suitable method to compensate absorption water of RCA. With this method, the self-compacting behavior can be maintained for all replacements until 45 min, and even until 90 min when the replacement percentage was less than 50%. The yield stress of concrete made with coarse RCA increased by increasing the replacement level due to the harshness of concrete caused by coarse RCA (Singh and Singh, 2018).

The difference in rheological behavior between SCC and SCRC is due to the intrinsic properties of RCA (shape, texture, and fines content) and its property of continuously absorbing water, which affects the effective w/c ratio. The texture and shape of aggregates mainly depend on the crushing machine type (Rashwan and AbouRizk, 1997). González-Taboada et al. (2017b) observed the increase in static yield stress and plastic viscosity by increasing the replacement level of coarse CRA. The authors believed that the fine content of recycled aggregates increased the quantity of fines in SCRC. The irregular shape and rough texture of these fine particles affect the rheology negatively. During mixing, fine particle content increased due to the loss of old adhered mortar, and some of them can present hydraulic activity. The influence of this effect is stronger in the case of a lower w/c ratio due to high friction forces. As a result, both the solid-phase factor and paste factor increased. González-Taboada et al. (2020) observed that the viscosity of SCRC was more sensitive than the viscosity of SCC due to the fine particles generated from recycled aggregate adhered cement mortar. Revilla-Cuesta et al. (2020) concluded from the literature review that the viscosity of a mixture increased and flowability decreased by increasing the percentage of coarse RCA substitution, because of the higher water absorption of coarse RCA. According to González-Taboada (González-Taboada et al., 2018), three parameters that influence the rheology and hence the robustness of SCRC are water absorption capacity, fine content, and morphology of RCA.

Singh and Singh (2018) investigated the rheology of different SCRC grades made with ternary blended powder (Portland cement+SF+FA). The results showed that yield stress increased with the increase of concrete grade due to the incorporation of SCMs. Singh et al. (2017) revealed the rheology of two-grade SCRC with the addition of FA, SF, and metakaolin. The authors observed increase in shear-thickening behavior by increasing coarse RCA within the grades. Moreover, a decrease in shear thickening with an increase in concrete grade was observed, which was attributed to the use of metakaolin or SF as part of the ternary cement. Revilla-Cuesta et al. (2020) reviewed the properties of SCC having RCA and other waste, including FA, SF, ground granulated blast furnace slag (GGBFS), rubber, and recycled asphalt pavement (RAP). The authors reached the conclusion that the addition of different wastes in the concrete mix produced SCC with adequate flowability for large-scale use if the dosage is adjusted to the flowability requirements, to the amounts of added residues, and to their percentage substitution.

The absorption capacity of fine RCA (RCFA) is greater than that of coarse RCA. The water absorption (%) of coarse RCA is lying in the range of 4.53–6.27, while for RCFA, this value ranges from 7.76 to 11.06 (Revilla-Cuesta et al., 2020). Carro-López et al. (2015) studied the rheology of SCC mortar by using RCFA with replacement ratios (in volume) of 0%, 20%, 50%, and 100%. The water was adjusted by the addition of extra water equivalent to the absorption of aggregates at 10 min. The results indicate the increase in the static yield stress with time for all levels of RCFA replacements. Moreover, increasing the fraction of RCFA increased the yield stress over time, indicating the loss of the filling ability of concrete. By increasing the RCFA fraction, a rapid increase in plastic viscosity was observed, and loss of passability after 45 min was also found for large replacement ratios. The loss of SCC characteristics at 90 min was reported with 50% and 100% replacement of fine recycled aggregates (Carro-López et al., 2017). According to Behera et al. (2020), the increase in the RCFA addition results in an increase in static yield stress with time, and this observation was more rapid in the case of 100% RCFA replacement. This behavior was due to the friction induced by the rough surface texture, high water absorption capacity, and interlocking of irregularly shaped of RCFA. Moreover, the increase in viscosity was observed by increasing RCFA into mixes containing SF, while it decreased without SF. However, Güneyisi et al. (2016a) investigated the rheology of SCC with saturated surface dry fine RCA.

The reduction in shear thickening and the increase in the flowability of SCC were observed by increasing the percentage of fine RCA. According to Nasrollah et al. (2020), the quantity of fine or coarse RCA should be limited to 25% in order to meet the rheological requirements of SCC.

8.2.1.10 Binary and ternary binder system

High cement content usually generates high heat of hydration, high cost, and high autogenous shrinkage. In addition, cement production causes serious environmental problems through the consumption of natural resources and the emission of carbon dioxide. To reduce the cost of SCC, mineral admixtures or waste materials are used. Researchers used these waste materials in SCC in binary, ternary, and quaternary blends and investigated the effect on fresh properties. Ahari et al. (2015b) observed a decrease in plastic viscosity by using SF and BFS in SCC mixtures and an opposite effect in the case of MK, class C fly ash (FAC), and class F fly ash (FAF). A similar result was also found by Ahari et al. (2015c). Moreover, the influence of MK and FAC was observed to have a greater influence on plastic viscosity and yield stress as compared to other admixtures.

Güneyisi et al. (2015) studied the relationship between torque and rotational speed of cement paste incorporating FA (0–100 wt.%) and nano-silica (NS) (0–6 wt.%). The results showed a higher torque value by increasing NS content while lowered torque value by increasing FA content. Moreover, shear-thickening behavior was observed, which was increased by increasing NS and decreased by increasing FA content.

Diamantonis et al. (2010) investigated the rheology of cement paste with the addition of FA, SF, limestone, and pozzolan. It was observed that the addition of limestone showed the best rheological properties. The yield stress and plastic viscosity decreased with the addition of limestone, and the value of plastic viscosity was observed to be even lower than pure cement paste. For improving rheological properties, limestone was proved to be the best additive at 40% content, and this paste could serve as a base of SCC production.

8.2.1.11 Other constituents

Heikal et al. (2013) studied the behavior of ground clay bricks (GCB) on the rheology of SCC. The results showed that the value of shear stress increased by increasing GCB content. To improve fluidity, the authors used polycarboxylate-based superplasticizers and observed a decrease in the efficiency of polycarboxylate at 12.5% GCB, whereas the efficiency increased with GCB content up to 37.5%.

An increase in the content of TiO_2 nanoparticles generally increased the rheological properties of SCC, which could be due to the filler effect of the ultra-fine particles in the cement paste (Jalal et al., 2013). Durgun and Atahan (2017) used three different colloidal nano SiO_2 (CNS) particles, having an average particle diameter of 35, 17, and 5 nm, respectively, to reduce the total fine materials content of SCC. The FA was also used together with CNS at different proportions. For the CNS having the size of 35, 17, and 5 nm, the average threshold dosages to obtain the best rheological properties were 1.5%, 1%, and 0.3%, respectively. Independent of the size of CNS, the static and dynamic yield stresses increase by increasing CNS content but decrease by increasing FA amount. A slight increase in plastic viscosity was observed in mixtures with CNS and high amount of FA.

Porcelain polishing residue (PPR) is generated from the ceramic tiles industry, and it has pozzolanic activity due to a high level of amorphous silica and alumina and high fineness ($D_{50} \sim 1$–10 μm). De Matos et al. (2018) used the PPR to replace cement (up to 30 wt.%) in SCC and investigated the rheological properties of the mixture. The incorporation of PPR enhanced the shear-thinning behavior of the paste. This can be attributed to the introduction of finer particles between the cement grains, which decreases the interparticle friction due to the bearing effect. The reduction of flow index due to the incorporation of finer minerals than cement and the opposite behavior in the case of minerals coarser than the cement were also reported by other researchers, for example, SF (Yahia, 2010) and limestone powder (Ma et al., 2016). In addition, the yield stress is also affected by the fineness of materials. Finer materials decrease the distance and increase the interaction between particles. The yield stress is exponentially inversely proportional to the distance between the particles (Guo et al., 2017). This can be verified by the results of Matos et al. (2018). Besides, plastic viscosity increased by increasing PPR content (Matos et al., 2018). Sua-Iam et al. (2016) showed that SCC with a ternary blend (Type 1 Portland cement+limestone powder+residual risk husk ash) performed better workability than conventional SCC.

Ouldkhaoua et al. (2019) replaced the sand with catodique ray tube (CRT) glass and metakaolin in the SCC mix and measured the rheological properties by a modified slump test. The addition of 50% of CRT sand glass improved the rheological properties significantly and minimized the dosage of SP. Güneyisi et al. (2016b) incorporated crumb rubber and tire chips as a partial replacement (0%–25%) of natural aggregates and measured the rheological properties of SCC mixtures with the replacement of cement by FA (30 wt.%). By increasing the replacement level of natural aggregates with rubber particles, a higher value of torque was observed at the same rotational speed, indicating the deterioration of rheology at fresh-state rubberized SCC.

Benabed et al. (2012) used various types of sand (river sand, crushed sand, and dune sand) in the SCC mix and measured the viscosity. The addition of dune sand deteriorates the SCC rheology due to its fineness which requires high water and high cement to achieve high fluidity. Crushed sand with limestone fines (10%–15%) showed the best rheological properties in SCC mixtures. Silva et al. (2017) employed river gravel having a smooth surface texture as coarse aggregates instead of crushed aggregate and found improved rheological properties compared to crushed aggregates.

Blankson and Erdem (2015) investigated the effect of organic (migrating corrosion inhibitor (MCI)) and inorganic (calcium nitrite inhibitor) corrosion inhibitors in SCC and measured their rheological properties. Both organic and inorganic corrosion inhibitors reduced the viscosity of SCC, but this effect was more prominent in the case of organic MCI. Both types of inhibitors reduced the segregation resistance. Moreover, high workability retention was observed in the case of an organic MCI inhibitor.

8.2.2 Special rheological behaviors

8.2.2.1 Thixotropy

Thixotropic is an intrinsic property of fresh cementitious materials, playing a significant role in pumpability, stability and formwork filling of concrete, and buildability in 3D printing (Jiao et al., 2021). The mixture is said to be thixotropic if it shows a continuous decrease of viscosity with time when applying a shear stress and then subsequent recovery of viscosity over time when the shear stress is removed. The physical origin of this rebuilding might be the colloidal interactions between solid particles and the nucleation of early hydration products at their contact points. Structuration rate A_{thix}, which is defined as the increased rate of the static yield stress when a material is at rest, is 0 Pa/s for non-thixotropic concrete and 2 Pa/s for most thixotropic concrete.

The phenomenon of thixotropy has an important influence on the various applications of SCC, such as formwork pressure development, pumpability, and multilayer casting (interface behavior of the successive layers). The design and construction of the formwork are critical aspects of erecting a concrete structure. Due to the high fluidity of SCC, it is assumed that the actual pressure of SCC on the formwork will be higher than the hydrostatic pressure. Hence, a stronger formwork needs to be designed, which increases the cost of design and construction of formwork. The thixotropy of cement-based materials plays an important role in limiting the magnitude of high lateral pressure on formwork. In the case of a low casting rate, the materials build up an internal structure that improves its ability to withstand the additional load cast above it (Ovarlez and Roussel, 2006).

In case of a high rate of flocculation, the apparent yield stress of SCC increases above the critical value, resulting in an already-cast layer of SCC not mixed with the layer cast subsequently. Due to the absence of vibration in concrete, the negative impact of this phenomenon is the development of a weak interface between two adjacent layers in the final product. As a result, in addition to poor physical appearance and mechanical strength, the porosity and permeability of concrete increase, causing severe durability problems (Roussel, 2007).

The choice of structure type is affected by the thixotropy of SCC. It is not advisable to use highly thixotropic material ($A > 0.5$ Pa/s) for cast-in-situ concrete formwork due to its longer dimensions and shapes. This is because its high structuration rate impedes the free flow of the material. It is recommended to use SCC having a high flocculation rate in precast concrete components (Rahman et al., 2014).

To control the structuration rate without greatly affecting the slump flow, Roussel and Cussigh (2007) proposed five main factors to adjust the mix proportions:

- The total amount of the powders in the mixture.
- The weight ratio of water/powders (w/p) ratio.
- The fineness of the powders.
- The amount of superplasticizer.
- The amount of viscosity agent (VA).

Rahman et al. (2014) added fly ash and limestone powder (LSP) in different percentages into the SCC mixture and concluded that increasing the FA and LSP content increased the thixotropy of the mix. The results can be attributed to the increase in the structuration rate due to the increase in the fineness of the powders. Ahari et al. (2015a) investigated the thixotropic behavior of SCC by incorporating FA, SF, metakaolin (MK), class C fly ash (FAC), class F fly ash (FAF), and granulated BFS. The results showed that the increase in thixotropic behavior with time was higher in the case of SF and MK as compared with reference mixes and other minerals. In addition, recycled coarse aggregates also influence the thixotropy of SCC. Bir (2018) noted an increase in the degree of thixotropy of SCC mixture by increasing the replacement of natural aggregates with recycled aggregates. This was attributed to the high internal friction associated with the recycled aggregates.

8.2.2.2 Shear-thinning or shear-thickening behavior

Shear thinning is the decrease in the apparent viscosity with shear rate, while shear thickening is the increase of apparent viscosity by increasing the shear rate.

Generally, shear thickening is considered a potential industrial problem in terms of concrete production and casting. The popular theories to explain shear-thickening behavior include order–disorder transition theory, grain inertia theory of large particles, and clustering theory of fine particles. For the order–disorder theory, the flow is easy if particles are ordered into layers, while the disordered structure consumes more energy to flow due to the jamming of particles, increasing the viscosity. In the case of grain inertia theory of large particles, the momentum is transferred between the suspended particles and the dominance of grain inertia, depending on the particle's Reynolds number. For the clustering theory, shear thickening could occur at certain critical shear stress where hydrodynamic forces due to flow become larger than the interparticle repulsive forces. De Larrard et al. (1998) and Feys et al. (2008b) believed that SCC showed a shear-thickening phenomenon. Chia and Zhang (2004) studied the shear-thickening phenomenon in SCC with AEA. The results showed that SCC mixtures exhibited a transition from shear thickening to shear thinning at 8.3% of air content. By increasing the air content, the adsorption of the bubble on cement particles could increase, thus hindering the formation of the cluster. The addition of air bubbles also enhances the deformation capacity of the mixture and reduces the hydrodynamic forces. Therefore, the mixture tends to show shear-thinning behavior at high air content.

Huang et al. (2018) observed the shear-thickening behavior of SCC with the addition of SP. They proposed three reasons for this behavior: (1) SP disperses the fine particles in the solution. According to the clustering theory, the shear-thickening phenomenon is attributed to the change of particles from dispersion to cluster. As a result, the addition of SP facilitates the occurrence of shear thickening. (2) The increase of shear rate results in the disorder degree of fine particles and polymer chains, which leads to shear thickening following the order–disorder theory. (3) The adsorbed SP can be torn off from fine particles due to the shear stress, and the desorption state is more likely to form a cluster.

Güneyisi et al (2016b) used tire chips and crumb rubber to prepare concrete, and they found that the exponent 'n' (the Herschel–Bulkley) values and 'c/μ' (the modified Bingham) coefficients increased by increasing the replacement level of natural aggregates with crumb rubber and tire chips, indicating the shear-thickening behavior. The highest 'c/μ' coefficients and exponent 'n' values were observed in the case of replacement of natural coarse aggregates with tire chips, and the lowest values were achieved when fine aggregates were replaced with crumb rubber.

Shear-thickening behavior was observed when natural aggregates were used in SCC, but this behavior decreases when natural aggregates were replaced with lightweight aggregates (LWA) made with FA (Gesoglu et al., 2015). Due to the spherical shape of LWA, decreased yield stress and plastic viscosity were observed.

Le et al. (2015) pointed out a decrease in the degree of shear thickening of SCC with the incorporation of RHA (MPS 5.7 μm) and SF. It was found that the effect of SF was much stronger than that of RHA.

8.3 FORMWORK PRESSURE OF SCC

In comparison with conventional concrete, SCC has very high workability, which may result in large lateral pressure on formwork and cause economic and safety problems. If contractors and engineers design the formwork of SCC based on the full hydrostatic pressure without considering the formwork pressure variations during casting and pressure decay following placement, this will lead to increased costs for formwork systems—compromising profitability by offsetting the cost savings associated with SCC due to the rapid placement and labor savings.

Form or molds receive SCC in different shapes and sizes. The fluidity and stability performance of the SCC mixture are influenced by the formwork characteristics. To avoid the honeycombing and surface defects, the formwork should be watertight and grout-tight, especially when the SCC has a low viscosity.

8.3.1 Factors affecting formwork pressure

The lateral pressure exerted by concrete on formwork is influenced by the mixture composition, formwork characteristics, and placement conditions. Mixture composition includes the binder type and content, water-to-cementitious ratio, SCMs, fillers, paste volume, characteristics and content of coarse aggregates, chemical admixtures, concrete consistency, concrete unit weight, and temperature (Omran and Khayat, 2014). The most important factors that affect the formwork pressure are casting rate, concrete yield stress, and formwork height (Billberg, 2012, Geiker and Jacobsen, 2019). In the case of pumping concrete from the bottom, the full height of concrete will be in motion, and the resulting pressure will be the cumulative pressure of hydrostatic and pump pressure. If the concrete is not in motion, then the pressure can be reduced due to the thixotropic properties of SCC (Billberg, 2012).

The phenomenon of thixotropy has an important influence on the various applications of SCC, such as formwork pressure development, pumpability, and multilayer casting (interface behavior of the successive layers). There could be a relationship between the lateral pressure of SCC and thixotropy. The greater the thixotropy, the lower the lateral pressure and the faster the rate of pressure drop with time. This is attributed to the faster re-gaining of the yield stress of the mixture when left at rest without any shearing action. The sharper drop of lateral pressure can be obtained by increasing the structural buildup. Omran and Khayat (2014) concluded that the greater the level of thixotropy, the greater the structural buildup at rest, and thus the lower the lateral pressure during placing and the more rapid the drop in pressure with time. The structuring rate of concrete has a considerable influence on the distribution of lateral pressure on the formwork, regardless of the time. A faster degree of structuring leads to the speedy development of cohesiveness soon after casting, and thus, a cohesive mixture exerts lower lateral pressure on formwork than its full hydrostatic pressure. At a faster flocculation rate (a fast structural buildup rate) and a lower casting rate, the maximum value of lateral stress remains lower than the hydrostatic pressure during casting (Ovarlez and Roussel, 2006).

The structural buildup of cement-based materials has been extensively studied. It is found that mineral admixtures can increase the structure buildup rate by accelerating the hydration rate. However, a high hydration rate did not always lead to a high structural buildup rate (Yuan et al., 2017b). Kim et al. (2012) incorporated FA and limestone powder in SCC mixes and measured the relationship between flowability and lateral pressure on formwork. The results showed that the incorporation of both minerals increased the lateral pressure on formwork. The formwork pressure can be reduced by adding a small amount of processed clay, MK, and alumino-silica (Kim et al., 2010).

8.3.2 Formwork pressure prediction

Due to the high casting rate of SCC, the formwork should be adequately designed to accommodate the expected liquid head formwork pressure. It is recommended to design the SCC formwork for a full liquid head, especially at a high casting rate. However, ACI 347R-14 includes new provisions and recommendations for casting SCC based on the rate of concrete placement relative to the rate of development of concrete stiffness/strength. The measure of stiffening characteristics of the SCC was included in the methods, and the methods are capable of being easily checked on-site using some easy measurements. The details of three typical methods for measuring the lateral pressure for SCC are described in the following.

8.3.2.1 Method proposed by Gardner (Gardner et al., 2012)

Based on some field testing results, Gardner et al. (2012) proposed a parameter—t_{400} (time for the drop of slump flow to 400 mm)—to characterize the SCC. The time to reach zero slump flow (t_0) is defined as follows:

$$t_0 = t_{400} \times \frac{\text{inital slump flow}}{\text{initial slump flow} - 400 \text{ mm}} \tag{8.1}$$

Using t_0, a simple equation was developed to estimate the development of lateral pressure (P_{max}) with time. After the placement, the limiting value of lateral pressure P_{max} (kPa) with time can be calculated as follows:

$$P_{max} = wR\left(t - \frac{t^2}{2t_0}\right), \quad t < t_0 \tag{8.2}$$

$$P_{max} = wRt_0/2, \quad t \geq t_0 \tag{8.3}$$

where w is the unit weight of concrete (kN/m³), h is the height of placement (m), and R is the rate of placement (m/h).

The equation cannot be used for the cases when the casting time is greater than that required to achieve P_{max}. For $t > t_0$, the pressure is assumed to remain constant at the maximum value.

If the time to fill the form, t_h (= height of form/R), is less than t_0, $t = t_h$ is used in the equation:

$$P_{max} = wR\left(t - \frac{t_h^2}{2t_0}\right) \tag{8.4}$$

It is worth mentioning that the experimental results in Gardner et al. (2012) are limited in that the maximum concrete head available to the authors was 4 m, implying the hydrostatic pressure of 96 kPa.

8.3.2.2 Method proposed by Khayat (Khayat and Omran, 2010)

Khayat proposed several methods for measuring and predicting the formwork pressure. A portable measurement device, referred to as the UofS2 pressure column, was developed to evaluate the lateral pressure exerted by plastic concrete. The UofS2 pressure column is a polyvinyl chloride (PVC) cylindrical pressure vessel with flat-plate flange closures. During the test, the vessel is initially filled with concrete (from the top) to a height of 0.5 m above the centerline of the bottom sensor, at a given placement rate and without any vibration. The top of the pressure vessel is then closed and sealed, and the internal air pressure is gradually increased to simulate the hydrostatic head of the placement of additional SCC at the given placement rate. The corresponding lateral pressure exerted on the vessel wall was recorded using the sensor and plotted against the hydrostatic pressure.

Another method is based on the structural buildup of SCC. In this empirical method, the on-site shear strength of SCC is measured by either the inclined plane (IP) test or the portable vane (PV) test. The static yield stress measured from IP$\tau_{0res@15 \text{ min}}$ and PV$\tau_{0res@15 \text{ min}}$ is measured after 15 min of rest, indicating the structural buildup of the concrete at rest. These values are used to calculate P_{max} with the pressure envelope being hydrostatic from the free surface to P_{max}. Khayat et al. (2011) carried out a comprehensive testing program to evaluate the key mixture parameters affecting the formwork pressure exerted by SCC. The investigated parameters included mixture proportions, concrete constituents, concrete temperature, casting characteristics, and the minimum formwork dimension. The UofS2 pressure column and the empirical test methods were employed to evaluate the lateral pressure characteristics and related them to the SCC rheological properties.

Approximately 780 data points were used to derive equations for predicting from pressures exerted by SCC, employing analyses of multiple parameters using special statistical software. The equations used to calculate the P_{max} are as follows:

$$P_{max} = \frac{wh}{100}\left[112.5 - 3.8h + 0.6R - 0.6T + 10D_{min} - 0.021 PV\tau_{0rest@15min@T=22°C}\right] \times f_{MSA} \times f_{WP} \tag{8.5}$$

$$P_{max} = \frac{wh}{100}\left[112 - 3.83h + 0.6R - 0.6T + 10D_{min} - 0.023 IP\tau_{0rest@15min@T=22°C}\right] \times f_{MSA} \times f_{WP} \tag{8.6}$$

where w is the unit weight of concrete (kN/m³), h is the height of placement (m), R is the rate of placement (m/h), T is the actual concrete temperature (°C), and D_{min} is the equivalent to the minimum formwork dimension (d). Use $D_{min} = d$, in case of $0.2 < d < 0.5$ m, and use $D_{min} = 0.5$ m, in case of $0.5 < d < 1$ m.

$PV\tau_{0rest@15min@T=22°C}$ is the value of static yield stress after 15 min rest at 22°C, and $PV\tau_{0rest@15min@T}$ is the value of static yield stress at actual concrete temperature measured by the PV test.

$IP\tau_{0rest@15min@T=22°C}$ is the value of static yield stress after 15 min rest at 22°C, and $IP\tau_{0rest@15min@T}$ is the value of static yield stress at actual concrete temperature measured by the IP test.

f_{MSA} is the factor for maximum aggregate size (MSA). For MSA = 10 mm and low thixotropic SCC ($PV\tau_{0rest@15min@T=22°C} \leq 700$ Pa), use $1.0 \leq f_{MSA} \leq 1.10$ for $4 \leq H \leq 13$ m. For SCC of various thixotropic levels with an MSA of 14–19 mm, $f_{MSA} = 1$ for any H.

f_{WP} is the factor accounting the delay between successive lifts and varies linearly with the SCC thixotropy: $f_{WP} = 1$ for continuous casting for any thixotropic SCC. $f_{WP} = 1.0–0.85$ for a placement that is interrupted for 30 min waiting period in the middle of casting for very low to very high thixotropic SCC $(PV\tau_{0rest@15min@T=22°C} = 50 - 1000$ Pa$)$, respectively.

Besides, many other models or methods are proposed to predict the formwork pressure of SCC. For example, Ovarlez and Roussel (2006) proposed a physical model and compared it with previous work in the literature. They demonstrated an excellent match and acceptable prediction. Their results showed that the lateral stress was equal to the hydrostatic pressure based on the assumption that the material is not able to flocculate and, therefore, is maintained as a fluid. They assumed that the yield stress at rest increases linearly with time:

$$P_{max} = \rho g H - \frac{H^2 A_{thix}}{eR} \tag{8.7}$$

where A_{thix} is the increase rate of static yield stress at rest (Pa/s), H is the height (m), e is the width (m), R is the casting rate (m/h), and ρ is the concrete density (kg/m³).

Perrot et al. (2009) further developed the equation considering the effect of the cross-sectional area of steel bars on the maximum pressure. As a result of this change, the maximal horizontal pressure can be calculated according to the following equation:

$$P_{\max} = \left[\rho g H - \left(\frac{\phi_b + 2S_b}{(e - S_b)\phi_b} \right) \frac{A_{\text{thix}} H^2}{R} \right] \tag{8.8}$$

where S_b is the horizontal steel section per linear meter of the form length (m²) and ϕ_b is the average diameter of the vertical rebars (m).

It is worth mentioning that all methods presented above have some limitations. Thus, it is recommended to evaluate the lateral pressure on the basis of more than one method until the confirmation of the performance based on the range of parameters associated with the project.

Due to the thixotropic nature of SCC, any agitation of already-placed concrete in the form will increase the formwork pressure of SCC. There are site and placement conditions that will increase formwork pressure. Site conditions that can transmit vibrations to freshly placed concrete can cause it to lose its internal structure and increase pressure. Heavy equipment operating close to the forms, or continued work on the forms, will also transmit vibration. Dropping concrete from the pump hose or placing a bucket will also agitate the already-placed concrete. It is worth mentioning that pumping concrete from the bottom will generate higher pressure than the full liquid head.

8.4 STABILITY OF SCC

SCC is a multi-phase material, including air, cement paste, and fine and coarse aggregates which have different densities. SCC is designed with a relatively high content of cement paste and high plastic viscosity and low yield stress to achieve higher flowability and passability compared to conventional concrete. Such high fluidity cannot lead to the excellent workability of concrete in all cases and may have an adverse effect on its stability. The stability of concrete is its ability to maintain the homogeneous distribution of its various constituents during its flow and setting. The stability of SCC comprises two aspects, i.e., static stability and dynamic stability. The former refers to the ability to resist the separation of paste and aggregate in concrete at rest, while the latter refers to the ability to maintain uniform distribution of aggregates in concrete during the process of mixing, pumping, casting, and vibration. Poor stability of SCC leading to segregation may cause pipe blockage during pumping, blocking around reinforcement, holes and honeycomb, etc., which negatively affect the workability, mechanical properties, and durability of concrete.

8.4.1 Static stability

Due to the density difference between different components in SCC, coarse aggregate can settle down, and water and air bubbles can go up when the SCC is at rest. Gravity-induced segregation of aggregate or bleeding has significant detrimental effects on the mechanical properties, top-bar effect, appearance, and durability of concrete structures. Thus, it is essential to understand the mechanisms of static segregation or bleeding phenomenon and identify the influencing mixture parameters. This can be useful in improving the mixture proportions to achieve static stability and targeted mechanical properties.

Under the condition of no external force, the single aggregate particle suspended in the paste is subjected to three vertical forces, i.e., gravity, buoyancy, and resistance generated by the yield stress of the cement paste. If the particle remains stable in the paste, the following equation shall be satisfied:

$$F_{grav} - F_{buoy} \leq F_{res} \tag{8.9}$$

where $F_{grav} - F_{buoy}$ represents the combined force of gravity and buoyancy on the aggregate in N, and F_{res} represents the restoring force exerted by the cement paste in N.

By applying Stokes' law and considering the rheological behavior of cement paste, the moving velocity of the aggregate in concrete can be expressed as follows:

When considering cement paste as the Bingham flow,

$$v = \frac{d^2(\rho_s - \rho_l)g - 18d\tau}{18\eta_{pl}} \tag{8.10}$$

When considering cement paste as the Herschel–Bulkley model,

$$v = \sqrt{\frac{d^{n+1}(\rho_s - \rho_l)g - 18d^n\tau}{18\eta_{pl}}} \tag{8.11}$$

where v is the constant terminal settling rate of the particle (m/s); ρ_s and ρ_l represent the density of aggregate and paste (kg/m³), respectively; g represents the gravitational constant (m/s²); d represents the diameter of the particle (m); η_{pl} is the plastic viscosity of the paste (Pa.s); τ represents the yield stress of the paste (Pa); and n represents the rheological index.

The above equations are developed for spherical particles, which is not accurate for the aggregate particle which is irregular. The drag force of a non-spherical aggregate exerted by paste depends on its shape and orientation with respect to the direction of movement. The non-spherical aggregate not only bears drag forces parallel to the stream velocity but also bears the lateral force, which may result in the rotation and wobble of the particle. It is very complicated to consider the effect of aggregate shape during the settlement.

The above equations are also developed for a single aggregate. Bethmont et al. (2009) believed that the contribution of the multiple aggregates to the stability mainly depends on the solid fraction of the granular skeleton and reported that the ratio of aggregate particle size to spacing can be used to consider the solid fraction. Roussel et al. (2010) found that the direct contact force is dominant over other interparticle forces in the suspension when the solid volume fraction is greater than $0.85\phi_{div}$ which can be approximated as the maximum packing fraction.

In the absence of segregation, the relative velocity of the settling particle with respect to suspending fluid must be zero. Therefore, the critical diameter "d_c" below which ($d < d_c$) the particles are stable can be estimated as follows:

$$d_c = \frac{K \cdot \tau_0}{|\rho_s - \rho_f|g} \tag{8.12}$$

where the shape-dependent factor "K" is equal to 18 for spheres.

If the particles' diameter is lower than d_c ($d < d_c$), concrete is stable. If $d > d_c$, it is still possible to prepare stable SCC. The stable system depends on the aggregate fraction and rheological parameters of paste. The critical solid fraction (ϕ_c), below which the suspension is unstable (i.e., $\phi < \phi_c$), can be obtained by the following equation:

$$\phi_c = \frac{\phi_{max}}{\sqrt[3]{\frac{6M \cdot \tau_0}{\tau|\rho_s - \rho_f| \cdot d \cdot g} + 1}} \tag{8.13}$$

where M is a shape-dependent parameter which equals to $3\pi/4$ for the identical spherical particles, and ϕ_{max} is the maximum random packing fraction of aggregate.

Therefore, the static stability criteria of concrete are as follows (Roussel, 2006):

$$\text{Static stability criteria} \begin{cases} d < d_c, & \text{Regardless of solid fraction } \phi \\ \phi > \phi_c, & \text{if } d > d_c \end{cases} \tag{8.14}$$

The static stability of SCC can be evaluated by different methods adopted by several researchers, including the Visual Stability Index (VSI), Column Test, and Sieve Segregation Test. A summary of commonly used test methods is shown in Table 8.1.

8.4.2 Dynamic stability

Compared with static stability, dynamic stability is more complex and needs to consider the stability of concrete during different processes including agitation, pumping, casting, and vibration. Generally, shear-induced particle migration that can occur during SCC casting is referred to dynamic segregation. This is attributed to the developed shear-rate gradients during flow. Aggregate in high-shear-rate zones migrates toward the lower-shear-rate ones. Assessment of dynamic segregation SCC under different casting and transportation processes simulating high-velocity regimes and flow distances (e.g. pumping) is of particular interest to understand the particle migration phenomenon. Due to their relatively low plastic viscosity and yield stress values, the mortar matrix in SCC is not able to provide enough drag force and carry the coarse aggregate. This causes the segregation of coarse particles,

Table 8.1 A summary of evaluation methods of static stability of SCC (Zhang et al., 2021)

Test methods	Evaluation parameters (basis)	Characteristics of test methods
Visual Stability Index (ASTM International, 2014b)	The appearance of the mixture after the slump flow test	Largely depends on the human factors
Hardened Visual Stability Index (Shen et al., 2005)	The distribution of coarse aggregate on cut surface of SCC	An easy quality control method but still affected by human factors to some extent
Image Analysis Test (Nili, 2018, Shen, 2007)	The distribution of coarse aggregate on cut surface of SCC	A quantitative and accurate method
Column Test (ASTM International, 2014b)	The difference in the quantity of upper and lower aggregates in the device after aggregate settlement	The test result is reliable, while the testing device is bulky, time-consuming, and laborious
Rapid Penetration Test and Segregation Probe (ASTM International, 2014a, Shen et al., 2005)	The penetration depth in SCC (the thickness of the upper mortar layer)	The test is convenient and quick, but the test result is easily affected by the aggregate content, shape, and its distribution in SCC
Static Sieve Segregation Test (JGJ/T 283-2012, 2012, Nili and Razmara, 2020)	The ratio of the mass of mortar flowing through the total mass of SCC poured into the sieve	The initial height of the poured mixture has a great influence on the test result
Electrical Conductivity Method (Khayat et al., 2003)	The difference in electrical conductivity in different parts of SCC	A non-destructive and accurate test, but only suitable for laboratory A
Wave Analysis Method (Naji et al., 2017)	The change in wave velocity along with the SCC height	A non-destructive test and capturing the data for the whole period, but only suitable for laboratory

especially in the case of higher flow distances. Therefore, lower coarse aggregate content can be observed in the flow front compared to the casting point due to dynamic segregation.

Although the theoretical models of rheological properties of cement-based materials have been extensively studied, these developments have not been applied to the evaluation of the dynamic stability of coarse aggregate. Most knowledge is still based on experimental observations. Cai et al. (2021) developed a 3D multi-phase numerical model for fresh concrete to better understand the coarse aggregate settlement under vibration. The settlement rate of the coarse aggregate in vibrated concrete is considered based on the Stokes law, and the calibrated rheological parameter of mixtures is determined by the segmented sieving method.

Several methods have been proposed to evaluate the dynamic stability of SCC during different processes. For example, the pressure bleed test mainly assesses stability during pumping. VSI and the Flow-through Test (JGJ/T 283-2012, 2012) mainly assess the stability during flowing induced by casting. The vibration with sieve segregation test mainly assesses the stability during vibration. A summary of evaluation methods of dynamic stability is shown in Table 8.2.

Table 8.2 A summary of evaluation methods of dynamic stability of SCC (Zhang et al., 2021)

Dynamic stability	Test methods	Evaluation parameters (basis)	Characteristics of test methods
The stability of SCC during agitation	–	–	–
The stability of SCC during casting	The pressure bleed test (Browne and Bamforth, 1977)	The amount of water emitted from concrete under pressure	Estimating the ability of concrete to maintain homogeneity under a pressure gradient
	The vibrating table test (Chalimo et al., 1989)	The non-segregation characteristic, represents the resistance to deposition of sand	Assessing the ability of concrete to withstand dynamic stresses
	The experiment on colored concrete (Jacobsen et al., 2009)	The shape of interfaces between uncolored and colored concrete	Limiting to the short flow lengths and low pressure
	The concrete shear test (Bin, 2019)	The migration and redistribution of aggregate in the shear layer	Only the coarse aggregate migration in laminar flow was studied
	The full-scale pumping (Choi et al., 2014, Kaplan et al., 2005, Secrieru et al., 2018)	The rheological properties of concrete and pumping pressure	A most effective and reliable method, but the test equipment and cost are too large
The stability of SCC during casting	Visual stability index (ASTM International, 2014c)	The VSI value, the appearance of the mixture after slump flow test	Largely depends on human factors
	Slump flow test (Tregger et al., 2011)	The radial aggregate distribution of aggregate after the test	The flowing distance is short, and the test is not sensitive enough to the reveal segregation
	Flow-through test (Shen et al., 2015)	The change in coarse aggregate content in a concrete mixture flowing through the trough	Largely depends on the friction between the concrete and the bottom of the device

(Continued)

Table 8.2 (Continued) A summary of evaluation methods of dynamic stability of SCC (Zhang et al., 2021)

Dynamic stability	Test methods	Evaluation parameters (basis)	Characteristics of test methods
	Tilting-box test (Behrouz et al., 2014, Esmaeilkhanian et al., 2014)	The volume difference of aggregate between the tilt-down section and tilt-down section and tilt-up section of the box	The results are not accurate enough for concrete with very low slump flows and high flow time
	Dynamic sieve segregation test (Alami et al., 2016)	The ratio of the mass of mortar flowing through the 5 mm square sieve and the total mass of concrete poured into the sieve at the rest time of 15 min	Not suitable for site use, and the repeatability of the test is still questionable
	3-Compartment sieve test (Gökçe and Andiç-Çakır, 2018)	The distribution of the weight of coarse aggregate in 3-compartment Sieve	The validity of the method needs further research
The stability of SCC during vibration	The vibration with sieve segregation test (Chia et al., 2005, Safawi et al., 2004)	The degree of separation of the aggregate after the vibration stops	The accuracy of the method is still questioned
	The test monitors the settlement of aggregate in concrete (Petrou et al., 2000)	The spatial displacements and settlement velocity of aggregate	A real-time monitoring method, but only suitable for laboratory

Generally, increasing the consistency of mortar matrix by lowering its paste volume, water-to-binder ratio, and the dosage of superplasticizers, as well as the incorporation of viscosity-modifying agents, can enhance the stability of SCC (Koura et al., 2020, Ley-Hernández and Feys, 2019). On the other hand, Koura et al. (2020) reported the positive effect of higher values of viscoplastic (i.e., yield stress and plastic viscosity) and viscoelastic properties of mortar on the dynamic segregation of SCC. More uniform and coarser particle size distributions of aggregate (Esmaeilkhanian et al., 2014) and lower sand content (Ley-Hernández and Feys, 2019) were shown to increase the dynamic segregation of SCC. Koura et al. (2020) reported that the ratio of the volumetric content-to-packing density (ϕ/ϕ_{max}) of coarse aggregate showed a dominant effect on the dynamic stability of SCC. The effect of morphological characteristics of aggregate on dynamic stability of SCC and shear-induced variation in the particle size distribution of the granular skeleton were investigated by Hosseinpoor et al. (2021).

8.5 SUMMARY

Since the invention of SCC, it has been extensively studied throughout the world widely. Due to its high flowability and high stability, SCC has been widely used in the precast industry to save manpower. However, SCC has not been widely used in on-site construction because of its sensitivity to materials and the environment. The study and application of SCC promote the use of the rheological tool in concrete science. SCC is completely different from conventional vibrated concrete in terms of fresh properties and shares the same hardened properties. The fresh properties of SCC are closely related to rheology. SCC with high viscosity and low yield stress requires the use of a high-range water reducer or superplasticizer and viscosity-enhancing agent. The mixture design of SCC is also mainly based on its flowability. In terms of application, the formwork pressure of SCC and the stability control of SCC should be paid more attention, which are determined by its rheological properties.

REFERENCES

Ahari, R. S., et al. (2015a). "Time-dependent rheological characteristics of self-consolidating concrete containing various mineral admixtures." *Constructions and Building Materials*, 88, 134–142.

Ahari, R. S., et al. (2015b). "Effect of various supplementary cementitious materials on rheological properties of self-consolidating concrete." *Constructions and Building Materials*, 75, 89–98.

Ahari, R. S., et al. (2015c). "Thixotropy and structural breakdown properties of self consolidating concrete containing various supplementary cementitious materials." *Cement Concrete Composites*, 59, 26–37.

Alami, M. M., et al. (2016). "Development of a new test method to evaluate dynamic stability of self-consolidating concrete." *SCC 2016 - 8th International RILEM Symposium on Self-Compacting Concrete*, Washington D.C., United States, Khayat, K. H., ed., 113–122.

Alberti, M. G., et al. (2019). "The effect of fibres in the rheology of self-compacting concrete." *Constructions and Building Materials*, 219, 144–153.

Aleksandra, K.-S., and Gołaszewski, J. (2016). "Rheological properties of high performance self-compacting concrete: Effects of composition and time." *Constructions and Building Materials*, 115, 705–715.

Andersen, P. J., et al. (1988). "The effect of superplasticizer molecular weight on its adsorption on, and dispersion of, cement." *Cement and Concrete Research*, 18(6), 980–986.

Ashish, D. K., and Verma, S. K. (2019). "An overview on mixture design of self-compacting concrete." *Structural Concrete*, 20(1), 371–395.

ASTM International. (2014a). "Standard test method for static segregation of self-consolidating concrete using column technique." ASTM International, West Conshohocken, PA.

ASTM International. (2014b). "Standard test method for rapid assessment of static segregation resistance of self-consolidating concrete using penetration test." ASTM International, West Conshohocken, PA.

ASTM International. (2014c). "Standard test method for slump flow of self-consolidating concrete." ASTM International, West Conshohocken, PA.

Behera, M., et al. (2020). "Rheology study of fresh self-compacting concrete made using recycled fine aggregates." *Proceedings of Rheology and Processing of Construction Materials*, Springer International Publishing, Berlin, Germany, 467–475.

Behrouz, E., et al. (2014). "New test method to evaluate dynamic stability of self-consolidating concrete." *ACI Materials Journal*, 111(3), 299–308.

Benabed, B., et al. (2012). "Properties of self-compacting mortar made with various types of sand." *Cement and Concrete Composites*, 34(10), 1167–1173.

Benaicha, M., et al. (2015). "Influence of silica fume and viscosity modifying agent on the mechanical and rheological behavior of self compacting concrete." *Constructions and Building Materials*, 84, 103–110.

Benaicha, M., et al. (2019). "Dosage effect of superplasticizer on self-compacting concrete: Correlation between rheology and strength." *Journals of Materials Research and Technology*, 8, 2063–2069.

Bethmont, S., et al. (2009). "Contribution of granular interactions to self compacting concrete stability: Development of a new device." *Cement and Concrete Research*, 39(1), 30–35.

Billberg, P. (2012). "Understanding formwork pressure generated by fresh concrete", in *Understanding the rheology of concrete*, Roussel, N., (ed.) Woodhead Publishing, pp. 296–330.

Bin, W. (2019). "Effect of rheological properties on stability of concrete." *China Building Materials Academy*, Master thesis, Beijing, China, 1–102.

Bingham, E. C. (1922). *Fluidity and plasticity*. McGraw-Hill, New York, United States.

Bir, S. R. (2018). "Thixotropy of self-compacting concrete containing recycled aggregates." *Magazine of Concrete Research*, 71(1), 1–12.

Blankson, M. A., and Erdem, S. (2015). "Comparison of the effect of organic and inorganic corrosion inhibitors on the rheology of self-compacting concrete." *Construction and Building Materials*, 77, 59–65.

Boukendakdji, O., et al. (2012). "Effects of granulated blast furnace slag and superplasticizer type on the fresh properties and compressive strength of self-compacting concrete." *Cement and Concrete Composites*, 34(4), 583–590.

Browne, R. D., and Bamforth, P. B. (1977). "Tests to establish concrete pumpability." *ACI Materials Journal*, 74(5), 193–203.

Cai, Y., et al. (2021). "An experimental and numerical investigation of coarse aggregate settlement in fresh concrete under vibration." *Cement and Concrete Composites*, 122, 104153.

Carro-López, D., et al. (2015). "Study of the rheology of self-compacting concrete with fine recycled concrete aggregates." *Constructions and Building Materials*, 96, 491–501.

Carro-López, D., et al. (2017). "Proportioning, fresh-state properties and rheology of self-compacting concrete with fine recycled aggregates." *Hormigon Acero*, 69(286), 213–212.

Chalimo, T., et al. (1989). "Osobennosti trouboprovodnogo transporta betonnikh cmeceiy betonona-çoçami." *Minsk*.

Chia, K. S., et al. (2005). "Stability of fresh lightweight aggregate concrete under vibration." *ACI Materials Journal*, 102(5), 347–354.

Chia, K. S., and Zhang, M. H. (2004). "Effect of chemical admixtures on rheological parameters and stability of fresh lightweight aggregate concrete." *Magazine Concrete Research*, 56(8), 465–473.

Choi, M. S., et al. (2014). "Effect of the coarse aggregate size on pipe flow of pumped concrete." *Constructions and Building Materials*, 66, 723–730.

Concrete, S.-C. (2005). "The European guidelines for self-compacting concrete." *BIBM*, et al., 22, 563.

De Larrard, F., et al. (1998). "Fresh concrete: A Herschel-Bulkley material." *Materials and Structures*, 31(7), 494–498.

De Matos P R, et al. (2018). "Rheological behavior of Portland cement pastes and self-compacting concretes containing porcelain polishing residue." *Construction and Building Materials*, 175, 508–518.

Diamantonis, N., et al. (2010). "Investigations about the influence of fine additives on the viscosity of cement paste for self-compacting concrete." *Constructions and Building Materials*, 24(8), 1518–1522.

Domone, P. (2006). "Mortar tests for self-consolidating concrete." *Concrete International*, 28(4), 39–45.

Du, L., and Folliard, K. J. (2004). "Mechanisms of air entrainment in concrete." *Cement and Concrete Research*, 35(8), 1463–1471.

Durgun, M. Y., and Atahan, H. N. (2017). "Rheological and fresh properties of reduced fine content self-compacting concretes produced with different particle sizes of nano SiO_2." *Constructions and Building Materials*, 142, 431–443.

Esmaeilkhanian, B., et al. (2014). "Effects of mix design parameters and rheological properties on dynamic stability of self-consolidating concrete." *Cement and Concrete Composites*, 54, 21–28.

Ferrari, L., et al. (2010). "Interaction of cement model systems with superplasticizers investigated by atomic force microscopy, zeta potential, and adsorption measurements." *Journal of Colloid and Interface Science*, 347(1), 15–24.

Feys, D., et al. (2008b). "Fresh self compacting concrete, a shear thickening material." *Cement and Concrete Research*, 38(7), 920–929.

Gang, L., et al. (2018). "Rheological behavior and microstructure characteristics of SCC incorporating metakaolin and silica fume." *Materials*, 11(12), 2576.

Gardner, N. J., et al. (2012). "Field investigation of formwork pressures using self-consolidating concrete." *Concrete International*, 34(1), 41–47.

Geiker, M., and Jacobsen, S. (2019). "Self-compacting concrete (SCC)", in *Developments in the formulation and reinforcement of concrete*. Edited by Mindess, M., Woodhead Publishing, Sawston, United Kingdom, pp. 229–256.

Gesoglu, M., et al. (2015). "Shear thickening intensity of self-compacting concretes containing rounded lightweight aggregates." *Constructions and Building Materials*, 79, 40–47.

Gökçe, H. S., and Andiç-Çakır, Ö. (2018). "A new method for determination of dynamic stability of self-consolidating concrete: 3-Compartment sieve test." *Constructions and Building Materials*, 168, 305–312.

Gonz, I., et al. (2018). "Thixotropy and interlayer bond strength of self-compacting recycled concrete." *Constructions and Building Materials*, 161, 479–488.

González-Taboada, I., et al. (2017a). "Tools for the study of self-compacting recycled concrete fresh behaviour: Workability and rheology." *Journal of Cleaner Production*, 156, 1–18.

González-Taboada, I., et al. (2017b). "Analysis of rheological behaviour of self-compacting concrete made with recycled aggregates." *Constructions and Building Materials*, 157, 18–25.

González-Taboada, I., et al. (2018). "Robustness of self-compacting recycled concrete: Analysis of sensitivity parameters." *Materials and Structures*, 51(1), 8.

González-Taboada, I., et al. (2020). "Self-consolidating recycled concrete: Rheological behavior over time." *ACI Materials Journal*, 117(1), 3–14.

Güneyisi, E., et al. (2015). "Fresh and rheological behavior of nano-silica and fly ash blended self-compacting concrete." *Constructions and Building Materials*, 95, 29–44.

Güneyisi, E., et al. (2016a). "Rheological and fresh properties of self-compacting concretes containing coarse and fine recycled concrete aggregates." *Constructions and Building Materials*, 113, 622–630.

Güneyisi, E., et al. (2016b). "Evaluation of the rheological behavior of fresh self-compacting rubberized concrete by using the Herschel-Bulkley and modified Bingham models." *Archives of Civil and Mechanical Engineering*, 16(1), 9–19.

Guo, Y., et al. (2017). "Evaluating the distance between particles in fresh cement paste based on the yield stress and particle size." *Constructions and Building Materials*, 142, 109–116.

Hassan, A., et al. (2010). "Effect of metakaolin on the rheology of self-consolidating concrete", in *Design, production and placement of self-consolidating concrete*. Edited by Khayat, K., Feys, D., Springer, Dordrecht, pp. 103–112.

Hayakawa, M., et al. (1993). "Development and application of superworkable concrete", in *Special concretes-workability and mixing*. Edited by Bartos, P.J.M., CRC Press, London, England, pp. 185–192.

Heikal, M., et al. (2013). "Mechanical, microstructure and rheological characteristics of high performance self-compacting cement pastes and concrete containing ground clay bricks." *Constructions and Building Materials*, 38, 101–109.

Hosseinpoor, M., et al. (2021). "Rheo-morphological investigation of static and dynamic stability of self-consolidating concrete: A biphasic approach." *Cement and Concrete Composites*, 121, 104072.

Huang, F., et al. (2018). "The rheological properties of self-compacting concrete containing superplasticizer and air-entraining agent." *Construction and Building Materials*, 166, 833–838.

Jacobsen, S., et al. (2009). "Flow conditions of fresh mortar and concrete in different pipes." *Cement and Concrete Research*, 39(11), 997–1006.

Jalal, M., et al. (2013). "Effects of fly ash and TiO_2 nanoparticles on rheological, mechanical, microstructural and thermal properties of high strength self compacting concrete." *Mechanics of Materials*, 61, 11–27.

JGJ/T 283-2012. (2012). *Technical Specification for Application of Self-Compacting Concrete*. China Architecture and Building Press, Beijing.

Jiao, D., et al. (2017). "Effect of constituents on rheological properties of fresh concrete-A review." *Cement and Concrete Composites*, 83, 146–159.

Jiao, D., et al. (2021). "Thixotropic structural build-up of cement-based materials: A state-of-the-art review." *Cement and Concrete Composites*, 122, 104152.

Kaplan, D., et al. (2005). "Avoidance of blockages in concrete pumping process." *ACI Materials Journal*, 102(3), 183–192.

Khayat, K. H. (1998). "Viscosity-enhancing admixtures for cement-based materials — An overview." *Cement and Concrete Composites*, 20(2), 171–188.

Khayat, K. H., et al. (1999). "Factorial design model for proportioning self-consolidating concrete." *Materials and Structures*, 32(9), 679–686.

Khayat, K. H., et al. (2003). "Analysis of variations in electrical conductivity to assess stability of cement-based materials." *ACI Materials Journal*, 100(4), 302–310.

Khayat, K. H., et al. (2010). "Evaluation of SCC formwork pressure." *Concrete International*, 32(6), 30–34.

Khayat, K. H., and Omran, A. F. (2011). "Field validation of SCC formwork pressure prediction models." *Concrete International*, 33(6), 33–39.

Kim, J. H., et al. (2010). "Effect of mineral admixtures on formwork pressure of self-consolidating concrete." *Cement and Concrete Composites*, 32(9), 665–671.

Kim, J. H., et al. (2012). "Effect of powder materials on the rheology and formwork pressure of self-consolidating concrete." *Cement and Concrete Composites*, 34(6), 746–753.

Koura, I. O., et al. (2020). "Coupled effect of fine mortar and granular skeleton characteristics on dynamic stability of self-consolidating concrete as a diphasic material." *Construction and Building Materials*, 263, 120131.

Kuroiwa, S., et al. (1993). "Application of super workable concrete to construction of a 20-story building." *ACI Symposium Publication*, 140, 147–162.

Larrard, D., et al. (1998). "Fresh concrete: A Herschel-Bulkley material." *Materials and Structures*, 31(7), 494–498.

.Le, H. T., et al. (2015). "Effect of macro-mesoporous rice husk ash on rheological properties of mortar formulated from self-compacting high performance concrete." *Constructions and Building Materials*, 80, 225–235.

Le, H. T., et al. (2016). "Effect of rice husk ash and other mineral admixtures on properties of self-compacting high performance concrete." *Materials & Design*, 89, 156–166.

Ley-Hernández, A. M., and Feys, D. (2019). "How rheology governs dynamic segregation of self-consolidating concrete." *ACI Materials Journal*, 116(3), 131–140.

Ling, G., et al. (2018). "Rheological behavior and microstructure characteristics of SCC incorporating metakaolin and silica fume." *Materials*, 11(12), 2576.

Lu, C., et al. (2015). "Relationship between slump flow and rheological properties of self compacting concrete with silica fume and its permeability." *Constructions and Building Materials*, 75, 157–162.

Ma, B., et al. (2013). "Rheological properties of self-compacting concrete paste containing chemical admixtures." *Journal of Wuhan University of Technology-Mater. Sci. Ed.*, 28(2), 291–297.

Ma, K., et al. (2016). "Effects of mineral admixtures on shear thickening of cement paste." *Constructions and Building Materials*, 126, 609–616.

Mardani-Aghabaglou, A., et al. (2013). "Effect of different types of superplasticizer on fresh, rheological and strength properties of self-consolidating concrete." *Constructions and Building Materials*, 47, 1020–1025.

Matos, P. R. D., et al. (2018). "Rheological behavior of Portland cement pastes and self-compacting concretes containing porcelain polishing residue." *Constructions and Building Materials*, 175, 508–518.

Naji, S., et al. (2017). "Assessment of static stability of concrete using shear wave velocity approach." *ACI Materials Journal*, 114(1), 105–116.

Nasrollah, B., et al. (2020). "Optimum recycled concrete aggregate and micro-silica content in self-compacting concrete: Rheological, mechanical and microstructural properties." *Journal of Building Engineering*, 31, 101361.

Nili, M. (2018). "Automatic image analysis process to appraise segregation resistance of self-consolidating concrete." *Magazine of Concrete Research*, 70(8), 390–399.

Nili, M., and Razmara, M. (2020). "Proposing a new apparatus to assess the properties of self-consolidating concrete." *Journal of Testing and Evaluation*, 48(4), 3188–3201.

Okamura, H., and Ozawa, K. (1994). "Mix design method for self-compactable concrete." *Japan Society Civil Engineers*, 496, 1–8.

Omran, A. F., and Khayat, K. H. (2014). "Choice of thixotropic index to evaluate formwork pressure characteristics of self-consolidating concrete." *Cement and Concrete Research*, 63, 89–97.

Ouldkhaoua, Y., et al. (2019). "Rheological properties of blended metakaolin self-compacting concrete containing recycled CRT funnel glass aggregate." *Epitoanyag: Journal of Silicate Based and Composite Materials*, 71(5), 154–161.

Ovarlez, G., and Roussel, N. (2006). "A physical model for the prediction of lateral stress exerted by self-compacting concrete on formwork." *Materials and Structures*, 39(2), 269–279.

Pal, S. C., et al. (2003). "Investigation of hydraulic activity of ground granulated blast furnace slag in concrete." *Cement and Concrete Research*, 33(9), 1481–1486.

Perrot, A., et al. (2009). "SCC formwork pressure: Influence of steel rebars." *Cement and Concrete Research*, 39(6), 524–528.

Petrou, M. F., et al. (2000). "A unique experimental method for monitoring aggregate settlement in concrete." *Cement and Concrete Research*, 30(5), 809–816.

Rahman, M. K., et al. (2014). "Thixotropic behavior of self compacting concrete with different mineral admixtures." *Constructions and Building Materials*, 50, 710–717.

Rashad, A. M. (2013). "Metakaolin as cementitious material: History, scours, production and composition - A comprehensive overview." *Constructions and Building Materials*, 41, 303–318.

Rashwan, S., and AbouRizk, S. (1997). "The properties of recycled concrete." *Concrete International*, 19(7), 56–60.

Revilla-Cuesta, V., et al. (2020). "Self-compacting concrete manufactured with recycled concrete aggregate: An overview." *Journal of Cleaner Production*, 262, 121362.

Roussel, N. (2006). "A thixotropy model for fresh fluid concretes: Theory, validation and applications." *Cement and Concrete Research*, 36(10), 1797–1806.

Roussel, N. (2007). "Rheology of fresh concrete: From measurements to predictions of casting processes." *Materials and Structures*, 40(10), 1001–1012.

Roussel, N., et al. (2010). "Steady state flow of cement suspensions: A micromechanical state of the art." *Cement and Concrete Research*, 40(1), 77–84.

Roussel, N., and Cussigh, F. (2007). "Distinct-layer casting of SCC: The mechanical consequences of thixotropy." *Cement and Concrete Research*, 38(5), 624–632.

Safawi, M. I., et al. (2004). "The segregation tendency in the vibration of high fluidity concrete." *Cement and Concrete Research*, 34(2), 219–226.

Safiuddin, M., et al. (2010). "Flowing ability of the mortars formulated from self-compacting concretes incorporating rice husk ash." *Constructions and Building Materials*, 25(2), 973–978.

Schmidt, W., et al. (2014). "Influences of superplasticizer modification and mixture composition on the performance of self-compacting concrete at varied ambient temperatures." *Cement and Concrete Composites*, 49, 111–126.

Secrieru, E., et al. (2018). "Changes in concrete properties during pumping and formation of lubricating material under pressure." *Cement and Concrete Research*, 108, 129–139.

Sedran, T., and Delarrard, F. (1999). "Optimization of SCC thanks to packing model." *First International RILEM Symposium on Self-Compacting Concrete*, Stockholm, Sweden, 321–332.

Sethy, K. P., et al. (2016). "Utilization of high volume of industrial slag in self compacting concrete." *Journal of Cleaner Production*, 112, 581–587.

Shen, L., et al. (2005). "Testing static segregation of SCC." *Proceedings of the 2nd North American Conference on the Design and Use of SCC*, Chicago, USA.

Shen, L. (2007). *Role of aggregate packing in segregation resistance and flow behavior of self-consolidating concrete*. University of Illinois at Urbana-Champaign.

Shen, L., et al. (2015). "Testing dynamic segregation of self-consolidating concrete." *Constructions and Building Materials*, 75, 465–471.

Silva, M. A. D, et al. (2017). "Rheological and mechanical behavior of high strength steel fiber-river gravel self compacting concrete." *Construction and Building Materials*, 150, 606–618.

Singh, R. B., et al. (2017). "Effect of supplementary cementitious materials on rheology of different grades of self-compacting concrete made with recycled aggregates." *Journal of Advanced Concrete Technology*, 15(9), 524–535.

Singh, R. B., and Singh, B. (2018). "Rheological behaviour of different grades of self-compacting concrete containing recycled aggregates." *Construction and Building Materials*, 161, 354–364.

Sua-iam, G., et al. (2016). "Novel ternary blends of Type 1 Portland cement, residual rice husk ash, and limestone powder to improve the properties of self-compacting concrete." *Construction and Building Materials*, 125, 1028–1034.

Tregger, N., et al. (2011). "Correlating dynamic segregation of self-consolidating concrete to the slump-flow test." *Construction and Building Materials*, 28(1), 499–505.

Vejmelková, E., et al. (2010). "Properties of self-compacting concrete mixtures containing metakaolin and blast furnace slag." *Construction and Building Materials*, 25(3), 1325–1331.

Wallevik, O., and Nielsson, I. (1998). "Self-compacting concrete-a rheological approach." *Proceedings of the International Workshop on Self-Compacting Concrete*, Kochi University of Technology, Kami, Japan, 23–26.

Yahia, A. (2010). "Shear-thickening behavior of high-performance cement grouts — Influencing mix-design parameters." *Cement and Concrete Research*, 41(3), 230–235.

Yuan, Q., et al. (2016). "Sealed-space-filling SCC: A special SCC applied in high-speed rail of China." *Construction and Building Materials*, 124, 167–176.

Yuan, Q., et al. (2017a). "Coupled effect of viscosity enhancing admixtures and superplasticizers on rheological behavior of cement paste." *Journal of Central South University (Engl. Ed.)*, 24(9), 2172–2179.

Yuan, Q., et al. (2017b). "On the measurement of evolution of structural build-up of cement paste with time by static yield stress test vs. small amplitude oscillatory shear test." *Cement and Concrete Research*, 99, 183–189.

Chapter 9

Rheology of other cement-based materials

9.1 RHEOLOGY OF ALKALI-ACTIVATED MATERIALS (AAMs)

9.1.1 Introduction

Alkali-activated materials (AAMs) have been gradually accepted by academia and industry as an effective and promising alternative to Portland cement-based materials (PCMs). AAMs consist of various precursors and activators, and these materials can be manufactured by activation of a reactive solid aluminosilicate. In addition, the development of AAMs has reduced the adverse impact on the environment and avoided energy consumption and carbon dioxide emission due to the production of raw materials (Duxson et al., 2007). However, the use of sodium silicate will lead to high carbon dioxide emissions and acidification of surface water, which will have serious impacts on environmental benefits (Habert et al., 2011). Therefore, a better mixture design is required. The effect of composition on the rheological properties of AAMs must be fully understood in order to control its workability. This part introduces the effects of the properties and dosage of activator and precursor on the rheological properties of AAMs. This provides a reference for future research, and the application of this material on the construction site should be better controlled.

The Bingham model, modified Bingham model, and Herschel–Bulkley (H–B) model are the most common models that describe the rheological behaviors of cement-based materials (Jiao et al., 2017). Table 9.1 summarizes the rheological models and features of AAMs made with various precursors, activators, and different additions/admixtures. It is obvious that the Bingham model, modified Bingham model, and H–B model can empirically characterize the rheological properties of AAMs. However, the applicability of different systems is different. It is generally believed that the alkalinity of the alkali activator controls the dissolution behavior of precursor particles and the alkali cations in the activator play the role of charge balance cations. In addition, the polymerization state and the adsorption of anionic groups on the surface of precursors in activator solution, especially silicate species, may affect the interaction between particles. These mechanisms determine the effect of activators on the rheological behavior of AAMs. Generally speaking, the Bingham model is suitable for sodium hydroxide-activated slurry, while the H–B model is more suitable for sodium silicate-activated slurry (Palacios et al., 2008). Furthermore, the Bingham model seems to suit both NaOH- and sodium silicate-activated mortar/concrete, and this behavior is similar to Portland cement (PC)-based mortar/concrete (Puertas et al., 2018). The modified Bingham model not only describes the non-linear behavior deviating from the Bingham fluid but also yields a variable dimension parameter like in the H–B model without mathematical limitation in the low-shear-rate region (Jiao et al., 2017). However, it is rarely used for describing the rheological behavior of AAMs (Zhang et al., 2018a).

Table 9.1 Rheological models and features for AAMs manufactured with various precursors, activators, and different additions/admixtures

Models	Activators	Precursors	Admixtures/additions	Correlation coefficient (R^2)	Curve features	Refs.
Bingham model	Na silicates	Slag (concrete)	–	0.965–0.997	High thixotropy; $60 \leq \tau_0 \leq 220$; $88 \leq \mu \leq 99.5$	Puertas et al. (2018)
	NaOH				Low thixotropy; $175 \leq \tau_0 \leq 375$; $57 \leq \mu \leq 84$	
	Na silicates	Fly ash (concrete)	Two high-range water reducer agents	0.973–0.987	Thixotropy decreased as admixture doses increased; $25 \leq \tau_0 \leq 120$; $80 \leq \mu \leq 115$;	Laskar and Bhattacharjee (2011)
H-B model	Na silicates	Slag	–	–	High thixotropy; $5 \leq \tau_0 \leq 87.5$	Puertas et al. (2014)
		Fly ash+slag	–	> 0.99876	$9.43 \leq \tau_0 \leq 96.86$; $0.81 \leq n \leq 1.38$ $0.21 \leq K \leq 3.66$ $659 \leq$ area of thixotropy ≤ 7113	Dai et al. (2020)
Modified Bingham model	Na silicates	Slag+steel slag	Stabilizer	0.968–0.998	$0.339 \leq \tau_0 \leq 3.439$; $0.004 \leq c/\mu \leq 0.02$	Zhang et al. (2018a)
		Slag+glass powder+calcium aluminate cement	–	–	$1.47 \leq \tau_0 \leq 12.72$ $1.18 \leq \mu \leq 13.85$	Li et al. (2020)
	NaOH	Fly ash	–	–	$3.3 \leq \tau_0 \leq 37$; $-0.0087 \leq c/\mu \leq 0$	Rifaai et al. (2019)

τ_S here means static yield stress (Pa); τ_0 means (dynamic) yield stress (Pa); μ means plastic viscosity (Pa.s); area of thixotropy (Pa/s).

In addition to the activator used, the source and composition of the initial aluminosilicate also have a great influence on the rheological behavior. Due to their plate-like particles and large specific surface area, alkali-activated MK systems usually exhibit higher viscosity and apparent yield stress (Romagnoli et al., 2012), which means that suitable activators and sufficient water are required to ensure good workability (Xiang et al., 2019). As reported, alkali-activated fly ash pastes behaved like a Bingham fluid (Vance et al., 2014), while the rheological behavior of the fly ash-slag blended systems was more consistent with the H–B model (Dai et al., 2020). These rheological shapes are related to the differences in particle size, dissolution behavior and reactivity of different precursor particles, and the initially precipitated gels. The effects of composition factors on the rheology of AAMs are discussed in detail in the following sections.

9.1.2 Effect of alkaline activators on rheology of AAMs

It is well known that the properties and dosage of alkali have a significant impact on the reaction process, microstructure, and properties of AAMs. Therefore, the activator is the key component to control the rheological behavior of AAMs paste. Activators can be classified into six categories in terms of chemical composition: caustic alkalis, non-silicate weak acid salts, silicates, aluminates, aluminosilicates, and non-silicate strong acid salts. Among these activators, sodium-based activators are the most widely available and cost-effective, and potassium-based activators have been used in laboratory studies. This section discusses the effects of Na/KOH, $Na_2O \cdot nSiO_2$, and some unconventional activators on the rheology of AAMs.

9.1.2.1 Na/KOH

9.1.2.1.1 Ion nature

The most used alkali hydroxides in AAMs are NaOH and KOH. Published literature has shown that paste suspensions containing NaOH solution usually exhibit higher yield stress and plastic viscosity than KOH-based suspensions (Vance et al., 2014). The yield stress and plastic viscosity of fly ash paste activated by NaOH solutions are higher than those activated by KOH with the same dosage in terms of molarity (Vance et al., 2014). This is due to the lower charge density of K^+, which gives lower ion-dipole forces and viscosity of solution (Poulesquen et al., 2011). Similarly, irrespective of activator dosage, a lower concentration of Na^+ cations adsorbed on the surface of negatively charged particles and more adsorption of K^+ cations might reduce the van der Waals forces and increase the repulsive force of the double layer between charged particles, thus contributing to a decrease in yield stress. Therefore, Na^+ cations would rather combine with water and consume more free water.

9.1.2.1.2 Concentration

Concentration is another important factor that impacts the rheological properties of AAMs. Table 9.2 summarizes the effects of NaOH concentration on the rheological properties of AAMs prepared with various precursors. The value of zeta potential is more negative (−7.28 mV) due to the high dissolution and ionization degree of suspension for the paste with high NaOH concentration (Zhang et al., 2018b). It is generally believed that the dissolution rate of slag is faster than that of fly ash due to its higher reactivity. Therefore, the ionic bonds formed by modified cations (such as Ca^{2+}) and non-bridging oxygen in the calcium aluminosilicate glass structure are easier to dissolve, resulting in the release of Ca^{2+} ions in early slag (Newlands et al., 2017). Due to the re-adsorption of Ca^{2+}, the zeta potential

Table 9.2 Effects of NaOH concentration on the rheological properties of AAMs prepared with various precursors

Precursors	w/b or L/S (by mass)	NaOH concentration (mol/L)	Critical concentration (mol/L)	Main features	Ref.
Fly ash	0.25	8; 10	–	τ_0 slightly decreased as concentration increased	Palacios et al. (2019)
	0.35	4; 8	–	τ_0 and μ increased as concentration increased	Vance et al. (2014)
Slag	0.45	1; 2	–	τ_0 increased as concentration increased	Kashani et al. (2014)
	0.55	2.57; 3.37; 4.13	3.37	τ_0 decreased as concentration (<3.37 mol/L) increased, and then increased; μ increased as concentration increased	Puertas et al. (2014)
Slag+fly ash	0.6	0.83; 1.67; 2.5; 3.33; 4.17; 5; 5.83; 6.67	4.17	The growth rate of G' increased as concentration (<4.17 mol/L) increased	Zhang et al. (2019)
	0.5	1; 2; 3; 4; 5; 6; 7; 8	–	τ_0 and μ decreased as concentration increased	Zhang et al. (2020)

w/b = water/binder ratio; L/S = liquid/solid ratio.
Per Alonso et al. (2017), τ_0 and μ are calculated based on the speed-torque curve.

becomes less negative. However, silicates and aluminates may be partially replaced by low-charge cations. These two mechanisms may be related to the measurement of zeta potential and dilution time. In addition, at the early stage, the water-binding capacity of the dominant N-A-S-H in the fly ash-slag mixing system was weaker than that of the C-A-S-H gel in the slag system, so the water freedom of the hybrid system was higher (Zhang et al., 2020). However, this effect may not be as significant as that of static repulsion.

More importantly, by reviewing the published literature, there seems to be a critical value for NaOH concentration to change the rheological characteristics of AAMs, especially the development of viscoelasticity. It is mainly because OH⁻ ions play a chemical driving role to stimulate the dissolution of solid aluminosilicates. However, Na⁺ ions are necessary to house in the voids in tetrahedral coordination structure to compensate for the electrical charge due to the substitution of Si^{4+} by Al^{3+} (Shi et al., 2011). Therefore, ion concentration affects the competition between dissolution and polycondensation kinetics, thus affecting the viscoelastic behavior (Rifaai et al., 2019).

Figure 9.1 shows that NaOH concentration of 7 mol/L is a critical value, below which the yield stress, storage modulus (G'), and hardening rate of alkali-activated fly ash (AAFA) pastes increase as the concentration increases, while the rheological properties show a downward trend (Rifaai et al., 2019). The reasonable explanation for this phenomenon is the competition between dissolution and polycondensation. On the one hand, a low concentration of NaOH can accelerate the kinetics of dissolution and polycondensation. On the other hand, the dissolution rate is dominant in the extremely alkaline environment, which will lead to more negatively charged monomers, resulting in significant repulsion between particles (Rifaai et al., 2019). However, the critical value of NaOH concentration may also be related to the properties of precursors.

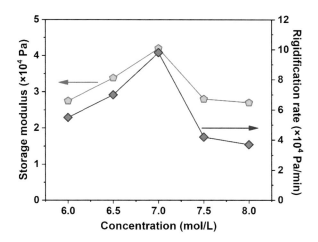

Figure 9.1 Effects of NaOH concentration on storage modulus and rigidification rate of alkali-activated fly ash pastes. (Data from Rifaai et al., 2019.)

As discussed earlier, the higher CaO content and lower degree of glass polymerization in the slag are more conducive to dissolve, thus generating a higher reaction rate than fly ash (FA) (Newlands et al., 2017). This means that the introduction of slag can accelerate the dissolution rate of overall raw materials, compared with the pastes containing FA as a single precursor. Therefore, the critical NaOH concentration is reduced. Furthermore, although an increase in NaOH concentration promotes the formation of Si-O-Na bond, excessively high concentration is unfavorable due to early hydration product precipitation on surfaces of particles. Xie et al. (2020) pointed out that 8.87 mol/L of NaOH (the mass ratio of NaOH/phosphorous slag was 16%) was a threshold and the influence of NaOH concentration changed near this threshold on viscoelasticity. Below this value, the viscoelastic transition (the transition from the viscous behavior to the elastic behavior served as the dominant feature) of alkali-activated phosphorous slag was delayed with the increase of NaOH concentration. Low hydration rate and good dispersion because of the enhanced electrostatic repulsive force may be responsible for this phenomenon. Above this threshold, the viscoelastic transition accelerated with the increase of NaOH concentration, which was attributed to the rapid generation of cross-linked gels and the decreased electrostatic repulsive force (Xie et al., 2020).

9.1.2.2 Na/K-silicates

9.1.2.2.1 Ion nature and ways of adding activators

It is well known that AAMs activated with sodium silicates usually exhibit greater workability, i.e., with a lower yield stress than that of PCMs and alkali hydroxide-activated systems due to the plasticizing effect (Alonso et al., 2017). Figure 9.2 shows that the yield stress of the suspensions containing sodium silicates is lower than that of the NaOH-activated systems. For example, Kashani et al. (2014) reported that the initial zeta potentials of Na/KOH-slag system and sodium silicate-slag system are both negative due to the deprotonation of slag silanol groups. However, as activator concentration increases, the zeta potential of Na/KOH-slag system changes from negative to positive, which means that repulsive double-layer forces decrease and then increase. By contrast, the zeta potential of the sodium silicate-slag system becomes more negative (Kashani et al., 2014). This is because silicate

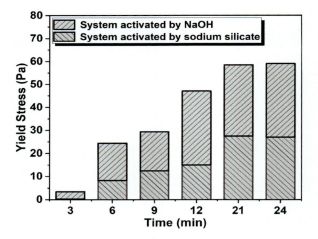

Figure 9.2 The comparison of yield stress evolution in slag AAMs suspensions activated by NaOH and sodium silicate (Palacios et al., 2008).

oligomers are adsorbed on the surface of slag particles, resulting in insufficient dissolved Ca^{2+} in the solution and greater double-layer repulsion. Therefore, the yield stress of Na silicate-based FA pastes is lower than that of NaOH-FA (Vance et al., 2014). The silicate adsorption in the activator solution increases the overall negative surface charge of FA, enhances the repulsion between particles, and reduces the flocculation of particles (Palacios et al., 2019).

The previous studies of sodium silicates-slag suspensions also showed obvious thixotropy characteristics (Palacios et al., 2008). This feature is mainly reflected in the destruction and reconstruction of thixotropic materials, which is largely related to the flow history and static time (Yahia and Khayat, 2001). Thixotropy can be described by a thixotropic hysteresis loop (qualitative) (Yahia and Khayat, 2001) and a thixotropic index (quantitative) (Hou et al., 2021). Figure 9.3 shows the thixotropy of slag pastes activated with sodium silicates and NaOH, respectively. It can be seen that sodium silicates-activated slag paste shows a larger area of the hysteresis loop and a higher shear stress than that of NaOH system,

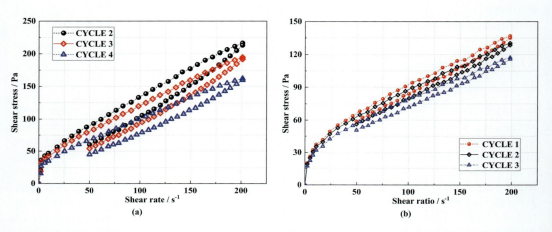

Figure 9.3 Hysteresis cycles for AAS pastes (Palacios et al., 2008): (a) slag+sodium silicate solution and (b) slag+NaOH solution.

especially in the first cycle (Palacios et al., 2008). This is mainly attributed to the initial and rapid formation of primary C-A-S-H gels, which generate colloidal forces between particles. It is important to note that this colloidal force can be interfered by continuous shearing. Therefore, a prolonged mixing will improve the workability of sodium silicate-activated slag concrete (Puertas et al., 2018).

The spherical particle feature of FA could be another important factor. Interestingly, Poulesquen et al. (2011) conducted a strain sweep test to investigate the influence of silicate-activating solutions synthesized by Na/KOH and two types of silica on the viscoelastic parameters (storage modulus G', loss modulus G'' and $\tan\delta = G''/G'$) of metakaolin (MK) systems during the polymerization process. Although Na-based activator has weak alkalinity, it can better dissolve MK and show a faster geological polymerization effect. K^+ shows a weaker association with water (Poulesquen et al., 2011). In other words, K^+ ions are more likely to combine with negatively charged silicates, exhibiting an interfering effect on the condensation process. Smaller $\tan\delta$ in K-based system was observed due to stronger interactions between the dissolved Si and Al tetrahedral monomers, which indicated a more rigid structure (Steins et al., 2012). However, smaller Na^+ with higher charge density strongly attracts and retains its hydration layer of water molecules.

Solid sodium silicate has higher yield stress and plastic viscosity than liquid sodium silicate. The dissolution reaction of solid sodium silicate leads to less free water acting as a lubricant, and the additional dissolution heat can improve the formation and size of gels in slag-FA systems (Xiang et al., 2018). Such rheological characteristics seem to make solid sodium silicate more suitable for preparing extrusion-based 3D printing mortars than liquid sodium silicate. In addition, some studies focus on the rheological behavior and fluidity of single-component AAMs and different raw materials.

9.1.2.2.2 Modulus ($nSiO_2/nNa_2O$)

Mehdizadeh et al. (2018) found that the yield stress of AAPS declined as the modulus of sodium silicate activator increased at a constant Na_2O/Al_2O_3 ratio of 1.0, which may be caused by the plasticizing effect of silicate (Kashani et al., 2014). It is found that increasing modulus can reduce the yield stress, thickening rate and plastic viscosity of the fresh paste. Therefore, in order to obtain satisfactory thixotropic properties, the modulus of 3D-printed AAM concrete based on slag is reduced to ensure the stability between layers of printing materials. However, some recent studies have shown that increasing the modulus of sodium silicate will increase the apparent viscosity and yield stress of alkali-activated fly ash slurry (Vance et al., 2014).

In the slag-FA blended pastes (fly ash/slag = 9:1), when the modulus increases from 1.4 to 1.8, the plastic viscosity decreases (Yin et al., 2019). Similar results were also confirmed in the slag-FA blended system (fly ash/slag = 5:5). When the modulus increased from 0.4 to 1.6, the yield stress and thixotropic area decreased by about 76% and 83%, respectively (Dai et al., 2020). This might be attributed to the high alkalinity and high degree of ionization in suspensions of lower modulus (1.4) at a high liquid–solid ratio, which promoted the dissolution and consequent polymerization, thus resulting in higher plastic viscosity. The effects of low modulus ($0 < Ms \leq 0.4$) and high modulus ($0.4 < Ms \leq 1.4$) on the formation and evolution of structures are very different (Dai et al., 2020). The mixture with a low modulus showed a higher initial storage modulus G', and then the structure formation rate increased slowly. However, a high modulus will delay the formation of the structure and then increase rapidly (Dai et al., 2020). In contrast, low-modulus activators exhibit higher alkalinity and promote early dissolution precipitation in the mixture, resulting in higher initial modulus G'. Consequently, precipitated gels on the surface of precursor particles might hinder further reactions (Dai et al., 2020).

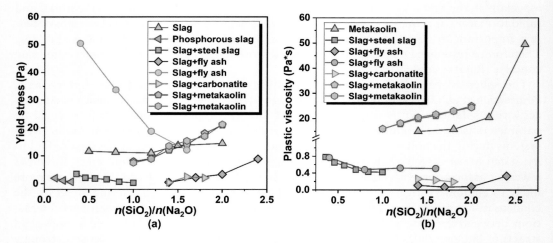

Figure 9.4 Effect of $n(SiO_2)/n(Na_2O)$ molar ratio of sodium silicate activator on yield stress (a) and plastic viscosity (b) of AAMs at the first shear cycle (Mehdizadeh et al., 2018).

Figure 9.4 shows the relationship between modulus, yield stress, and plastic viscosity of AAMs in different literature. In general, the increase of modulus is helpful to reduce yield stress of AAMs thanks to the plasticizing effect when the modulus of activator is below around 1.2 with a relatively high degree of ionization. However, a higher modulus (>1.4) may not be conducive to reducing yield stress. In addition to the strong increase of ion-dipole forces, a rise of modulus also induced more colloidal HO-Si-OM complexes in activator solution forming clusters with several nanometers, and thus the polymerization degree of silicate anions raises. Further, it must consider the effects of early C-(A)-S-H gels precipitate by Ca^{2+} ions and silicate species on rheology in calcium-containing systems (Palacios et al., 2021). Higher modulus-induced more active $[SiO_4]^{4-}$ reacts with dissolved Ca^{2+} ions, producing more hydration products (Yin et al., 2019). Therefore, the viscosity of suspension increased and the complex network structure of highly polymerized silicates is not easy to be damaged by shearing (Yin et al., 2019). Theoretically, the viscosity of interstitial fluid (silicates) has an immediate impact on the viscosity of AAMs. Figure 9.4b confirms this viewpoint and shows that the plastic viscosity would ordinarily increase as modulus, due to more colloidal clusters.

9.1.2.2.3 Concentration

The effect of sodium silicate concentration on the rheological properties of high-calcium AAMs has been extensively studied, which provides guidance for applications requiring better processability and fluidity. Figure 9.5 illustrates sodium silicate at lower concentrations (2.2 and 4.4×10^{-4} mole Na_2O/g slag), and the increase of repulsive force reflects the plasticization and flocculation effect of $[SiO_4]^{4-}$ groups. However, the yield stress will further increase with the content of Na_2O (6.6×10^{-4} mole Na_2O/g slag). This indirectly shows that the electric double-layer force may show attraction at this time, which promotes flocculation and leads to higher yield stress (Palacios et al., 2019). The effect of Na_2O% on yield stress was complex in sodium silicates-slag systems. At a relatively low modulus (<0.8), the fact that the rise in Na_2O% led to an increase in pastes yield stress was consistent with the discussion above (Puertas et al., 2014). Suspensions, however, did not show similar behavior when increased Na_2O% at a higher modulus (>1.2) of Na silicates (Puertas et al., 2014). This might be ascribed to the fact that the modulus needed for primary C–S–H gel to form and break down in dynamic shearing conditions declined as Na_2O% rose.

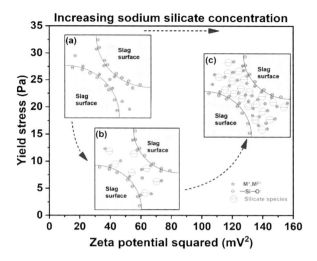

Figure 9.5 A schematic sketch of electrostatic interparticle forces with different dosages of sodium silicate in slag paste. (Adapted from Kashani et al., 2014.)

The effects of Na/K-silicate concentration (Na/K_2O%) on the rheological properties of some mixed systems and calcium-free systems, especially the viscoelasticity, were also studied. Dai et al. (2020) reported that increasing sodium silicate dosage (3%–6% of N_2O%) in slag-FA paste caused a delayed structural build-up followed by an abrupt increase, which was related to the longer initial setting time. Interestingly, they observed humps in the loss factor evolution curve during the rapidly increasing structural build-up stage for pastes with higher Na silicate concentration, which indicated that the loss modulus (G'') increases faster than the storage modulus (G') for a given time. This phenomenon could be explained by the release of extra water from the condensation of $Si(OH)_4$ species, leading to an increase in loss factor and a more liquid-like behavior (Dai et al., 2020). These humps were also related to gelation (Steins et al., 2012). Furthermore, the dynamic yield stress, thixotropy, and shear-thickening response decreased as Na silicate dosage increased, and even behaved like a Bingham fluid.

9.1.3 Effect of precursors on the rheology of AAMs

AAMs are reported to cover a variety of materials, and their differences depend on the physical and chemical properties of the starting aluminosilicates. Many researchers have extensively studied the alkali activation reactions and products of different precursors under different curing conditions. In particular, the rheological behaviors of different precursors (such as slag and FA) and their mixtures were studied (Yang et al., 2018). When different precursors are used, the rheological behavior and parameters of AAMs are very different.

9.1.3.1 Chemical and physical properties of precursors

They are usually used as a single precursor alone in AAM manufacturing because of their high content of reactive components. Gonzalez et al. (2019) used 0.25 M NaOH solutions to activate a modified basic-oxygen-furnace slag incorporated by 13 wt.% Al_2O_3 (reactive amorphous: 55.5 wt.%) and found that the plastic viscosity of the paste was about two times higher than that of the modified slag induced by 11 wt.% Al_2O_3 and 5 wt.% SiO_2 (reactive amorphous: 65.5 wt.%). The results mean that the total amount of reactive component of a specific calcium-rich precursor cannot determine its reactivity. The chemistry of the reactive component seems more important. In a class C fly ash-waste brick powder blended high-calcium system, it was found that increasing the content of calcium-rich waste brick powder tended to increase yield stress and consistency coefficient of pastes activated by Na silicate (Guo et al., 2016), due to the fact that more divalent cations (Ca^{2+}) could lead to attractive ion correlation forces and more primary gels (Vyšvařil et al., 2018). Poor dispersion of brick powder (which contained some concrete powder) also led to a higher solid volume fraction in the suspension, which induced an increase in rheological responses (Guo et al., 2016).

Class F FA is commonly used as calcium-poor precursor, and their reactivity lies on content and composition of vitreous phase(s). The spherical and fine FA particles exhibited beneficial impacts on the rheological properties in the alkali-activated blended system dominated by low-calcium FA. Aboulayt et al. (2018) studied the alkali-activated metakaolin-fly ash (MK-FA) blended system. The maximum packing volume fraction (ϕ_M) of MK-FA blended AAMs grout increased with the replacement level of MK by FA, which was due to the dense packing of spherical FA particles. In addition, the replacement of FA makes more surface-wetting water flow freely. Therefore, it was concluded that the viscosity (consistency) decreased and the setting time and bleeding risk increased.

Yang et al. (2018) proposed to use fly ash microspheres (FAM) as a novel inorganic dispersant based on its "ball-bearing" effect. When the fly ash replacement ratio by FAM increased, the plastic viscosity for all samples declined, irrespective of the modulus (Ms) of sodium silicate (in Figure 9.6a). This phenomenon was ascribed to the "ball-bearing" effect, together with the lubricating effect of FAM particles which breaks down the interlocking between flocs and fragmentation (Vance et al., 2013). Apart from the packing density, the yield stress appeared to be more susceptible to the activator modulus (in Figure 9.6b). In the Ms = 1.4 system, only 40% FAM reduced the yield stress of suspensions, which could be inferred that the "ball bearings" of FAM destroyed the interlocking between fly ash and slag particles to resist the increased particle packing (Yang et al., 2018). By contrast, "ball-bearing" effect in the Ms = 1.8 system seems less important, since more colloidal silicate species might react with Ca^{2+} released from slag to produce more primary gels, and thus increasing the attraction between suspended particles. Again, it should be noted that particles with larger specific

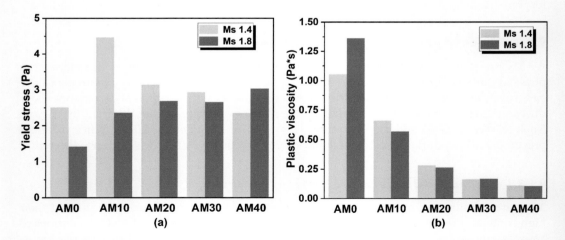

Figure 9.6 The plastic viscosity (a) and yield stress (b) of the alkali-activated fly ash-slag blended pastes prepared by Ms = 1.4 and 1.8 sodium silicate solutions. At constant slag content of 40 wt.%, the 60 wt.% fly ash is replaced by 0, 10, 20, 30, and 40 wt.% fly ash microspheres, denoted as AM0, AM10, AM20, AM30, and AM40, respectively. (Data from Yang et al., 2018.)

surface area need to consume more water. Therefore, there is a balance between the three effects of bulk density, surface water absorption, and "ball bearing".

The ratio of fly ash/slag has a great influence on the rheology of alkali slag-FA composite system. Different content of slag directly affects the initial dissolution behavior, while higher calcium content will lead to the change of precipitation gel system. Compared with FA paste activated by a composite activator ($NaCO_3$ and Na silicate), a replacement of slag to fly ash (fly ash/slag = 3:1; 4:1) reduced the yield stress of pastes. Similar to this finding, Dai et al. (2020) increased the replacement amount of slag to FA to 70 wt.% in Na silicate-based pastes, and the yield stress and consistency coefficient could be reduced by 51% and 81%, respectively. This is attributed to the reduction of flocculation caused by very fine FA particles and a faster dissolution of slag resulting in low solid content (Dai et al., 2020). Different w/b ratios and particle sizes may be responsible for these differences. Dai et al. (2020) designed the w/b ratios to decrease with the increase of FA content, and the fly ash used was finer than slag. This enhances attractive interparticle forces leading to more flocculation structures that are more difficult to be disturbed by shearing. However, it must be considered that the introduction of more slag means faster dissolution and higher structure build-up rate. These results are due to the formation of more early C-A-S-H gels (Palacios et al., 2021).

The introduction of metakaolin (MK) with very high specific surface area generally leads to higher rheological response of pastes (Rovnaník et al., 2018), which indicates that suitable activators, dispersing agents, and sufficient water are required to achieve good workability. Moreover, compared to cement-based suspensions, MK-based pastes usually exhibit high viscosity due to the fact that plate-shaped MK grains have low ϕ_M. In addition to the negative effect of irregular spiny plate-shaped particles of MK, the viscosity-enhancing effect of MK was attributed to the larger specific area, requiring more mixing water for wetting. Moreover, MK with distorted aluminum layer structure had higher reactivity than brick powder; hence, the suspensions containing more MK led to faster growth of yield stress and storage modulus due to rapid gelation (Rovnaník et al., 2018). Table 9.3 summarizes the effects of precursor nature on the rheological properties of alkali-activated blended systems. It can be seen that the physical properties of precursors, morphology in particular, jointly affect the surface water absorption and packing density to alter the rheological properties of AAMs. However, the correlation between these film thicknesses and the rheological properties of AAMs has not been proposed quantitatively. In addition, due to the changes of polymer structure and polymer particle interaction, when the precursors with different reactivity are mixed in different proportions, the rheological parameters will also be affected.

9.1.4 Effects of chemical admixtures on the rheology of AAMs

9.1.4.1 Water-reducing admixtures

As we all know, water-reducing admixtures are widely used to improve the rheology and workability of PCMs. Commonly used water-reducing admixtures include (1) lignosulfonate-based type; (2) naphthalene-based, aminosulfonate-based and melamine-based water-reducing admixtures; and (3) polycarboxylate-based superplasticizers. A large number of literatures have studied the working mechanism of superplasticizer and its compatibility with PC. The effective adsorption of admixture on the surface of cement particles leads to electrostatic repulsion and steric resistance between particles, so as to avoid flocculation (Alonso et al., 2007, Ren et al., 2013). Thus, to some extent, such admixtures can reduce the water demand and improve durability and mechanical properties (Puertas et al., 2005). Table 9.4 summarizes and compares effects of various water-reducing admixtures with different dosages on the rheological properties of AAMs in literatures.

Table 9.3 Effects of precursor nature on the rheological properties of sodium silicate-activated blended systems.

Precursors	L/S ratio (by volume /mass)	Controlling factors	Rheological parameters τ_0 (Pa)	K or μ (Pa*s^n/ Pa*s)	Decisive factors	Refs.
Metakaolin+fly ash (Grout)	L/S 0.75	MK/FA ratio 0:10; 2:8; 4:6; 6:4; 8:2	Decreased from 0.40 to 0.02	Increased from 0.01 to 0.10	The higher specific surface area of MK; dense packing of FA	Aboulayt et al. (2018)
Fly ash+silica fume (Grout)	w/b 0.75	SF/FA ratio 0:10; 1:9; 2:8; 3:7; 4:6	Increased from 1.81 to 9.51	Increased from 0.0105 to 0.066	Higher specific surface area and reactivity of SF	Güllü et al. (2019)
Slag+fly ash+fly ash microspheres (Paste with constant slag of 40 wt.%)	L/S 0.5	FA/FAM ratio 6:0; 5:1; 4:2; 3:3; 2:4	Increased from 2.51 to 4.47, then decreased to 2.35	Decreased from 1.05 to 0.11	The "ball-bearing" effect, dense packing, and higher specific surface area of FAM	Yang et al. (2018)
Class C fly ash+waste brick powder	w/b 0.3	FA/waste brick powder ratio 4:6; 3:7; 2:8; 1:9; 0:10	Decreased from 693 to 130	Decreased from 675 to 88	The waste brick powder contains higher CaO content and shows poor solubility	Guo et al. (2016)
Metakaolin+waste red brick powder (Paste)	L/S 0.71	The brick powder/MK ratio 0:10; 2.5:7.5; 5:5; 7.5:2.5; 10:0	no significant effect	Decreased from 2.18 to 0.09	Higher specific surface area and reactivity of MK	Rovnanik et al. (2018)

L/S = liquid/solid ratio; w/b = water/binder ratio; L/B = liquid/binder ratio.

Table 9.4 The comparison of effects of various water-reducing admixtures with different dosages on the rheological properties of AAMs

Plasticizers	Dosage (by mass of binder)	Precursors	Activators	Main performance	Ref.
Polycarboxilate (PC); naphthalene sulfonate (NaS)	1.0%–5.0%	Fly ash (paste)	Na silicates	PC slightly reduced the viscosity; NaS not only increased the viscosity, but also induced flash set	Montes et al. (2012)
N and PC	1.0%–4.0%	Fly ash+slag (paste)		PC improved workability more effectively than N	Jang et al. (2014)
Polycarboxylic ethers (PCE); LS	1%; 1.5%	Fly ash (concrete)		The minimum μ and τ_0 from mixtures with LS (undergo segregation with 1.5%)	Laskar and Bhattacharjee (2013)
PCE; melamine formaldehyde derivative (M); naphthalene formaldehyde derivative (NF); vinyl copolymer (V)	0.3%; 0.5%; 1.0%; 1.5%; 2.0%	Slag (paste and mortar) Slag (paste and mortar)	NaOH	Effects of all plasticizer admixtures are negligible Mixtures with NF reduced τ_0	Palacios et al. (2008)
M; NF; V	~0.4%	Slag (paste)		Mixtures with 1.26 mg NF significantly reduced τ_0	Palacios et al. (2009)
HPEG PCEs with various side chain lengths and anionic charge density	0.05%			HPEG PCEs with high anionicity, high molecular weight and short side chain length showed a better dispersion	Lei and Chan (2020)

9.1.4.1.1 Water-reducing admixtures in alkali-activated slag (AAS) systems

The applicability of commercial water-reducing admixtures to control the rheological properties of AAS has been studied by many researchers. In NaOH-AAS system, the plasticizer with appropriate dosage shows a certain plasticizing effect in the initial stage, but it may become unstable in the later stage, which is related to the molecular structure under high-pH conditions. Palacios et al. (2008) investigated the influences of four different high-range water-reducing admixtures (HRWRAs) on the rheological behavior of NaOH-activated slag pastes/mortars. Only the naphthalene-based HRWRA could lower the yield stress prominently due to its chemical stability in the alkaline environment. Though the lower-molecular-weight fragments of the polycarboxylate-based, melamine-based and vinyl copolymer HRWRAs may still adsorb on the particle surface and cause electrostatic repulsion, its contribution to space is not obvious (Palacios et al., 2008).

In addition, the authors also reported the similar behavior of these admixtures in the experiment of 13.6-pH NaOH-AAS paste (Palacios et al., 2009). As shown in Figure 9.7, the yield stress and plastic viscosity of AAS pastes with different amounts of the melamine-based admixture were all higher than those of the blank control group, when the shearing regime was implemented for at least 10 min. The plasticizing effect of vinyl copolymer seems less prominent in AAS pastes. Only the naphthalene-based superplasticizer with 1.26 mg substantially reduced both yield stress and plastic viscosity in NaOH-AAS pastes (in Figure 9.7c and d). This is due to the fact that the adsorption-mediated decrease in the size and number of agglomerated particles further promotes the chemical stability of the mixture. This reflects the suitability and plasticizing effectiveness of different superplasticizers in the alkaline environment of NaOH-AAS pastes.

In addition, the effective adsorption of admixture on precursor particles has a great impact on its plasticization efficiency. Generally, with the increase of adsorption of anionic superplasticizer molecules (admixtures containing sulfonic acid groups) on the particle surface, the zeta potential of blast furnace slag and FA suspension gradually decreases (more negative), indicating that electrostatic repulsion may play a great role in the dispersion mechanism (Nägele and Schneider, 1989). The working mechanism of these superplasticizers in NaOH-AAS was further studied by Palacios et al. later, and it turned out that steric hindrance was dominant in the working mechanism of both the sulfonate and polycarboxylate superplasticizers in NaOH-AAS (Palacios et al., 2012). Among them, melamine admixture contributes the most to the electrostatic repulsion. The naphthalene derivative admixture with the lowest molecular weight has the smallest spatial repulsion range. Compared to the melamine derivative and vinyl copolymer, polycarboxylate superplasticizer with a lower molecular weight showed the highest steric hindrance, which might be attributed to different adsorption conformations related to molecular structures (Palacios et al., 2012).

More importantly, in addition to chemical stability, the solubility in alkaline media (Conte and Plank, 2019) and the competitive adsorption between anionic activator and anionic superplasticizer (Lei and Chan, 2020) will also affect the adsorption efficiency and dispersion effect of these superplasticizers in AAMs. The solubility difference of PCEs with different molecular structures in NaOH and Na_2CO_3 solutions induced different dispersing efficiency (Conte and Plank, 2019). All the PCEs tested in the study had no important impact on improving the flow of Na_2CO_3-slag paste due to the extremely low solubility in Na_2CO_3 solution. Moreover, allyl ether-based PCEs with a high molecular weight (Mw > 30,000 Da) and a short side chain ($n_{EO} = 7$) exhibited the best dispersing performance in NaOH-slag system, which might be attributed to their good solubility in NaOH solution and stronger chelating ability with Ca^{2+} ions. It should be noted that better solubility of these polymers in alkaline solutions does not always mean superior dispersing ability. Lei and Chan (2020)

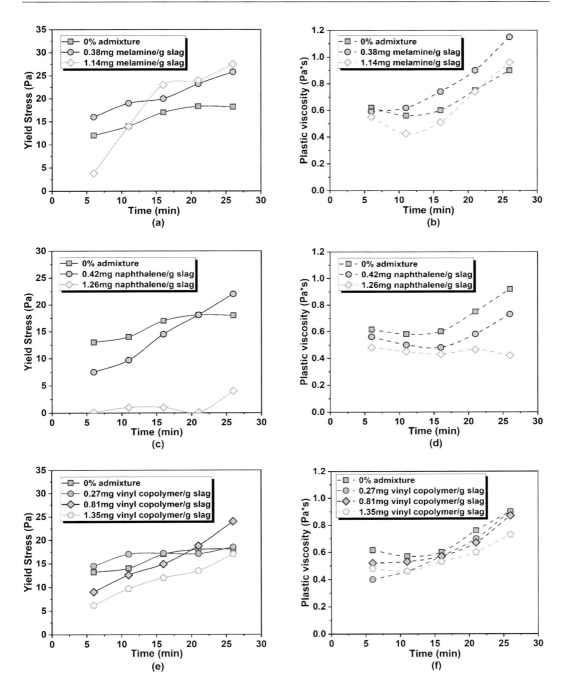

Figure 9.7 (a–f) The yield stress and plastic viscosity evolution in 13.6-pH NaOH-AAS pastes containing three types of superplasticizers: melamine-based, naphthalene-based, and vinyl copolymer at different dosages. (Data from Palacios et al., 2009).

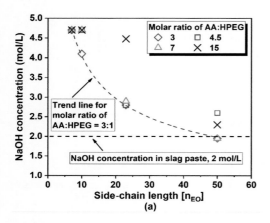 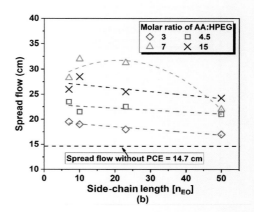

Figure 9.8 NaOH concentrations (a) needed to achieve the cloud point of different HPEG PCEs, and spread flow (b) of NaOH-slag pastes containing 0.05 wt% of different HPEG PCEs (Lei and Chan 2020).

synthesized a series of PCEs with different anionic charge densities or side chain lengths, based on α-methallyl-ω-hydroxy poly (ethylene glycol) ether (HPEG). As shown in Figure 9.8, even 23-HPEG-15 (Mw = 42,000 Da; n_{EO} = 23) showed higher solubility (4.50 mol/L) in NaOH solutions than 23-HPEG-7 (Mw = 410,000 Da; n_{EO} = 23; 2.96 mol/L NaOH), and it presented poor dispersing performance in NaOH-slag pastes. Therefore, the solubility of PCEs is a limit factor of dispersing efficiency of such polymers in AAMs, but is not the decisive factor.

It is well known that many commercial anionic superplasticizers improve adsorption efficiency by the strong chelation with Ca^{2+} ions. Due to the deprotonation of hydroxyl, the negatively charged slag particles attract some Ca^{2+} ions in an AAS system, which provides a possibility for the adsorption of anionic superplasticizers on the slag particles' surface. However, current research shows that the adsorption capacity of superplasticizers on slag surfaces is still far less than that in PCMs. To some extent, the low adsorption efficiency might be generally grouped under competitive adsorption between such polymer molecules and anionic species in activators (Conte and Plank, 2019). Marchon et al. (2013) considered that the fast flow loss of PCEs in the alkali-activated system was probably due to the competitive adsorption between PCE molecules with the hydroxyls. As previously discussed, the low adsorption performance of PCEs in Na_2CO_3-slag pastes could also be accounted in the almost complete removal of Ca^{2+} ions in solutions resulting from precipitation by the CO_3^{2-} ions (Conte and Plank 2019). It could lead to an insufficient adsorption of Ca^{2+} ions on slag particles, lowering adsorption efficiency of PCEs.

9.1.4.1.2 Water-reducing admixtures in alkali-activated fly ash (AAFA) systems

The influence of the third-generation water-reducing agent on the fluidity of AAFA paste and concrete was also studied. Compared with the blank control slurry, AAFA-containing water-reducing agents (including lignin sulfonate, melamine, and polycarboxylate) showed the effect of reducing yield stress and plastic viscosity after mixing. However, this phenomenon only existed in the early stage (Criado et al., 2009). As mentioned earlier, the chemical stability and solubility of superplasticizer in alkali solution and the degree of adsorption on particles should be considered as possible reasons for the low plasticization efficiency. Interestingly, Laskar and Bhattacharjee (2013) pointed out that the concentration of 4 M for NaOH solution was a critical value, below which increasing dosages of both lignosulphonate-based and polycarboxylate-based admixtures could improve workability of fly ash paste activated by sodium silicate. Furthermore, unlike in AAS suspensions, Criado et al. (2009) reported that polycarboxylate superplasticizer in AAFA system shows the most effective plasticizing performance among the three types of superplasticizers. Most admixtures, including the third-generation superplasticizers, are designed to form complexes with dissolved Ca^{2+} during the early hydration of PC (Uchikawa et al., 1995).

Slag and high-calcium (class C) FA can also release large amounts of Ca^{2+}, but there is almost no calcium dissolution in the low-calcium (class F) FA system. Therefore, the steric hindrance effect is prohibited or not trigged in class F FA systems. As reported by Montes et al., the polycarboxylate-based admixture only slightly reduced the viscosity of FA paste activated by Na silicate (Montes et al., 2012). In another study, however, polycarboxylates were more effective to induce an apparent drop in shear stress and viscosity in sodium silicate–class C FA, and this higher adsorption efficiency was attributed to the presence of positively charged Ca^{2+} (Xie and Kayali, 2016). In contrast to PC, the drop in the plastic viscosity and yield stress of AAFA pastes containing plasticizers is not always accompanied by the improvement of fluidization performance. There was hardly a correlation between a slump and the rheological parameters in the research by Criado et al. (2009). This is adverse to the report from Laskar and Bhattacharjee (2013), which concluded that good correlation between rheological parameters and slump test results in AAFA concretes containing different plasticizers. It is evident that the properties of FA, the interaction between water-reducing admixtures and FA, and the type of activator and the addition method should be considered in future research for different AAM systems.

9.1.4.2 Other chemical admixtures

In addition to water-reducing agents, the effects of other common admixtures, such as shrinkage-reducing agents, retarder, and pour point depressants, on the rheological properties of AAMs were also studied. Palacios et al. (2008) studied the rheological modification effect of polypropylene glycol derivative shrinkage reducer on AAS. A recent publication showed that fluidity of AAS pastes with polypropylene glycol (PPG)-based shrinkage-reducing agent could be increased (PPG1000), reduced (PPG400), or maintained (PPG2000) at a similar level, which depended on the dosages and molecular weight of PPG (Ye et al., 2020). The difference in fluidity can be attributed to the effect of PPG on the viscosity of suspensions and the dissolution process of slag grains (Ye et al., 2020). Hence, the chemical stability and rheological properties of shrinkage-reducing agents with different structures and molecular weights in high-alkaline AAMs need more research. Xu et al. (2018) found that citric acid can work not only as a setting retarder but also as a potential plasticizer in AAMs.

As shown in Table 9.5, the NaOH-activated FA (class C)–lime kiln dust paste (marked as AAFA-LKD) with the addition of 3 wt.% citric acids showed a low plastic viscosity and comparable yield stress to PC, and these rheological parameters could be maintained

Table 9.5 Plastic viscosity and yield stress of PC paste and alkali-activated fly ash (class C)–lime kiln dust (marked as AAFA-LKD) pastes without and with 3 wt.% citric acid at different times after reaction (Xu et al., 2018)

Pastes	PC	AAFA-LKD		AAFA-LKD + 3% citric acid		
Time (min)	10	5	10	5	10	60
Plastic viscosity (Pa.s)	12.4	12.6	13.6	4.4	3.9	6.5
Yield Stress (Pa)	105	69.3	432·	91.8	72.5	64.7

In the table of Xu et al. (2018), the unit of yield stress is MPa. This might be wrong, so here it is corrected as Pa.

at viable levels in the first 1 h. The introduction of citric acid reduces the alkalinity of suspension and thus delays the dissolution process. In addition, the formation of carboxylate complexes between the carboxylic acid groups of citric acid and Ca^{2+} ions released from calcium-containing precursors would also affect the gelation process and then reduce yield stress.

Sodium polyacrilate with low molecular weight (Mw = 2000 au) was used to work as a deflocculant to reduce the yield stress and shear viscosity of aqueous suspensions of kaolinite (Romagnoli et al., 2012). However, sodium polyacrylate has a limited effect on reducing yield stress and viscosity of metakaolin-based AAMs, or even for the aqueous suspensions of metakaolinite (Romagnoli et al., 2012). Therefore, the admixture, in this case, behaves more like a flocculant than deflocculant. Grout materials usually require a good balance between flowability and stability. Viscosity-modifying agents or stabilizers are usually added to prevent segregation and bleeding. Aboulayt et al. (2018) applied polysaccharide-based stabilizer (xanthan gum) in metakaolin-FA blended AAM grouts and found a beneficial effect on the stability. For a given FA ratio, both the yield stress and plastic viscosity at different shear rates increased with the increase of stabilizer dosage.

9.1.5 Effects of mineral additions on the rheology of AAMs

Whether inert or latent hydraulic mineral additions with small particle sizes, they all have a physical filling effect in the hydration process, not only in PCMs but also in AAMs. The high surface area of the fines can provide abundant nucleation sites for hydration. Accordingly, both the fresh and hardened properties will be affected to varying degrees.

9.1.5.1 Reactive mineral additions

Silica fume (SF) can be used as a precursor in AAMs to mix with slag or FA. Here, in order to better explain the influence of SF, it will be discussed in this part. The rheological modification of AAMs by SF is closely linked with the water to binder ratio (w/b) and its physicochemical properties. The amount of SF that caused a significant increase in the yield stress and plastic viscosity of AAFA grout decreased as w/b ratio. For example, the critical content of SF in the suspension with a w/b of 0.75 is 20%, while 50% is the critical content of SF in the suspension with a w/b of 1.25. Beyond this critical content, it was observed that the yield stress and plastic viscosity of AAFA grout increased with the increase of SF proportion (Güllü et al., 2019). This is similar to the impact of adding SF in the conventional grouts. On the other hand, the "ball-bearing" effect of this spherical ultra-fine powder (the passing rate of 0.01 mm is 93% for SF, while 18% passing for FA) could play a role as a lubricator (Yang et al., 2018). Below this critical content, the lubricating effect of SF was dominant due to spherical particles. In a recent study, ultra-high-performance concrete based on AAS with a very low w/b ratio of 0.175 retained good workability by incorporating with 12.5 wt.% SF (Wetzel and Middendorf 2019).

Steel slag (SS) is a metallurgical slag containing a large number of low reactive or inert ingredients, and its pozzolanic reactivity is low (Luxán et al., 2000). Increasing the replacement ratio of SS in AAS-based 3D printing resulted in the reduction of thixotropy (Zhang et al., 2018b). Raising SS content reduces the degree of hydration, and therefore, the connection between particles is weakened due to fewer initial gels formed after mixing, thus reducing the reversible connection of particles (Zhang et al., 2018a). Besides, the addition of SS reduced the plastic viscosity of AAM due to the smaller specific surface area of SS (Papo et al., 2010).

9.1.5.2 Inert mineral additions

The addition of a proper amount of inert mineral additions, such as limestone (LS) powder, can improve rheological properties by working as fillers to optimize the particle packing model of the blend system, and more water for lubrication. This can compensate for the negative effects of high specific surface area and nucleation effect. According to Xiang et al. (2018), the incorporation of 5–20 wt.% LS powder in sodium silicate-activated slag-FA blended grout did not alter the rheology model (Bingham model), but Figure 9.9 shows a positive effect on reducing yield stress and plastic viscosity, especially for the grouts activated by solid sodium silicate. The particle size range and d_{50} of LS are lower than those of slag and FA, and its specific surface area (1602 m²/kg) is also lower than that of FA (2428 m²/kg). To a certain extent, LS can therefore play a filling role and reduce the wetting water demand.

Similarly, to modify the very sticky AAFA binders for 3D-printing purposes, Alghamdi et al. (2019) used 15–30 wt.% limestone powder (d_{50} =1.5 µm) to replace FA (d_{50} =17.9 µm). As expected, a partial replacement by limestone powder can effectively address this issue and improve print quality, due to the consistency and water retention. Moreover, Xiang et al. (2019) also found that a similar positive effect could be achieved in sodium silicate-activated MK grout by replacing MK with different percentages of Fuller-fine sand powders (by 10, 20, 30, and 40 wt.%). For example, in the sample group synthesized with 70 wt.% MK (specific surface area = 16,646 m²/kg) and 30 wt.% Fuller-fine sand (specific surface area = 8886 m²/kg), the yield stress was reduced by 73% at the first shear cycle. However, it should be noted that adding different volume fractions (10%, 30% and 60 wt.%) of fine silica sand as filler to Na silicate-based MK pastes caused an increase in viscosity, but less than if the water content of pastes was reduced to obtain the same solid volume fraction (Kuenzel et al., 2014).

9.1.6 Effect of aggregates on the rheology of AAMs

Currently, most research efforts are being focused on the rheology AAM paste while only very few studies regarding the rheology of AAM mortar/concretes have been reported. For AAMs paste, the Bingham model is normally recommended to describe the rheological behavior of NaOH-activated pastes, and the H–B model is more suitable for sodium silicate-activated pastes (Puertas et al., 2014). However, for alkali-activated mortar/concrete, the

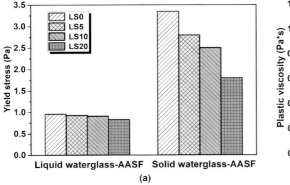
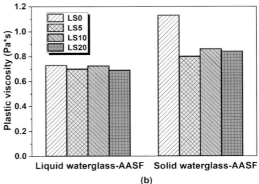

Figure 9.9 The 27 min yield stress (a) and plastic viscosity (b) of the alkali-activated slag-fly ash (AASF) grouts using liquid waterglass and solid waterglass as the activator, after adding 0–20 wt.% limestone (LS) (Xiang et al., 2018).

Bingham model suits both two systems, and this behavior is similar to PC-based mortar/concrete (Alonso et al., 2017).

Furthermore, compared with PC mortars, it was reported that the workability of both alkali-activated fly ash and AAS mortars was more sensitive to changes of liquid/solid ratio (Alonso et al., 2017). You et al. (2020) found that using copper slag (CS) as fine aggregate to replace natural sand (NS) could alter the rheological characteristics of the alkali-activated fly ash-slag mortar. It was observed that a low replacement by 20% and 40% (by volume) CS resulted in higher yield stress and consistency, due to the characteristics of CS by more irregular shape and containing higher amounts of fine powders (Westerholm et al., 2008).

9.2 RHEOLOGY OF CEMENT PASTE BACKFILLING (CPB)

9.2.1 Introduction

Mining and mineral-processing operations are associated with a series of challenges such as the management of a large volume of underground and/or surface voids as well as a massive amount of mine wastes. The efficient management of the generated tailings has become a serious issue for sustainable mining operations since they contain heavy metals, mineral-processing reagents, and sulfur compounds. Due to the increasingly strict environmental rules, the changing milling practices, and the realization of profitable applications, there are some emerging new techniques that convert problematic tailings into useful materials. A common practice for tackling the aforementioned challenges relating to tailings management is to re-fill them into underground mine-out stopes or openings through cemented paste backfill (CPB) technique.

Backfilling is an environmentally friendly and safe tailings management method. The advantage of backfilling such stopes includes the permitted mining of the adjacent stopes, ground stabilization, increased ore recovery, and reduction in the surface waste disposal. It reduces up to 50% of harmful tailings that would otherwise be deposited on the surface (Ouattara, et al., 2017). Tailings are the residue generated after the extraction of recoverable metals and minerals from mined ore in a mine-processing plant. They are often composed of the finely grounded host rock, gangue minerals, and a small number of unrecovered ore minerals. The management of mine waste is expensive and presents serious problems to geotechnical (e.g., tailings dam failures) and environmental (e.g., groundwater pollution and acid mine drainage) hazards (Yin, et al., 2012). CPB is a non-segregated, complex composite composed of mine tailings, binder, and water (Ouattara et al., 2018b). In conventional paste backfilling, thickeners are used to densify the tailings slurry (20%–40% solid content, by mass) to produce thickened tailings (TTs) having 50–70 wt.% solid content. To improve the thickeners' performance, flocculent is added at a specific dosage. The TTs are then filtered to reach solid content $\geq 80\%$. The filtered tailing is then used for CPB preparation (Ouattara et al., 2018b). The solid content depends on the rheological properties of the paste and ranges from approximately 70% to 80% with about 15 wt.% particles < 20 μm. Such size distribution prevents settlement with low bleeding when transported in the pipeline (Panchal et al., 2018). The CPB composes of binder content ranging from 3 to 7 wt.% of total solids, and the content depends on the production cost and magnitude of stress expected around the excavated stopes (Simon and Grabinsky, 2013). The aim of adding cement is to provide sufficient cohesion to prevent liquefaction and provide mechanical strength (Huynh et al., 2006). The CPB has three major roles. First, it supports the ground, thus providing a safe work environment. Secondly, it can be used to construct a floor, wall, or roof/headcover for mining activities. Finally, it is an effective way for mine waste disposal (Yin et al., 2012).

After preparing fresh CPB in a backfilling plant located at the surface of the mine, the CPB makes a kind of slurry having relatively high density. The purpose of backfilling is to place the fresh CPB into underground open stopes to produce a hardened structure to strengthen adjacent ore bodies and ensure a safe mining environment. Transportability is one of the key properties of fresh CPB, which depends on fluidity or flowability. Fresh CPB is a kind of non-Newtonian fluid that flows under high shear stresses and stops under low shear stresses (Qi and Fourie, 2019). The fresh CPB must be flowable enough to ensure efficient pumping/delivery from the plant to underground open stopes. The upper limit for yield stress was reported to be 200 Pa to ensure effective centrifugal pumping of CPB (Ouattara et al., 2018b). Hence, the flowability of fresh CPB is crucial for cost-effective and efficient mine backfill operations, in which the cost represents 20% of the total cost of the mining process. Poor flowability not only affects the pumping/delivery efficiency but can also cause pipe clogging, leading to a significant delay in the production process and an increase in the cost of the mining process. Therefore, to avoid pipe clogging, fluent transportability of fresh CPB should be ensured (Wu et al., 2013). Depending on the physical characteristics and mineralogy of the tailings, slumps ranging from 12 to 25 cm generally perform well in transportation and backfilling (Ouattara et al., 2017). However, Sivakugan et al. (2015) believed that slumps ranging from 23 to 28 cm would perform better in pipeline transportation. The rheological characteristics of fresh CPB must be known for predicting the pressure gradient and flow velocity for pipeline flow (Ouattara, et al., 2017).

The pumping/delivery and placement depend on the rheological characteristics of fresh CPB. The rheological properties of fresh CPB depend on the physical properties of the mixture (e.g., density and CPB mixture concentration, characteristics of the CPB mixture constituents) (Wu et al., 2013) and type and magnitude of interparticle forces such as pH and ionic concentration of pore fluid (Simon and Grabinsky, 2013). It is known that the properties of tailings, like particle size, size distribution, mineralogical and chemical composition, from each mine, are quite different.

9.2.2 Factors affecting the rheological properties of CPB

9.2.2.1 Cement

Cement mainly determines the hydration reaction of CPB, and thus its time-dependent properties. The microstructure of CPB changes continuously during transportation due to the formation of hydration products, thus changing rheological characteristics. Panchal et al. (2018) observed an increase in yield stress and thixotropy by an increase in hydration age. The results showed that the yield stress increased at a faster rate during the initial stage of hydration and slowed down gradually after 105 min.

The rheological parameter of CPB changes with varying binder content. Panchal et al. (2018) observed an increase in yield stress, plastic viscosity, and thixotropy by increasing binder dosage from 4% to 8%. This variation in rheological properties can be due to the physical characteristics (e.g., packing density) and chemical reaction (e.g., hydration). Because of the different packing densities of tailing and binder, the packing density may be improved in the mixture. Hence, by increasing the binder content, the packing density was improved and the volume of capillary pores reduces, subsequently decreasing the flowability of CPB. Moreover, increasing cement content increases the hydration product, thus strengthening the bond between tailing particles which leads to an increase in rheological properties of CPB (Panchal et al., 2018).

Deng et al. (2018b) studied the influence of binder content on rheological properties of ultra-fine cement paste backfill (UCPB) (having 80% of particle size (d_{80}) < 10 μm), with cement content ranging from 0% to 25%. The Casson yield stress exhibited the highest value at 6% cement content, followed by a decrease by further increasing binder dosage. The possible reason for this behavior is as follows. The particle size of cement is much larger than that of ultra-fine simulated tailings (UST). After mixing, due to the difference in the particle size distribution of cement and UST, packing density was improved and reached a peak value, at which the mixture showed the highest yield stress. Moreover, it was observed that at higher shear rates (25, 50, 100, 200, and 300 s^{-1}), the apparent viscosity and shear stress of CPB with cement first decreased and then increased by increasing shearing time (shearing for a long period at the constant shear rate), regardless of the content of cement. The initial shear stress (τ_0), minimum shear stress (τ_{min}), and final shear stress after 3600 s ($\tau 3600$) increased first and then decreased by increasing the cement/tailings ratio (c/t) (Deng et al., 2018c). Peng et al. (2019) noted that yield stress and plastic viscosity increased by an increase of cement content from 3% to 6% in CPB (initial sulfate concentration 25,000 ppm) with the curing time from 0 to 4 h. Di et al. (2014) revealed that shear stress and viscosity increased by increasing the c/t ratio and water/cement (w/c). According to Lang et al. (2015), the tailing/cement (t/c) ratio showed no influence on rheological properties, but the slump was affected by the ratio. A slight increase in the slump was observed by increasing the t/c ratio. However, by increasing the solid concentration, this relationship was weakened.

9.2.2.2 Solid concentration

As a suspension, solid concentration has a significant influence on the rheological parameters of the cement paste. In the mineral industry, the most common effect of solid concentration on the viscosity is a shift from Newtonian behavior at low solid concentrations, through Bingham behavior at intermediate concentrations to shear-thinning behavior at a higher concentration (Sofra, 2017). The rheological properties of the CPB system are also affected by the solid concentrations.

The initial shear stress of UCB increased from 8.6 to 76.5 Pa by increasing solid content from 50% to 60%. Moreover, for all solid content percentages, the apparent viscosity and shear stress at steady-state shear testing were observed to decrease first and then increase slowly over prolonging shearing time at a constant shear rate (Deng et al., 2018b). This transient flow of UCB may be due to the following reasons. The initial decrease of both properties may be due to the destruction of microstructure by the shearing, while the gradual increase is due to the rebuild-up of microstructure induced by cement hydration and sedimentation of particles during testing. In the hydration process, the formation of calcium silicate hydrates (C–S–H), calcium hydroxide (CH), and ettringite causes the structural build-up and increases the apparent viscosity of UCB. On the other hand, due to the sedimentation process, the solid content in the bottom of the rheological measurement cup increases, which may cause an increase in measured values (Deng et al., 2018b). Increasing the solid content from 60 to 70 wt.%, the Casson yield stress was observed to increase from 6.16 to 61.70 Pa (Deng et al., 2018c). The increase in rheological properties by increasing solid content was also observed by Lang et al. (2015). Yang et al. (2020) investigated the rheological properties of cemented unclassified tailings backfill (CUTB) and observed obvious shear-thinning behavior. Moreover, the increase in viscosity and yield stress was observed by increasing cement-to-tailing ratio and solid content as quadratic. Fresh CPB showed shear-thinning property which is more pronounced in the mixture having higher yield stress (Deng et al., 2018c). Cao et al. (2018b) studied the effect of solid content (60%–72%) on the viscosity of CPB. The results showed that the lowest viscosity (8.26 Pa.s) was observed at a solid content of 60% and a c/t ratio of 1:10. Besides, viscosity was observed to increase by increasing solid content and the c/t ratio.

9.2.2.3 Mixing intensity

The mixing process homogenizes paste constituents and plays an important role in determining the rheological properties of fresh CPB. Reducing the mixing time and mixing intensity without damaging the quality and productivity of the mix is widely pursued. Therefore, a well-designed mixing technique is important for the preparation of CPB (Yang et al., 2019a).

Different mixing intensity results in different rheological properties. This phenomenon can be affected by the interparticle forces, which is explained by the Particle Flow Interaction (PFI) theory. The PFI theory demonstrates the processes of coagulation, dispersion, and re-coagulation of the particles. Coagulation refers to the contact between two or more particles. The particles should be separated from each other to ensure improved rheological properties in CPB. These particles are forced together due to the total potential energy of particles, which is generated from the combined forces of van der Waals' attraction, electrostatic repulsion, and steric hindrance. To separate the particles, the shear force should be greater than the total potential energy. In this way, the rheological properties of CPB decrease. The other phenomenon responsible for the change in mixture rheology is the structural breakdown. The structural breakdown is attributed to the shrinking electric double layer of cement particles and/or connection breaking between particles formed during the hydration process. When cement in the mixture reacts with water, it forms hydration products that surround the surface of particles in the form of a membrane, causing particles to be bound together. During the mixing, connections between particles are broken and particles are separated. Higher mixing intensity increases the structural breakdown, dissolving more hydration products into pore solution, which may affect the rheological properties (Yang et al., 2019a).

Yang et al. (2019a) observed a proper fluidity and decrease in viscosity at a shear rate between 360 and 600 s^{-1}. For energy saving and high productivity, the authors proposed an optimal mixing speed of 300–400 rpm. At a higher shear rate (>500 rpm), numerous ions and early hydration products dissolve into the pore solution, which enhances the particle agglomeration and increases the yield stress and Bingham viscosity of CPB, and this effect is higher in the case of higher cement content. Yang et al. (2019b) observed lower yield stress and shear-thinning behavior at a low shear rate, and higher yield stress and shear-thickening behavior at a high shear rate. At a constant shear rate (300 s^{-1}), both apparent viscosity and yield stress of CPB decrease first and then steadily increase over long periods of shearing (Deng et al., 2018c). The shear rate around 100 s^{-1} promotes the breaking of loose inter-connection of particles or random collisions of particles, thus lowering the viscosity (shear-thinning). A higher shear rate (around 400 s^{-1}) enhances the particle aggregation, called the formation of hydro clusters, which mildly increased the CPB viscosity (shear thickening) (Wang et al., 2019).

9.2.2.4 Particle size

In general, suspensions containing finer particles have higher yield stress and viscosity at a given solid content. This is due to the fact that finer particle has a greater particle surface area, and thus greater interparticle interactions and a higher amount of required water. By increasing solid content, finer slurries display a more gradual increase in yield stress, whereas a mixture with coarse particles exhibits a relatively sharp transition from liquid-like to solid-like behavior (Sofra, 2017).

Optimizing particle size is an important technique to tailor the rheology of the CPB slurry. Deng et al. (2018a) investigated the effect of particle size on the rheological properties of CPB. The result showed that the shear-thinning behavior of fresh CPB was more pronounced in the case of larger particle sizes and higher volume fractions to apparent maximum packing density (ϕ/ϕ_m) ratio. For the sample having (ϕ/ϕ_m) < 0.875, the CPB slurry with finer particle size resulted in a higher yield stress, but the time-dependent rheological behavior was less pronounced as compared with coarse particle size. Moreover, it was found that the apparent viscosity increases with the increase of particle sizes because of the higher content of hydration products and a higher rate of particle sedimentation. Ke et al. (2015) observed a decrease in the workability of fresh CPB slurry by increasing the proportion of fine particles.

9.2.2.5 High-range water reducer (HRWR)

The CPB technology is now moving toward a high-solid-content CPB to improve the mechanical properties. But high-solid-content CPB can be high viscous and undergo shear-thickening properties in high-solid granular suspensions. Moreover, the solid content above the critical value resulted in an exponential increase in the dynamic viscosity and yield stress in CPBs. So, the higher solid content can deteriorate the rheological properties of the CPB mixture and can cause pipeline blockage. This blockage can be avoided by using low solid concentration and high water content or by using a high-range water reducer (HRWR) to obtain a slump in the range of 15–25 cm (6–10 inches) (Ouattara et al., 2018b).

Ouattara et al. (2018a) studied the consistency of CPB by using different types of HRWR: a polymelamine sulfonate (PMS), a poly naphthalene sulfonate (PNS), and four polycarboxylates (PC). The PC-based HRWR was observed to be the most effective in increasing the consistency of CPB as compared with other types of HRWR. Ouattara et al. (2018b) also investigated the effect of PC-based, PMS-based and PNS-based HRWRs on the rheological properties of high-solid-content CPB made with tailings from polymetallic and gold ores and different types of a binder. The addition of HRWRs reduces the yield stress and infinite shear viscosity. The rheological behavior changed from shear thickening (for control CPB) to Bingham by increasing the HRWR dosage. The optimum PC dosage was reported to be 0.121% (by mass of total solid content of CPB) to ensure the yield stress < 200 Pa. Moreover, the better performance of PC-based HRWR for improving the rheology of CPB was observed as compared with PNS and PMS types. The binder content ranging from 3.5% to 6% showed a negligible effect on the rheology of CPB having PC-type HRWRs. Ercikdi et al. (2009) found that polynaptalene sulphonates (PNS) and polycarboxylate (PC)-based polymers improved the CPB rheology. Ercikdi et al. (2010) used the lignin, PNS, and PC-based water-reducing agents (WRA) and found that a slump of 17.8 cm (7 inches) can be achieved at the lowest dosage by using PC-based WRA as compared with the other two in CPB produced with sulfide-rich tailings. Guo et al. (2020) studied the effect of polycarboxylic acid and naphthalene serious WRA on the rheology of coal gangue-based cemented backfill material. The study revealed the reduction in yield stress and increase in plastic viscosity with the addition of WRA, and reasonable content was proposed to be 1.5% and 0.4% (by mass of binder) for naphthalene serious and polycarboxylic acid, respectively. Ouattara et al. (2017) studied the effect of superplasticizer (SP) on the rheology of CPB with a varying solid content and binder type. The results showed that the use of PC (46% solid concentration with a specific gravity of 1.10) at optimized dosage in blended cement (80% ground granulated blast slag furnace and 20% Portland cement (S-GU)) paste backfilling (75% solid content) showed better rheological properties. Improved rheological properties were observed in the case of superplasticized CPB having S-GU and GU-FA (50% PC and 50% class F fly ash) binder as compared with PC (GU) (Ouattara et al., 2018b). The S-GU binder seemed to be more compatible with PC-based HRWRs as compared with GU-FU and GU (Ouattara et al., 2018b). Panchal et al. (2018) concluded that the yield stress, plastic viscosity, and thixotropy of CPB (having carbonate-rich tailings) decreased by increasing the dosage of PC-based SP from 0% to 1%. Haruna and Fall (2020) investigated the influence of Master Glenium superplasticizer (carboxylic compounds) on the rheology of CPB made with different types of tailings. The study revealed that the addition of 0.125% (of total CPB weight) decreased plastic viscosity and yield stress by about 50% and suggested this amount as an optimum dosage. However, different SP dosages were required to obtain the same flowability when different types of tailings were used.

9.2.2.6 Temperature

When the fresh CPB is transported through the pipeline, regardless of the system used, that is, gravity, gravity/pump, or pump/gravity system, the heat is generated due to the friction of fresh CPB and the inner sidewall of the pipe. This happens due to the conversion of the kinetic energy of CPB into heat energy that leads to a temperature increase. Moreover, another source of temperature development in the CPB is the cement hydration while being transported deep underground opening cavities that take a relatively lengthy amount of transportation time (Wu, 2020). Moreover, in the low-temperature region, the backfilling process is carried out at sub-zero temperatures. Therefore, it is crucial to reveal the influence of high and sub-zero temperatures on the rheological properties of CPB.

Cheng et al. (2020) studied the effect of varying temperatures (from 5°C to 50°C) on the rheological characteristics of CPB. The study showed a gradual decrease in yield stress and plastic viscosity by increasing temperature. A greater number of floc network structures exist in a slurry that is stable to some extent. With an increase in temperature, the stable structure is destroyed, thus releasing free water that enhances the fluidity of the system. In this way, the fundamental rheological properties of CPB decreased. Moreover, the study observed that the thixotropy decreased by increasing temperature. However, Wu et al. (2013) observed an increase in yield stress and plastic viscosity of CPB by increasing temperature. By increasing the temperature of CPB, the rate of cement hydration is accelerated that leading to the generation of more hydration products. Due to an increase in hydration products, yield stress values of fresh CPB increase. A similar result was observed by Zhao et al. (2020). Moreover, they noticed that an increase in shear stress and apparent viscosity of CPB was observed to increase by increasing temperature from 2°C to 60°C. Both yield stress and plastic viscosity increased at elevated curing temperature, and a faster rate of gain of these properties was observed (Kou et al., 2020). As mentioned earlier, high temperatures increase hydration products and increase yield stress and plastic viscosity. Besides, the Bingham yield stress was noted to decrease from the temperature of 2°C to 20°C and then increased when the temperature rises further to 60°C (Wang et al., 2018).

Jiang et al. (2016) studied the yield stress of CPB at sub-zero temperatures (−1°C, −6°C, and −12°C). The yield stress at the studied temperature was much lower than that observed at room temperature. A decrease in the sub-zero curing temperature lowers the yield stress significantly. This is attributed to that at low temperatures, the early-age hydration of cement was inhibited and fewer hydration products were produced.

9.2.2.7 Other constituents

The effect of other constituents, including initial sulfate concentration, Minecem, flocculants, mined-process water, and alkali-activated cement (AAC) on the rheological properties of CPB, is briefly summarized as follows.

Kou et al. (2020) used AAC as a binder and investigated the rheological parameters with different binder dosage, curing time and temperature, slag fineness, and silicate modulus. Both yield stress and plastic viscosity increased by increasing curing time and temperature, binder content (up to 8%), and by lowering silicate modulus. The influence of silicate modulus on rheological properties is attributed to the higher negative zeta potential of solid particles and the lower pH value of pore solution at a higher silicate modulus. It was also observed that the rheological properties of AAC were more sensitive to the curing temperature as compared with OPC.

Peng et al. (2019) investigated the influence of different initial sulfate concentrations on the rheology of CPB. Lower yield stress and higher apparent viscosity were observed by increasing the dosage of initial sulfate concentration. Zhao et al. (2020) incorporated a new type of binder, Minecem (mainly a mixture of cement and slag), into CPB. The fundamental rheological properties were observed to be more sensitive to water content and temperature than CPB incorporating Minecem.

Dewatered tailings are required for shortening the time of preparing CPB. Therefore, some flocculants are used to accelerate the settling and dewatering rate of tailings. Flocculants bind the fine particles together to form dense aggregates that settle and dewater at a faster rate than individual particles. However, water remains entrapped in the flocs, creating a three-dimensional network with adjacent flocs which also retains water. This entrapped water is not fully expelled during the compressive force applied to the bed of the thickener, filter cake, etc. The network structure leads to higher yield stress. Rheology can be used to investigate the effect of shear on flocculated structures and can be used to identify and develop flocculants of desirable dewatering rates. Moreover, the rheological technique can be used to create flocs that are robust enough to remain intact during the dewatering phase, but then easily compress and/or sheared to facilitate the escape of inner and interfloc water and lowered the yield stress (Sofra, 2017). A better understanding of the rheological parameters of CPB-containing flocculants is crucial for better transportation to underground stopes. Xu et al. (2020) studied the rheological properties of CPB-containing flocculants (anionic-type polyacrylamide). The results showed higher initial yield stress of fresh CPB with flocculants. Besides, with the increase in shear rate, the apparent viscosity decreases sharply and then a steady increase is achieved in a late period. Moreover, increasing curing time and binder content increased the apparent viscosity of CPB with flocculants.

The addition of mine-processed water increased the yield stress of CPB, thus decreasing the fluidity. Moreover, the increase in the salinity in the pore fluid also increased the yield stress (Zhao et al., 2020).

9.3 RHEOLOGY OF FIBER-REINFORCED, CEMENT-BASED MATERIALS

9.3.1 Introduction

With the increase of compressive strength, concrete becomes more brittle, leading to catastrophic failures. This brittle behavior of concrete can be overcome by using different types of fibers, which stitches the micro and macro cracks and takes subsequent tensile strains acting on concrete. Moreover, adding fibers increases the toughness or energy-absorbing capacity of hardened concrete. In short, fiber reinforcement improves ductility, toughness, shear, and flexure strength; arrests the growth and coalescence of cracks; reduces shrinkage, cracking, and permeability; and enhances fatigue and impact resistance (Boulekbache et al., 2010).

Fibers have been applied in cement-based systems to replace rebar and enhance structural performance. The efficiency of fibers in improving the above properties depends on fibers' nature, quantity, aspect ratio, orientation, dispersion, and flowability of the fresh concrete (Zollo, 1997). As the shape of fibers is elongated, their effect depends on the orientation and position in the structure and relative to principal stresses (Grünewald, 2012). The efficiency of fibers in a hardened state depends on the optimized mixture composition and controlled rheological behavior. To achieve this, the interlocking and mechanical interaction of fibers should be minimized during the flow and production stage.

Though fiber reinforcement can improve the mechanical properties and durability of concrete, the inclusion of fibers can affect the fresh state properties of materials negatively, and in turn, can affect the efficiency of the reinforcement. Fibers can make concrete difficult to mix and place, possibly leading to excessive voids or poor fiber dispersion in the cement-based system. For the effective use of fiber, the influence on the fresh state or rheological properties of cement-based materials must be known.

Fibers with needle-like shapes increase flow resistance and play a part in the formation of an internal structure of fibers and aggregate grains. The fiber effect on the workability is mainly due to four reasons: first, fibers are more elongated in shape as compared to aggregate, and thus the surface area of fibers is higher than the aggregates at the same volume. Secondly, stiff fibers change the structure of the granular skeleton, and flexible fibers fill the space between them. Stiff fibers can push the particles with sizes relatively larger than the fiber length, thus increasing the granular skeleton porosity. Thirdly, the behavior of the surface of fibers is different from that of cement and aggregates, i.e., plastic fibers might be hydrophobic or hydrophilic. Finally, fibers can be deformed (i.e. crimped, hooked ends, or wave-shaped) for anchorage improvement between them and with surrounding matrix (Grünewald, 2012).

The incorporation of fibers in concrete changed the rheology negatively. The suspension with fibers is non-Newtonian fluid, which can display normal stress differences. Flow resistance arises due to the opposite movement of the particles. During the flow, the fibers rotate and orientate due to their slender shape. This motion is affected and/or counteracted by neighboring fibers/particles (Grünewald, 2012). At moderate to high fiber content, the use of plasticizing admixtures may be useful for obtaining desired rheological properties of concrete. At the higher dosage of fiber, higher paste volumes for the proper placing of fiber-reinforced concrete may be required. Therefore, a mixture like self-consolidating concrete may be used as a practical solution for placing fiber-reinforced concrete.

The rheological properties of fiber-reinforced concrete depend on various parameters, i.e., fiber content, fiber aspect ratio, fiber geometry, fiber type, wall effect generated by the geometry of formwork, interaction of different types of fibers, interaction of fibers with cement paste and aggregates, packing density, and maximum aggregate size. The influence of particle shape on the plastic viscosity of mortar follows the sequence of spherical particle < grain shape particle < tabular particle < needle-like particle.

Swamy and Mangat (1974) pointed out that once the critical fibers concentration is above a critical threshold, concrete could not flow anymore. This result holds even for concrete with high flowability, such as self-compacting concrete. Fibers form clumps or balls above this concentration. This critical volume fraction decreases by an increase in coarse particle volume fraction in the mixture. The decrease was also observed by increasing the aspect ratio. The value of fiber factor (product of aspect ratio and fiber volume fraction) was observed to be 0.2–2 when this critical concentration is reached (Martinie et al., 2010a). Depending on the mixture composition, Grünewald (2004) proposed this value between 0.3 and 1.9. Martinie et al. (2010a) proposed that the fiber factor should not exceed 4, because above this critical value fibers tend to form clumps or balls. Moreover, they postulated that fiber-reinforced concrete having a fiber factor below 3.2 should have less influence on the rheological properties of concrete. By increasing the fiber factor, the relative increase in the plastic viscosity was reported. The workability of concrete decreased by increasing the volume fraction of fibers and aspect ratio (Martinie et al., 2010a). Fibers contribute to the formation of internal structure and thus increase the yield stress (Grünewald, 2012).

Normally, no need to make changes in the mix design of conventional concrete when using fibers at low and moderate dosages, that is, up to approximately 18 kg/m^3 for steel fibers and 2.4 kg/m^3 for synthetic macro fibers (macro fibers refer to a fiber having diameters > 0.3 mm). At higher dosage, some adjustments are necessary for the mix design, depending on the fiber type. This includes increasing the dosage of a water-reducing agent to maintain workability without changing the water–cement ratio (w/c). When the dosage of fiber increases further, it is necessary to increase paste volume (cementitious materials) and incorporate more fine aggregates to ensure proper accommodation and dispersion of the fibers in the concrete (ACI 544.4R). However, the rheological parameters of fiber-reinforced cement paste increased by increasing sand content (Kuder et al., 2006).

Exceptional rheological properties can be obtained when blending normal grain-sized PC with ultra-fine cement (optimum content is 20% replacement), which makes the incorporation of fiber in cementitious materials possible (Kaufmann et al., 2004). Martinie et al. (2010a) proposed a simple mix design criterion for ultra-high-performance fiber-reinforced concrete and stated that fiber volume fraction can be increased by reducing the aspect ratio of fiber, and reducing the packing fraction of granular skeleton or using sand with a higher dense-packing fraction.

9.3.2 Influence of fiber on the rheology of FRCs

The effects of various factors on the rheology of fiber-reinforced cement-based materials are introduced in the following.

9.3.2.1 Fiber orientation

The rheology of concrete can be influenced by the orientation of fibers. It is commonly assumed that fibers can align along the flow direction, parallel to the anticipated stress, within a highly flowable cement-based system (Cao et al., 2016). The orientation of fibers in the matrix depends on many factors like fibers geometry and their interaction effects (cement paste–aggregates–fibers formwork), the flowability of concrete, and the concrete pouring and compacting process (Boulekbache et al., 2010). Due to the strain or flow of concrete, fibers can orient, resulting in the prediction of the structural behavior of fiber-reinforced cementitious materials more complex. The orientation and position of fibers in cementitious composites can deviate from the assumed isotropic distribution and orientation because of dynamic/static segregation or fibers floating, blocking of flow, and fibers orientation due to flow and wall effect (Grünewald, 2012). It was also observed that flexural strength was improved when fibers were oriented in the direction of tensile stress which is obtained when concrete is workable; on the other hand, the poor orientation of fibers decreased the workability of concrete and flexural strength (Boulekbache et al., 2010).

In flowable concrete with a low yield stress, fibers have high mobility within cement paste and can move and orient easily under light external vibration, while in stiff concrete, the risk of formation of fiber balls exists, which limits the movement of fibers and even blocks them (Boulekbache et al., 2010). The orientation of fibers in Newtonian fluid controlled by the yield stress and significant viscosity can, however, make the mix less homogeneous and affect the distribution of fibers (Sepehr et al., 2004).

Based on the mixture composition and rheological characteristics, cementitious materials with fibers can either be "rolling" or "flow as a plug". The former can happen in self-compacting concrete (with low yield stress), while the latter is a case of less-flowable concrete. Consequently, the flow pattern determines the orientation of fibers. Other parameters affecting the orientation of the fibers include slenderness, interaction with other particles, location of fiber (proximity to a wall), phase concentration, mixture composition, rheological characteristics, and the flow "history" during production. The closer the distance between fibers and wall, the more quickly they align in the flow direction.

The alignment of the fibers was affected by the wall effect and shearing body movement in a rheometer, so the rheological characteristics of fiber-reinforced concrete are also dependent on shear rate and shear strain (Alferes et al., 2016). It was pointed out that the distribution and orientation of fibers depend on the yield stress of fluid materials (Wang et al., 2017). A correlation between yield stress and orientation factor α ($\alpha = 0$, when fibers are not perpendicular to the cut surface: equivalent to concrete without fibers, and $\alpha = 1$ when all the fibers are perpendicular to the cut surface) was established. They observed that concrete with a yield stress of 36 and 45 Pa has an orientation factor value of 0.57 and 0.45, respectively, exhibiting the uniform distribution of fibers. On the other hand, for concrete with high yield stress (120 Pa), $\alpha = 0.3$, showing that fibers cluster and are badly dispersed (Boulekbache et al., 2010). The same authors observed a similar result on fiber orientation in fiber-reinforced, high-strength concrete (Boulekbache et al., 2012).

The fiber distribution in cement-based materials significantly affects its rheological properties, mechanical properties, and durability. The distribution of fibers depends on the size of the fibers relative to the aggregates. To enhance the efficiency of fibers in the hardened state, it is recommended to choose fibers longer than the maximum size of aggregates, normally, 2–4 times (Grünewald, 2012). The addition of fibers to a mixture that is not fluid or workable enough results in the fibers do not get mixed in a faster rate. Thus, they pile up on each other in the mixer and form balls (Boulekbache et al., 2012), and thus the properties of cement-based materials are deteriorated.

The fibers uniformly distributed without clumping or balling have a positive influence on the fluidity of fresh mortar (Kim et al., 2007). Wang et al. (2017) pointed out that moderate rheological properties of cement-based systems are necessary for the uniform dispersion of steel fibers in ultra-high-performance AAM concretes (UHPC). Too high plastic viscosity and yield stress can make fiber dispersion difficult during the mixing procedure, and too low rheological properties may cause remarkable segregation during the casting period. Li (2009) noted the poor fiber dispersion in ultra-high-performance, fiber-reinforced cementitious composite (UHPFRCC) when plastic viscosity falls below 5 Pa.s.

One way to obtain good fiber dispersion in the composite is high mixing energy by means of a high-speed mixer and longer mixing time (Choi et al., 2015). It is recommended to use the mixing speed of 10–12 rpm and the mixing time of 4–5 min (minimum 40 revolutions) after adding all the fibers to the trucks. For mixing in a central mixer, the mixing time and revolution rate are performed the same way as for plain concrete (ACI 544.4R).

9.3.2.2 Fiber length

Yield stress and plastic viscosity can be reduced by reducing fiber length (Vaxman et al., 1989). This can be due to the fact that longer fiber has more detrimental interference with aggregates and gives more resistance to flow than short fiber. Banfill et al. (2006) observed that both plastic viscosity and yield stress increased by increasing carbon fiber length. Zhang et al. (2019) used polypropylene fiber and concluded that yield stress and apparent viscosity gradually increased by increasing fiber length. Tattersall and Banfill (1983) pointed out that increasing fiber length mainly increased the yield value, but had little effect on the plastic viscosity. However, Ponikiewski (2011) noted that the length of steel fiber does not have a significant influence on the yield value and plastic viscosity on SCC, but in the case of hooked steel fiber, plastic viscosity is impaired by increasing fiber length. Long metallic fibers produced higher viscosity than short ones (Hossain et al., 2012).

Table 9.6 Fiber types and material characteristics (Grünewald, 2012)

Fiber type	Typical fiber diameter (μm)	Typical fiber length (mm)	Density (g/cm^3)	E-modulus (kN/mm^2)	Tensile strength (N/mm^2)	Elongation at break (%)
Steel fiber—hooked end	500–1300	30–60	7.85	160–210	>1000	3–4
Steel fiber—crimped	400	26–32	7.85	210	980	
AR-glass fiber	3–30	3–25	2.68–2.70	72–75	1500–1700	1.5–2.4
Polypropylene fiber—monofilament	18–22	6–18	0.91	4–18	320–560	8–20
Polypropylene fiber—fibrillated	50–100	6–19	0.91	3.5–10	320–400	5–15
Polyacrylnitrile fiber	18–104	4–24	1.18	15–20	330–530	6–20
Carbon fiber	5–10	6	1.6–2.0	150–450	2600–6300	0.4–1.6

9.3.2.3 Types of fiber

The most used fibers in the construction industry are steel fibers and polymeric synthetic fibers. Different types of fibers with their material characteristics are shown in Table 9.6.

9.3.2.3.1 Steel fiber

For long and stiff steel fibers (e.g., $L = 30$–$60\,\text{mm}$, $r = 45$–80), the mechanical interaction of aggregates and fibers dominates the flow behavior, probably due to the comparable surface area of these fiber types to the coarser sands. The addition of steel fibers affects the yield stress of cementitious composites but has minor effects on the characteristics of paste (Grünewald, 2012).

The yield stress and plastic viscosity of steel fiber-reinforced cement paste were observed to decrease until the critical volume fraction of fibers and then increase (Kuder et al., 2006). The stiff steel fibers might enhance the structural breakdown of cement paste during mixing, thus reducing yield stress and plastic viscosity initially, while increasing fiber dosage beyond a critical point could lead to mechanical interlocking and entangling of fibers that caused increased in both yield stress and plastic viscosity. According to Cao et al. (2018a), the addition of 2% steel fiber ($L = 13\,\text{mm}$, $D = 200\,\mu\text{m}$, and $r = 65$) gave a value of yield stress and plastic viscosity lower than that of plain mortar. Tattersall (1991) observed that both yield stress and plastic viscosity increased by increasing steel fiber content. Tattersall and Banfill (1983) also pointed out similar results. Figure 9.10 shows the effect of three types of steel fibers (the fibers are characterized by the aspect ratio/fiber length-content), and it clearly shows that fibers with different aspect ratios, lengths, and content increase plastic viscosity and yield stress as compared with the reference mixture. Choi et al. (2017) determined the effect of steel fiber ($\phi = 0.5$, $300\,\text{mm}$) on the rheology of wet-mix shotcrete. The torque viscosity markedly increased when a dosage of steel fibers > $50\,\text{kg/m}^3$ created the pumping problems during shotcreting. Filho et al. (2016) used hooked steel fibers ($L = 30\,\text{mm}$, $r = 47.6$) in SCC. They found that the addition of $20\,\text{kg/m}^3$ (0.25%) had no significant influence on the shear profile, but when fiber content goes above $80\,\text{kg/m}^3$ (1.02%), shear torque became proportional to the increase of fiber content. A direct relation between the growth of plastic viscosity and fiber increment was also observed.

9.3.2.3.2 Carbon fiber

Jiang et al. (2018) studied the effect of carbon fiber with a diameter of $7\,\mu\text{m}$ on the rheology of cement paste and concluded that yield stress increased by increasing fiber content,

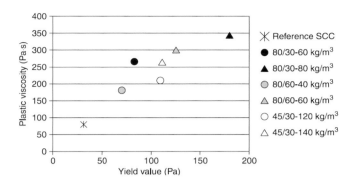

Figure 9.10 Increase of yield value and plastic viscosity of SCC with fibers. (Image obtained by Grünewald, 2012.)

but this increase was greater when using fiber of 6 mm length as compared to 3 mm length. The plastic viscosity also increases by increasing fiber content, and the plastic viscosity for 3 mm fiber (0%–1% fiber content) was greater than that of 6 mm fiber at the same dosage, but the effect was the opposite at 1.5% fiber dosage. Banfill et al. (2006) investigated the effect of carbon fibers ($V = 0.15$–0.5 vol.% and $L = 3$–12 mm) on the rheological properties of cement-based materials. It was found that both yield stress and plastic viscosity increased by increasing fiber content and fiber length, and by decreasing w/c.

Jiang et al. (2018) also studied the rheology of cement paste incorporating carbon nanofiber (CNF) and pointed out that CNF is sensitive to both yield stress and plastic viscosity, as dosage above 0.5% plays an obvious negative role in the rheology of cement paste. Meng and Khayat (2018) investigated the influence of CNF (0%–0.3%) on the rheological properties of UHPC and observed that plastic viscosity and yield stress first decreased and then increased and both give a minimum value when the content of CNF was 0.05%.

9.3.2.3.3 Cellulosic fiber

Kaci et al. (2011) studied the influence of cellulosic fibers (length = 1 mm and average diameter = 10 μm) on the rheology of mortar and concluded that below 0.55% fiber addition, the yield stress was only moderate dependent on this addition, but after this critical value, yield stress increases drastically. Moreover, the minimum viscosity was observed at 0.55% fiber addition at a high shear rate. Rapoport and Shah (2005) performed an experiment to correlate the dispersion of cellulose fiber (0.5% volume fraction) on the yield stress and plastic viscosity of cement paste and observed that the yield stress of cement paste and fiber dispersion do not influence each other. This means that for the given type and content of fiber, no correlation exists between fiber dispersion and matrix rheology. Moreover, they indicated that matrix viscosity does not influence the fiber dispersion.

9.3.2.3.4 Glass fiber

The yield stress of fresh mortar increases gradually by the addition of glass fiber. These fibers fill the gap between cement and sand particles, and adsorb free water in a mortar. Moreover, the agglomeration structure formed between fibers and grains increases the flow resistance. Thus, the addition of glass fibers results in higher yield stress. The addition of glass fiber also increases the equivalent plastic viscosity. The reason for this increase may be the friction forces among cementitious composites and fibers. In addition, it was reported that the incorporation of glass fiber increases the shear-thickening intensity of cement mortar (Jiao et al., 2019).

9.3.2.3.5 Other fibers

Calcium carbonate ($CaCO_3$) whiskers are inorganic crystals. These crystals have been used in cementitious materials as microscopic reinforcement to enhance toughness and reduce shrinkage. The length of the crystal ranges from 20 to 30 μm with a diameter of 0.5–2 μm and having an aspect ratio of 20–60. As compared with other types of fibers, like steel, glass, or PVA, $CaCO_3$ whiskers have a small particle size and large specific surface area. The addition of these fibers makes the paste more cohesive and hard to flow. Moreover, whiskers can easily agglomerate, thus reducing the reinforcing efficiency (Cao et al., 2016). Both yield stress and plastic viscosity increased by decreasing w/c and increasing $CaCO_3$ whisker (Cao et al., 2016).

Zhang et al. (2019) studied the influence of polypropylene fiber (equivalent diameter 0.033–0.048, length = 3, 6, and 12 mm) on the rheology of cement paste. They observed an increase in yield stress and apparent viscosity by increasing fiber length and dosage. According to their results, the incorporation of 2% short PVA fiber ($L = 6$ mm, and $D_{avg} = 14$ μm) in fiber-reinforced fly ash–geopolymer composites increased the initial yield stress from 12 to 67 kPa.

9.3.2.3.6 Hybrid fiber system

Ponikiewski and Szwabowski (2003) investigated the influence of steel, carbon, glass, and polypropylene fibers on the rheological properties of fresh mortars and pointed out that steel fiber volume fraction showed the significant influence on plastic viscosity, while volume fraction of the other three types of fibers had a significant influence on the yield stress. Cao et al. (2018a) performed an experiment to access the rheological properties of mortar by using steel fibers, PVA fibers, and $CaCO_3$ whiskers and concluded that yield stress and plastic viscosity increased by increasing the addition of PVA fiber for steel fiber and further $CaCO_3$ whisker for PVA fiber. This indicates that the addition of PVA fiber and $CaCO_3$ whisker influenced the flowability of mortars negatively. This may be due to the hydrophilic property of PVA fiber and the high specific surface area of $CaCO_3$ whisker. They also proposed that hybrid fiber-reinforced (1.75% steel fiber + 0.2% PVA fiber + 0.5% $CaCO_3$ whisker) cementitious materials possessed the best rheological properties. At lower fiber content (0.1%), both PVA and metallic fiber produced plastic viscosity lower than the control mix (0% fiber addition). Moreover, the yield stress of metallic fibers was higher than that of PVA at the specific fiber content. However, the increase of plastic viscosity by increasing PVA content is higher as compared to increasing metallic fibers content (Hossain et al., 2012).

9.3.3 Effect of fibers on the rheology of AAMs

Though AAMs have been shown to possess more compact microstructure and good mechanical properties than PCMs in numerous literature, large shrinkage property, particularly for sodium silicate-activated systems, remains a big challenge, particularly for potential application in aggressive environments. This is because that excessive shrinkage can cause microcracks. Therefore, in order to address this problem, adding fibers to AAMs becomes an approach and this leads to recent innovation of high-performance and ultra-high-performance AAM concretes (UHPC). As expected, the flowability of AAM-based UHPC decreased as steel fiber volume fractions increased (Liu et al., 2020).

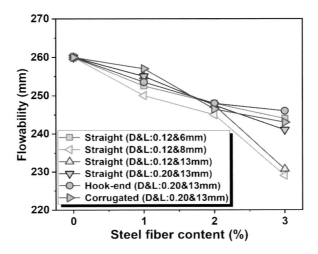

Figure 9.11 Flowability of fresh AAM-based UHPC with different steel fibers (Liu et al., 2020). Note: "D&L" means "diameter and length".

The mixtures with hooked-end fiber showed the best fluidity at high fiber content, while the effect of fiber shapes was negligible at low fiber content, as shown in Figure 9.11. Moreover, Liu et al. (2020) investigated the effects of various aspect ratios (length/diameter: 6/0.12, 8/0.12, 13/0.12, 13/0.2) of straight steel fiber on the flowability. They reported that the aspect ratio of 6/0.12 was the most optimal choice to ensure good flowability when the steel fiber volume fraction increased to 3% (Figure 9.12). The increase of the polyvinyl alcohol (PVA) fiber volume can also give rise to viscosity and yield stress of the AAS composite due to its poor dispersibility. Nevertheless, Choi et al. (2015) have successfully produced AAS-based composite with high ductility, low plastic viscosity, and yield stress, at the proportion of *w/b* of 0.4, superplasticizer of 0.27% and PVA fibers of 1.3 vol.%.

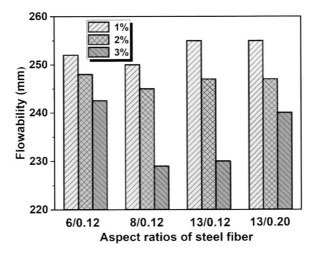

Figure 9.12 Flowability of fresh AAM-based UHPC with different aspect ratios of straight steel fibers (Liu et al., 2020).

9.3.4 Prediction of the yield stress of FRC

In the fresh state, fiber reinforcement can be defined as the suspension of solid particles in the fluid. Based on fiber volume fraction ϕ, some properties such as hydrodynamic or fiber–fiber interaction forces dominate the cement-based system. To design workable/flowable fiber-reinforced concrete, an important aspect to consider is that the fiber volume should be below the critical dosage. Three distinct regimes were proposed to correlate the inclusion of fibers and their influences on the flowability of cementitious composites. In the dilute regime ($\phi < 1/r^2$), where r is the aspect ratio (L/D, length divided by diameter), fibers can move freely without any interactions with the particles. In the case of a semi-dilute regime ($1/r^2 < \phi < 1/r$), some hydrodynamic interaction between fibers and some fibers' contact are possible. In dilute and semi-dilute regimes, fiber–fluid interaction is dominating. Finally, in the concentrated regime ($\phi > 1/r$), each fiber experiences contact with many neighbor fibers (Doi and Edwards, 1978). In order to predict the flow behavior of concentrated suspensions like cement-based systems, the relation between the phase volume and packing density should be considered (Grünewald, 2012).

According to Philipse (1996), for elongated particles ($r \gg 1$), the dense-packing fraction is defined as $\phi_{fm} = \alpha m/r$ and loose-packing fraction as $\phi_{fc} = \alpha_c/r$. The $\alpha_m = 4$ and $\alpha_c = 3.2$ were noted for random fiber orientation having the value of aspect ratio between 50 and 100. So, Martinie et al. (2010a) proposed three regions based on the degree of fiber suspension. In the case when $\phi < \phi_{fc}$, rheological behavior of the fiber suspension is close to the suspending fluid. While in the case of $\phi_{fc} < \phi < \phi_{fm}$, contact between the fibers happens that influences the rheological properties. Finally, cementitious composites are unable to flow when $\phi > \phi_{fm}$.

According to Martinie et al. (2010a) and Martinie and Roussel (2010b), after fibers' addition, for the yield stress to keep the same order of magnitude, and thus for concrete to display similar rheological behavior, the relative packing fraction should be lower than 0.8 $\left(\dfrac{V_f\left(\dfrac{L}{D}\right)}{4} + \dfrac{\phi_s}{\phi_m} \leq 0.8\right)$, as shown in Figure 9.13. A mixture having a relative packing fraction between 0.8 and 1 can be considered as optimized, while to obtain a high flowable mixture, the relative packing fraction should be limited to 0.8. They define a total relative volume fraction as the sum of the relative volume fraction of the fibers and the relative volume fraction of the granular skeleton $\left(\dfrac{V_f\left(\dfrac{L}{D}\right)}{4} + \dfrac{\phi_s}{\phi_m}\right)$, where ϕ_s and ϕ_m are the volume fraction and dense-packing fraction of sand (0.65 in the case of rounded sand), respectively. This proposed theory was confirmed by Boulekbache et al. (2012).

Wang et al. (2017) proposed that the optimal yield stress for fresh state UHPC ranged among 900–1000, 700–900, and 400–800 Pa for 1%, 2%, and 3% fiber volume fraction, respectively.

9.3.5 Prediction of plastic viscosity

The viscosity of suspension having spherical rigid particles in large concentrations including or excluding needle-shaped rigid steel fibers ($r \leq 85$, $V_f \leq 2$ vol.%) can be calculated by the following equation (Grünewald, 2012):

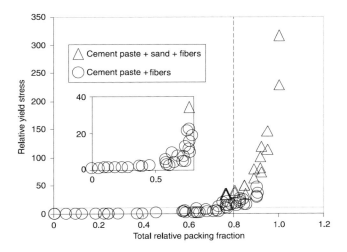

Figure 9.13 Relative yield stress as a function of the total packing fraction. The dashed line corresponds to the theoretical random loose packing (Martinie, et al., 2010a).

$$\eta_e = \eta \left\{ (1-\phi) + \frac{\pi \phi r^2}{3 \ln(2r)} \right\} \quad (9.1)$$

where η is the viscosity of the medium (Pa.s), ϕ is the volume fraction of fibers (vol.%), and η_e is the plastic viscosity of mix with fibers (Pa.s).

The local concentration of fibers (like fiber balling) and particle flocculation decrease the packing density and increase the viscosity (Grünewald, 2012). The critical volume fraction (ϕ_{cf}) for polymer melts ($r = 6\sim27$) can be predicted by the following equation:

$$\phi_{cf} = 0.54 - 0.0125(r) \quad (9.2)$$

9.4 SUMMARY

9.4.1 AAMs

The rheological behavior of AAMs can be described by different rheological models due to its complicated components. The type of activator seems to be the predominant factor controlling the determination of the rheological model. Generally, the Bingham model is recommended for NaOH-activated pastes, while the H–B model suits better the sodium silicate-activated pastes. Other component factors such as precursors and chemical admixtures also have important effects on rheological parameters. For example, Na silicate-activated FA pastes behaved like a Bingham fluid in some reports, while the rheological behavior of a fly ash-slag composite was more consistent with the H–B model.

The polymerization status of silicate species from activator solution and their adsorption mechanism on precursor surface plays an important role in affecting the rheological properties of AAMs. Furthermore, due to the different ion sizes, Na-based activators usually induce higher yield stress, higher viscosity, and faster viscoelastic evolution of suspensions compared to K-based activators. Silicates-based activators at proper dosages can have an excellent plasticizing effect. In general, there is a critical concentration of activators (modulus and $Na_2O\%$), which can reverse the trend of rheological properties by affecting the ion distribution and activation.

The physical properties of precursors affect their surface water absorption and packing density, whereas the chemical properties change the dissolution behavior of precursor particles and types of precipitated gels, and both altering the rheology of AAMs. In alkali-activated blended systems, the content of precursor with high specific surface area (such as MK) is usually the predominant factor that increases the yield stress and plastic viscosity.

Nowadays, most of the common superplasticizers for PC materials show lower dispersion efficiency in AAMs, which can be explained by three aspects: chemical stability in the extremely alkaline environment, the solubility in alkaline media, and the competitive adsorption between the anionic species of activators and anionic chemical admixtures. All these aspects depend on the rational design of polymers' molecular structure and charge. A few candidates show effectiveness for some specific systems: HPEG PCEs with high anionicity, high molecular weight, and short side chain length showed a good dispersion in high-calcium systems.

The addition of inert mineral additions, such as fine sand and limestone power, exhibited a positive effect on improving rheological properties due to the better particle dense packing, which can compensate for the negative effects of high specific surface area and nucleation effect. On the other hand, the impact of reactive mineral additions on the rheology of AAMs is complex, depending on their physicochemical characteristics and the water to binder ratio of suspension.

Although a great effort has been made in the field, there is still a large gap between recent progress and a full understanding of rheology of AAMs and the facilitation of engineering applications. In future research, the following concepts (concerns) can be considered:

More in-depth studies on the relationship between the composition factors and interparticle forces, as well as hydration kinetics in the AAMs should be conducted. This will provide a deeper understanding of the interaction mechanisms and a full picture of effects of various components.

Robust formulations and relevant chemicals to meet rheological and mechanical requirements of AAMs should be developed and engineered. For instance, the development of organic plasticizers tailored for AAMs or the appropriate addition of inorganic plasticizers to improve the poor workability of AAMs are required.

The rheology on AAM concrete needs to be further studied. There are some works on paste and mortar, but concrete study is limited, especially the effect of fibers and additives on rheology. Establishing empirical formulas to quantitatively describe the relationship between these materials and the rheological properties of AAMs is necessary, which promotes large-scale commercial applications of AAMs.

9.4.2 Cement paste backfilling

Cemented paste backfill (CPB), an engineered mix consisting typically of tailings, cement, and water, has become a regularly used structural material in most mines worldwide. To eliminate the tailings ponds and reduce the footprints of solid mining wastes, paste technology has been extensively discussed and applied. CPB pastes are also commonly applied in underground backfill operations to improve the stability of empty excavated areas. CPB pastes are often transported through pipelines and subsequently form a tailings stack in underground voids. High-density tailing slurries after thickening behaves as non-Newtonian flow and can flow over long distances before particle segregation occurs. Therefore, tailing pastes can be treated as homogenous and stable systems over long distances. The rheology of CPB is extremely important to paste technology. CPBs exhibit complex rheological behaviors such as yield stress and thixotropy. Time-dependent behavior of flocculated pastes or thickened tailings are getting extensively studied.

Although the rheology properties of CPB have been studied, there are still many aspects needed further investigation. The effect of chemical admixtures for CPB, such as superplasticizer and flocculation, on the rheology, pumpability, and stability needs in-depth research. New types of chemical admixtures which are less sensitive to the rheological properties are needed to be developed.

9.4.3 Fiber-reinforced, cement-based materials

Fiber reinforcement improves ductility, toughness, shear, and flexure strength; arrests the growth and coalescence of cracks; reduces shrinkage, cracking, and permeability; and enhances fatigue and impact resistance of cement-based system. The major drawback of fiber reinforcement is the negative influence of fibers on the rheological properties of cementitious composites.

The rheological properties of fiber-reinforced concrete depend on various parameters, i.e., fiber content, fiber aspect ratio, fiber geometry, fiber type, wall effect generated by the geometry of formwork, interaction of different types of fibers, mixture composition, interaction of fibers with cement paste and aggregates, packing density, and maximum aggregate size. There exists a critical value for fiber reinforcement, beyond which flowability and workability are severely deteriorated. The dosage of different types of fibers should be lower than the critical value for obtaining flowable mixture.

REFERENCES

Aboulayt, A., Jaafri, R., Samouh, H., et al. (2018). "Stability of a new geopolymer grout: Rheological and mechanical performances of metakaolin-fly ash binary mixtures." *Construction and Building Materials*, 181, 420–436.

Alferes, R., Motezuki, F., Romano, R., et al. (2016). "Evaluating the applicability of rheometry in steel fiber reinforced self-compacting concretes." *Revista IBRACON de Estruturas e Materiais*, 9, 969–988.

Alghamdi, H., Nair, S. A., and Neithalath, N. (2019). "Insights into material design, extrusion rheology, and properties of 3D-printable alkali-activated fly ash-based binders." *Materials & Design*, 167, 107634.

Alonso, M., Gismera, S., Blanco, M., et al. (2017). "Alkali-activated mortars: Workability and rheological behaviour." *Construction and Building Materials*, 145, 576–587.

Alonso, M., Palacios, M., Puertas, F., et al. (2007). "Effect of polycarboxylate admixture structure on cement paste rheology." *Construction and Building Materials*, 57, 65–81.

Banfill, P., Starrs, G., Derruau, G., et al. (2006). "Rheology of low carbon fibre content reinforced cement mortar." *Cement and Concrete Composites*, 28, 773–780.

Boulekbache, B., Hamrat, M., Chemrouk, M., et al. (2010). "Flowability of fibre-reinforced concrete and its effect on the mechanical properties of the material." *Construction and Building Materials*, 24, 1664–1671.

Boulekbache, B., Hamrat, M., Chemrouk, M., et al. (2012). "Influence of yield stress and compressive strength on direct shear behaviour of steel fibre-reinforced concrete." *Construction and Building Materials*, 27, 6–14.

Cao, M., Xu, L., and Zhang, C. (2016). "Rheology, fiber distribution and mechanical properties of calcium carbonate ($CaCO_3$) whisker reinforced cement mortar." *Composites Part A: Applied Science and Manufacturing*, 90, 662–669.

Cao, M., Xu, L., and Zhang, C. (2018a). "Rheological and mechanical properties of hybrid fiber reinforced cement mortar." *Construction and Building Materials*, 171, 736–742.

Cao, S., Yilmaz, E., and Song, W. (2018b). "Evaluation of viscosity, strength and microstructural properties of cemented tailings backfill." *Minerals*, 8, 352.

Cheng, H., Wu, S., Li, H., et al. (2020). "Influence of time and temperature on rheology and flow performance of cemented paste backfill." *Construction and Building Materials*, 231, 117117.

Choi, P., Yun, K.-K., and Yeon, J. H. (2017). "Effects of mineral admixtures and steel fiber on rheology, strength, and chloride ion penetration resistance characteristics of wet-mix shotcrete mixtures containing crushed aggregates." *Construction and Building Materials*, 142, 376–384.

Choi, S.-J., Choi, J.-I., Song, J.-K., et al. (2015). "Rheological and mechanical properties of fiber-reinforced alkali-activated composite." *Construction and Building Materials*, 96, 112–118.

Conte, T., and Plank, J. (2019). "Impact of molecular structure and composition of polycarboxylate comb polymers on the flow properties of alkali-activated slag." *Cement and Concrete Research*, 116, 95–101.

Criado, M., Palomo, A., Fernández-Jiménez, A., et al. (2009). "Alkali activated fly ash: Effect of admixtures on paste rheology." *Rheologica Acta*, 48, 447–455.

Dai, X., Aydın, S., Yardımcı, M. Y., et al. (2020). "Effects of activator properties and GGBFS/FA ratio on the structural build-up and rheology of AAC." *Cement and Concrete Research*, 138, 106253.

Deng, X., Klein, B., Hallbom, D., et al. (2018a). "Influence of particle size on the basic and time-dependent rheological behaviors of cemented paste backfill." *Journal of Materials Engineering and Performance*, 27, 3478–3487.

Deng, X., Klein, B., Tong, L., et al. (2018b). "Experimental study on the rheological behavior of ultra-fine cemented backfill." *Construction and Building Materials*, 158, 985–994.

Deng, X., Klein, B., Zhang, J., et al. (2018c). "Time-dependent rheological behaviour of cemented backfill mixture." *International Journal of Mining, Reclamation and Environment*, 32, 145–162.

Di, W., Cai, S.-J., and Huang, G. (2014). "Coupled effect of cement hydration and temperature on rheological properties of fresh cemented tailings backfill slurry." *Transactions of Nonferrous Metals Society*, 24, 2954–2963.

Doi, M., and Edwards, S. F. (1978). "Dynamics of rod-like macromolecules in concentrated solution: Molecular and Chemical Physics." *Journal of the Chemical Society*, 74, 560–570.

Duxson, P., Provis, J. L., Lukey, G. C., et al. (2007). "The role of inorganic polymer technology in the development of 'green concrete'." *Cement and Concrete Research*, 37, 1590–1597.

Ercikdi, B., Cihangir, F., Kesimal, A., et al. (2010). "Utilization of water-reducing admixtures in cemented paste backfill of sulphide-rich mill tailings." *Journal of Hazard Materials*, 179, 940–946.

Ercikdi, B., Kesimal, A., Cihangir, F., et al. (2009). "Cemented paste backfill of sulphide-rich tailings: Importance of binder type and dosage." *Cement and Concrete Composites*, 31, 268–274.

Gonzalez, P. L. L., Novais, R. M., Labrincha, J. A., et al. (2019). "Modifications of basic-oxygen-furnace slag microstructure and their effect on the rheology and the strength of alkali-activated binders." *Cement and Concrete Composites*, 97, 143–153.

Grünewald, S. (2004). *Performance-based design of self-compacting fibre reinforced concrete*, PhD Thesis, Delft University of Technology, The Netherlands.

Grünewald, S. (2012). *Fibre reinforcement and the rheology of concrete. Understanding the rheology of concrete*, edited by Roussel. Woodhead Publishing, Sawston, United Kingdom.

Güllü, H., Cevik, A., Al-Ezzi, K. M., et al. (2019). "On the rheology of using geopolymer for grouting: A comparative study with cement-based grout included fly ash and cold bonded fly ash." *Construction and Building Materials*, 196, 594–610.

Guo, X., Shi, H., Hu, W., et al. (2016). "Setting time and rheological properties of solid waste-based composite geopolymers." *Journal of Tongji University (Natural Science)*, 44, 1066–1070.

Guo, Y., Wang, P., Feng, G., et al. (2020). "Performance of coal gangue-based cemented backfill material modified by water-reducing agents." *Advances in Materials Science and Engineering*, 12, 1–11.

Habert, G., De Lacaillerie, J. D. E., and Roussel, N. (2011). "An environmental evaluation of geopolymer based concrete production: Reviewing current research trends." *Journal of Cleaner Production*, 19, 1229–1238.

Haruna, S., and Fall, M. (2020). "Time-and temperature-dependent rheological properties of cemented paste backfill that contains superplasticizer." *Power Technology*, 360, 731–740.

Hossain, K., Lachemi, M., Sammour, M., et al. (2012). "Influence of polyvinyl alcohol, steel, and hybrid fibers on fresh and rheological properties of self-consolidating concrete." *Journal of Materials in Civil Engineering*, 24, 1211–1220.

Hou, P., Muzenda, T. R., Li, Q., et al. (2021). "Mechanisms dominating thixotropy in limestone calcined clay cement (LC3)." *Cement and Concrete Research*, 140, 106316.

Huynh, L., Beattie, D., Fornasiero, D., et al. (2006). "Effect of polyphosphate and naphthalene sulfonate formaldehyde condensate on the rheological properties of dewatered tailings and cemented paste backfill." *Minerals Engineering*, 19, 28–36.

Jang, J. G., Lee, N., and Lee, H. K. (2014). "Fresh and hardened properties of alkali-activated fly ash/slag pastes with superplasticizers." *Construction and Building Materials*, 50, 169–176.

Jiang, H., Fall, M., and Cui, L. (2016). "Yield stress of cemented paste backfill in sub-zero environments: Experimental results." *Minerals Engineering*, 92, 141–150.

Jiang, S., Shan, B., Ouyang, J., et al. (2018). "Rheological properties of cementitious composites with nano/fiber fillers." *Construction and Building Materials*, 158, 786–800.

Jiao, D., Shi, C., Yuan, Q., et al. (2017). "Effect of constituents on rheological properties of fresh concrete-A review." *Cement and Concrete Research*, 83, 146–159.

Jiao, D., Shi, C., Yuan, Q., et al. (2019). "Effects of rotational shearing on rheological behavior of fresh mortar with short glass fiber." *Construction and Building Materials*, 203, 314–321.

Kaci, A., Bouras, R., Phan, V., et al. (2011). "Adhesive and rheological properties of fresh fibre-reinforced mortars." *Cement and Concrete Composites*, 33, 218–224.

Kashani, A., Provis, J. L., Qiao, G. G., et al. (2014). "The interrelationship between surface chemistry and rheology in alkali activated slag paste." *Construction and Building Materials*, 65, 583–591.

Kaufmann, J., Winnefeld, F., and Hesselbarth, D. (2004). "Effect of the addition of ultrafine cement and short fiber reinforcement on shrinkage, rheological and mechanical properties of Portland cement pastes." *Cement and Concrete Composites*, 26, 541–549.

Ke, X., Hou, H., Zhou, M., et al. (2015). "Effect of particle gradation on properties of fresh and hardened cemented paste backfill." *Construction and Building Materials*, 96, 378–382.

Kim, J.-K., Kim, J.-S., Ha, G. J., et al. (2007). "Tensile and fiber dispersion performance of ECC (engineered cementitious composites) produced with ground granulated blast furnace slag." *Cement and Concrete Research*, 37, 1096–1105.

Kou, Y., Jiang, H., Ren, L., et al. (2020). "Rheological properties of cemented paste backfill with alkali-activated slag." *Minerals*, 10, 288.

Kuder, K., Ozyurt, N., Mu, E., et al. (2006). "Rheology of fiber-reinforced cement systems using a custom built rheometer", in *Brittle matrix composites*, edited by Brandt, A. M., Li, V. C., and Marshall, I. H. Woodhead Publishing, Sawston, United Kingdom 8, 431–439.

Kuenzel, C., Li, L., Vandeperre, L., et al. (2014). "Influence of sand on the mechanical properties of metakaolin geopolymers." *Construction and Building Materials*, 66, 442–446.

Lang, L., Song, K.-I., Lao, D., et al. (2015). "Rheological properties of cemented tailing backfill and the construction of a prediction model." *Materials*, 8, 2076–2092.

Laskar, A. I., and Bhattacharjee, R. (2011). "Rheology of fly-ash-based geopolymer concrete." *ACI Materials*, 108, 536.

Laskar, A. I., and Bhattacharjee, R. (2013). "Effect of plasticizer and superplasticizer on rheology of fly-ash-based geopolymer concrete." *ACI Materials*, 110, 513–518

Lei, L., and Chan, H.-K. (2020). "Investigation into the molecular design and plasticizing effectiveness of HPEG-based polycarboxylate superplasticizers in alkali-activated slag." *Cement and Concrete Research*, 136, 106150.

Li, L., Lu, J.-X., Zhang, B., et al. (2020). "Rheology behavior of one-part alkali activated slag/glass powder (AASG) pastes." *Construction and Building Materials*, 258, 120381.

Li, M. (2009). *Multi-scale design for durable repair of concrete structures*, PhD Thesis. University of Michigan, United States.

Liu, Y., Zhang, Z., Shi, C., et al. (2020). "Development of ultra-high performance geopolymer concrete (UHPGC): Influence of steel fiber on mechanical properties." *Cement and Concrete Composites*, 112, 103670.

Luxán, M. P. D., Sotolongo, R., Dorrego, F., et al. (2000). "Characteristics of the slags produced in the fusion of scrap steel by electric arc furnace." *Cement and Concrete Research*, 30, 517–519.

Marchon, D., Sulser, U., Eberhardt, A., et al. (2013). "Molecular design of comb-shaped polycarboxylate dispersants for environmentally friendly concrete." *Soft Matter*, 9, 10719–10728.

Martinie, L., Rossi, P., and Roussel, N. (2010a). "Rheology of fiber reinforced cementitious materials: Classification and prediction." *Cement and Concrete Research*, 40, 226–234.

Martinie, L., and Roussel, N. (2010b). "Fiber-reinforced cementitious materials: From intrinsic isotropic behavior to fiber alignment", in *Design, production and placement of self-consolidating concrete*, edited by Khayat, K. H., and Feys, D. Springer, Montreal, Canada, 407–415.

Mehdizadeh, H., Kani, E. N., Sanchez, A. P., et al. (2018). "Rheology of activated phosphorus slag with lime and alkaline salts." *Cement and Concrete Research*, 113, 121–129.

Meng, W., and Khayat, K. H. (2018). "Effect of graphite nanoplatelets and carbon nanofibers on rheology, hydration, shrinkage, mechanical properties, and microstructure of UHPC." *Cement and Concrete Research*, 105, 64–71.

Montes, C., Zang, D., and Allouche, E. N. (2012). Rheological behavior of fly ash-based geopolymers with the addition of superplasticizers." *Journal of Sustainable Cement-Based Materials*, 1, 179–185.

Nägele, E., and Schneider, U. (1989). "The zeta-potential of blast furnace slag and fly ash." *Cement and Concrete Research*, 19, 811–820.

Newlands, K. C., Foss, M., Matchei, T., et al. (2017). "Early stage dissolution characteristics of aluminosilicate glasses with blast furnace slag-and fly-ash-like compositions." *Journal of the American Ceramic Society*, 100, 1941–1955.

Ouattara, D., Belem, T., Mbonimpa, M., et al. (2018a). "Effect of superplasticizers on the consistency and unconfined compressive strength of cemented paste backfills." *Construction and Building Materials*, 181, 59–72.

Ouattara, D., Mbonimpa, M., Yahia, A., et al. (2018b). "Assessment of rheological parameters of high density cemented paste backfill mixtures incorporating superplasticizers." *Construction and Building Materials*, 190, 294–307.

Ouattara, D., Yahia, A., Mbonimpa, M., et al. (2017). "Effects of superplasticizer on rheological properties of cemented paste backfills." *International Journal of Mineral Processing*, 161, 28–40.

Palacios, M., Alonso, M., Varga, C., et al. (2019). "Influence of the alkaline solution and temperature on the rheology and reactivity of alkali-activated fly ash pastes." *Cement and Concrete Composites*, 95, 277–284.

Palacios, M., Banfill, P. F. and Puertas, F. (2008). "Rheology and setting of alkali-activated slag pastes and mortars: Effect of organic admixture." *ACI Materials*, 105, 140.

Palacios, M., Bowen, P., Kappler, M., et al. (2012). "Repulsion forces of superplasticizers on ground granulated blast furnace slag in alkaline media, from AFM measurements to rheological properties." *Materiales de Construcción*, 62, 489–513.

Palacios, M., Gismera, S., Alonso, M. D. M., et al. (2021). "Early reactivity of sodium silicate-activated slag pastes and its impact on rheological properties." *Cement and Concrete Research*, 140, 106302.

Palacios, M., Houst, Y. F., Bowen, P., et al. (2009). "Adsorption of superplasticizer admixtures on alkali-activated slag pastes." *Cement and Concrete Research*, 39, 670–677.

Panchal, S., Deb, D., and Sreenivas, T. (2018). "Variability in rheology of cemented paste backfill with hydration age, binder and superplasticizer dosages." *Advanced Powder Technology*, 29, 2211–2220.

Papo, A., Piani, L., and Ricceri, R. (2010). "Rheological properties of very high-strength Portland cement pastes: Influence of very effective superplasticizers." *International Journal of Chemical Engineering*, 2010, 3–4.

Peng, X., Fall, M., and Haruna, S. (2019). "Sulphate induced changes of rheological properties of cemented paste backfill". *Minerals Engineering*, 141, 105849.

Philipse, A. P. (1996). "The random contact equation and its implications for (colloidal) rods in packings, suspensions, and anisotropic powders." *Langmuir*, 12, 1127–1133.

Ponikiewski, T. (2011). "The rheology of fresh steel fibre reinforced self-compacting mixtures." *ACEE Architecture Civil Engineering Environment*, 4, 65–72.

Ponikiewski, T., and Szwabowski, J. (2003). "The influence of selected composition factors on the rheological properties of fibre reinforced fresh mortar", in *Brittle Matrix Composites*, edited by Brandt, A. M., Li, V. C., and Marshall, I. H., Woodhead Publishing, Sawston, United Kingdom, 7, 321–329.

Poulesquen, A., Frizon, F., and Lambertin, D. (2011). "Rheological behavior of alkali-activated metakaolin during geopolymerization." *Journal of Non-Crystalline Solids*, 357, 3565–3571.

Puertas, F., González-Fonteboa, B., González-Taboada, I., et al. (2018). "Alkali-activated slag concrete: Fresh and hardened behaviour. *Cement and Concrete Composites*, 85, 22–31.

Puertas, F., Santos, H., Palacios, M., et al. (2005). "Polycarboxylate superplasticiser admixtures: Effect on hydration, microstructure and rheological behaviour in cement pastes." *Advances in Cement Research*, 17, 77–89.

Puertas, F., Varga, C., and Alonso, M. (2014). "Rheology of alkali-activated slag pastes. Effect of the nature and concentration of the activating solution." *Cement and Concrete Composites*, 53, 279–288.

Qi, C., and Fourie, A. (2019). "Cemented paste backfill for mineral tailings management: Review and future perspectives." *Minerals Engineering*, 144, 106025.

Rapoport, J. R., and Shah, S. P. (2005). "Cast-in-place cellulose fiber-reinforced cement paste, mortar, and concrete." *ACI Materials*, 102, 299.

Ren, J., Bai, Y., Earle, M. J., et al. (2013). "A preliminary study on the effect of separate addition of lignosulfonate superplasticiser and waterglass on the rheological behaviour of alkali-activated slags." *Third International Conference on Sustainable Construction Materials & Technologies (SCMT3)*, Kyoto, Japan, 1–11.

Rifaai, Y., Yahia, A., Mostafa, A., et al. (2019). "Rheology of fly ash-based geopolymer: Effect of NaOH concentration." *Construction and Building Materials*, 223, 583–594.

Romagnoli, M., Leonelli, C., Kamse, E., et al. (2012). "Rheology of geopolymer by DOE approach." *Construction and Building Materials*, 36, 251–258.

Rovnaník, P., Rovnanikova, P., Vyšvařil, M., et al. (2018). "Rheological properties and microstructure of binary waste red brick powder/metakaolin geopolymer." *Construction and Building Materials*, 188, 924–933.

Sepehr, M., Ausias, G., and Carreau, P. (2004). "Rheological properties of short fiber filled polypropylene in transient shear flow." *Journal of Non-Newtonian Fluid Mechanics*, 123, 19–32.

Shi, C., Jiménez, A. F., and Palomo, A. (2011). "New cements for the 21st century: The pursuit of an alternative to Portland cement." *Cement and Concrete Research*, 41, 750–763.

Simon, D., and Grabinsky, M. (2013). "Apparent yield stress measurement in cemented paste backfill." *International Journal of Mining, Reclamation and Environment*, 27, 231–256.

Sivakugan, N., Veenstra, R., and Naguleswaran, N. (2015). "Underground mine backfilling in Australia using paste fills and hydraulic fills." *International Journal of Geosynthetics and Ground Engineering*, 1, 1–7.

Sofra, F. (2017). *Rheological properties of fresh cemented paste tailings. Paste tailings management*, Edited by Yilmaz, E. and Fall, M. Springer, Switzerland.

Steins, P., Poulesquen, A., Diat, O., et al. (2012). "Structural evolution during geopolymerization from an early age to consolidated material." *Langmuir*, 28, 8502–8510.

Swamy, R., and Mangat, P. (1974). "Influence of fibre-aggregate interaction on some properties of steel fibre reinforced concrete." *Matériaux et Construction*, 7, 307–314.

Tattersall, G. H. (1991). *Workability and quality control of concrete*. CRC Press, Boca Raton, FL.

Tattersall, G. H., and Banfill, P. F. (1983). *The rheology of fresh concrete*, Pitman Books Ltd, London, England.

Uchikawa, H., Sawaki, D., and Hanehara, S. (1995). "Influence of kind and added timing of organic admixture on the composition, structure and property of fresh cement paste." *Cement and Concrete Research*, 25, 353–364.

Vance, K., Dakhane, A., Sant, G., et al. (2014). "Observations on the rheological response of alkali activated fly ash suspensions: The role of activator type and concentration." *Rheologica Acta*, 53, 843–855.

Vance, K., Kumar, A., Sant, G., et al. (2013). "The rheological properties of ternary binders containing Portland cement, limestone, and metakaolin or fly ash." *Cement and Concrete Research*, 52, 196–207.

Vaxman, A., Narkis, M., Siegmann, A., et al. (1989). "Short-fiber-reinforced thermoplastics. Part III: Effect of fiber length on rheological properties and fiber orientation." *Polymer Composites*, 10, 454–462.

Vyšvařil, M., Vejmelková, E., and Rovnaníková, P. Rheological and mechanical properties of alkali-activated brick powder based pastes: Effect of amount of alkali activator. IOP Conference Series: Materials Science and Engineering, IOP Publishing, 2018.

Wang, H., Yang, L., Li, H., et al. (2019). "Using coupled rheometer-FBRM to study rheological properties and microstructure of cemented Paste Backfill." *Advances in Materials Science and Engineering*, 2019, 6813929.

Wang, R., Gao, X., Huang, H., et al. (2017). "Influence of rheological properties of cement mortar on steel fiber distribution in UHPC. *Construction and Building Materials*, 144, 65–73.

Wang, Y.-S., Provis, J. L., and Dai, J.-G. (2018). "Role of soluble aluminum species in the activating solution for synthesis of silico-aluminophosphate geopolymers." *Cement and Concrete Composites*, 93, 186–195.

Westerholm, M., Lagerblad, B., Silfwerbrand, J., et al. (2008). "Influence of fine aggregate characteristics on the rheological properties of mortars." *Cement and Concrete Composites*, 30, 274–282.

Wetzel, A., and Middendorf, B. (2019). "Influence of silica fume on properties of fresh and hardened ultra-high performance concrete based on alkali-activated slag." *Cement and Concrete Composites*, 100, 53–59.

Wu, D., Fall, M., and Cai, S. (2013). "Coupling temperature, cement hydration and rheological behaviour of fresh cemented paste backfill." *Minerals Engineering*, 42, 76–87.

Wu, D. (2020). *Mine waste management in China: Recent development.* Springer, Singapore.

Xiang, J., Liu, L., Cui, X., et al. (2018). "Effect of limestone on rheological, shrinkage and mechanical properties of alkali–activated slag/fly ash grouting materials." *Construction and Building Materials*, 191, 1285–1292.

Xiang, J., Liu, L., Cui, X., et al. (2019). "Effect of Fuller-fine sand on rheological, drying shrinkage, and microstructural properties of metakaolin-based geopolymer grouting materials." *Cement and Concrete Composites*, 104, 103381.

Xie, F., Liu, Z., Zhang, D., et al. (2020). "Understanding the acting mechanism of NaOH adjusting the transformation of viscoelastic properties of alkali activated phosphorus slag." *Construction and Building Materials*, 257, 119488.

Xie, J., and Kayali, O. (2016). "Effect of superplasticiser on workability enhancement of Class F and Class C fly ash-based geopolymers." *Construction and Building Materials*, 122, 36–42.

Xu, L., Matalkah, F., Soroushian, P., et al. (2018). "Effects of citric acid on the rheology, hydration and strength development of alkali aluminosilicate cement." *Advances in Cement Research*, 30, 75–82.

Xu, W., Tian, M., and Li, Q. (2020). "Time-dependent rheological properties and mechanical performance of fresh cemented tailings backfill containing flocculants." *Minerals Engineering*, 145, 106064.

Yahia, A., and Khayat, K. (2001). "Analytical models for estimating yield stress of high-performance pseudoplastic grout." *Cement and Concrete Research*, 31, 731–738.

Yang, L., Wang, H., Li, H., et al. (2019a). "Effect of high mixing intensity on rheological properties of cemented paste backfill." *Minerals*, 9, 240.

Yang, L., Wang, H., Wu, A., et al. (2019b). "Shear thinning and thickening of cemented paste backfill." *Applied Rheology*, 29, 80–93.

Yang, S., Xing, X., Su, S., et al. (2020). "Experimental study on rheological properties and strength variation of high concentration cemented unclassified tailings backfill." *Advances in Materials Science and Engineering*, 2020, 6360131.

Yang, T., Zhu, H., Zhang, Z., et al. (2018). "Effect of fly ash microsphere on the rheology and microstructure of alkali-activated fly ash/slag pastes." *Cement and Concrete Research*, 109, 198–207.

Ye, H., Fu, C., and Lei, A. (2020). "Mitigating shrinkage of alkali-activated slag by polypropylene glycol with different molecular weights." *Construction and Building Materials*, 245, 118478.

Yin, S., Wu, A., Hu, K., et al. (2012). "The effect of solid components on the rheological and mechanical properties of cemented paste backfill." *Minerals Engineering*, 35, 61–66.

Yin, S., Guan, H., Hu, J., et al. (2019). "Rheological properties and fluidity of alkali-activated fly ash-slag grouting material." *Journal of South China University of Technology (Natural Science)*, 08, 120–128.

You, N., Liu, Y., Gu, D., et al. (2020). "Rheology, shrinkage and pore structure of alkali-activated slag-fly ash mortar incorporating copper slag as fine aggregate." *Construction and Building Materials*, 242, 118029.

Zhang, D.-W., Wang, D.-M., Lin, X.-Q., et al. (2018a). "The study of the structure rebuilding and yield stress of 3D printing geopolymer pastes." *Construction and Building Materials*, 184, 575–580.

Zhang, D.-W., Wang, D.-M., Liu, Z., et al. (2018b). "Rheology, agglomerate structure, and particle shape of fresh geopolymer pastes with different NaOH activators content." *Construction and Building Materials*, 187, 674–680.

Zhang, D.-W., Wang, D.-M., and Xie, F.-Z. (2019). "Microrheology of fresh geopolymer pastes with different NaOH amounts at room temperature." *Construction and Building Materials*, 207, 284–290.

Zhang, D.-W., Zhao, K.-F., Xie, F.-Z., et al. (2020). "Effect of water-binding ability of amorphous gel on the rheology of geopolymer fresh pastes with the different NaOH content at the early age." *Construction and Building Materials*, 261, 120529.

Zhao, Y., Taheri, A., Karakus, M., et al. (2020). "Effects of water content, water type and temperature on the rheological behaviour of slag-cement and fly ash-cement paste backfill." *International Journal of Environmental Science and Technology*, 30, 271–278.

Zollo, R. F. (1997). "Fiber-reinforced concrete: An overview after 30 years of development." *Cement and Concrete Composites*, 19, 107–122.

Chapter 10

Rheology and Pumping

10.1 INTRODUCTION

The concrete pumping technique is now one of the most common practices in the field of construction. This technique allows vertical and horizontal transportation of large quantities of materials to various locations, such as underground projects and high-rise building sites. Efficient placement of concrete is also a vital factor to ensure a construction project can be executed according to the schedule.

The pumping technique and its studies started decades ago, but the rapidly growing demands for the ultra-high structure and the ultra-deep substructure call for studies of longer pumping setups (Shah and Lomboy, 2014). Nowadays, the stunning Burj Kalifa project required concrete to be pumped 606 m high above the ground. At the same time, modern oil wells involve cement paste to be pumped as deep as 3000 m for fortification and safety purposes (ARTESIA). For such challenging pumping distance requirements, unless the changes in material properties can be clarified, unclear property changes due to the high pressure and long-term transportation doubt the placement efficiency and material stability. Therefore, the analysis of the effects of pumping on concrete properties should be further studied to meet with the fast development of modern concrete construction.

The scientific evaluation of concrete pumping performance, that is, the pumpability evaluation of concrete, is the basis of concrete pumping technology. Pumpability is a comprehensive concept, and there is no unified definition in the world. Moreover, with the progress of concrete technology and the increasing complexity of engineering structure, there are great differences in the performance between modern concrete and traditional concrete. A single evaluation index cannot adapt to the evaluation of the pumpability of modern concrete. To overcome the shortcomings of traditional evaluation methods and adapt to the complexity of modern concrete, researchers proposed to use a series of empirical indicators to evaluate the pumping performance of concrete, including pressure bleeding test, slump time, slump, and expansion. Although a series of indexes more comprehensively reflect the pumpability of concrete than a single index, they do not essentially characterize the pumpability of concrete and cannot be used to establish an accurate mathematical model to predict the pumping behavior of concrete. With the improvement of the elevation and distance of concrete pumping, it is more and more difficult to pump. The empirical series of indicators cannot meet the requirements of accurately predicting concrete pumping and controlling the quality of pumped concrete. For the super high-rise concrete pumping construction with certain technical difficulties, the construction unit mostly adopts the horizontal coil test to verify the pumpability of concrete, but this test is time-consuming and expensive. Therefore, rheology has become the main tool for researchers to study the working performance of fresh concrete. It has also been applied to predict the flow behavior of concrete in pump pipe under pressure and established the mathematical relationship between concrete rheological parameters, pump pipe diameter, pump pipe length, and concrete pumping pressure.

During the pumping process, the material undergoes various actions—strong shearing, high applied pressure, the temperature rises—which lasts from a few minutes to half an hour to cause the concrete flow and the fluctuation of material properties. Also, the pumping process can be very fast, so the maximum equivalent shear force exerted on the pumped concrete can be more than $35\,s^{-1}$ (Feys, 2019) and the maximum pumping pressure can reach tens of MPa (Jang and Choi, 2019). Each of these effects is large enough to induce the changes in concrete properties, which are associated with the time effect, which is a dependent variable of cement hydration, and a complex relationship with shearing and pressure effects. For a long-distance pumping job, such as the casting for an ultra-high-rise building with a moderate pumping rate, materials undergo these coupling effects for as much as half an hour, which is long enough to further influence the material property. Consequentially, the complex coupling effect of the pumping process leads to material changes that can be very hard to be described by a single equation or be explained by a simple mechanism.

What makes things complicated is the multi-phase nature, i.e., the granular fluid nature of concrete. The components of various modern concrete, such as self-compacting concrete and high-performance concrete, are even more intricate, when various supplementary cementitious materials and chemical admixtures are involved, which further adds to the complexity of the things. The contribution of each constituent to the material property can be altered by external conditions.

Some general trends have been observed in concrete properties after pumping, such as the increase in the yield stress and the decrease in the plastic viscosity (Feys et al., 2016a, Kwon et al., 2013a), but contradictory findings are reported, such as the variation in the air content (Jang et al., 2018, Secrieru et al., 2018a), making it necessary to sort out as well as to systematically analyze the data to clarify the mechanisms of the pumping-induced property changes.

10.2 CHARACTERIZATION OF PUMPABILITY

The movement of the concrete as fluid can be regarded as a hydrodynamic flow when the fluid is fully saturated (Browne and Bamforth, 1977, Feys et al., 2009b). In this case, the coarse aggregate inside the concrete mixture experiences little segregation and thus enjoys enough cement or mortar coverage. Under this situation, the pressure loss is linear to the length of the pipe, when the discharge rate remains constant, and the pipe has a straight configuration as well as a constant diameter. On the contrary, as a suspension, the concrete may contain less mortar or cement enough to lubricate coarse aggregates, leading to an "unsaturated" case, which makes the movement type of the fluid a frictional stress transfer (Feys et al., 2013). Under this circumstance, the straight-line relationship no longer exists, because frictional interactions of particles dominate and Coulomb's friction law explains the trend of pressure loss to the length of the pipe. Local pressure, instead of the overall pressure, of the fluid determines the pressure loss per unit length and the total pressure drops exponentially as the length increases (Coussot and Ancey, 1999).

10.2.1 Definition of pumpability

The definition of pumpability lacks universal standards, but it is well recognized that the term consists of two aspects: transportation without blockage and no significant alternation of material properties due to pumping (Jolin et al., 2006). Many countries have published

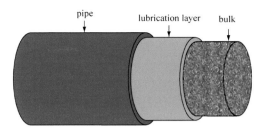

Figure 10.1 Schematic diagram of concrete flow in the pipe (Choi et al., 2014).

codes and standards to measure the pumpability of the concrete mixture. The first aspect of pumpability is well related to the level of saturation of the fluid, often related to coarse aggregate blockage, pressurized bleeding, or water mobility under pressure (Browne and Bamforth, 1977, Kaplan, 1999). On the other hand, the second part of the pumpability definition is also crucial for change in properties after pumping has been discovered and studied recently. However, detailed trends and mechanisms behind these effects are still under discussion. Multiple factors can affect the pumpability, including pipe dimensions, suspension fluidity, aggregate size distribution, and chemical admixtures.

10.2.2 Determination of pumpability

The most known equations describing the total pressure loss and the discharge rate are the Hagen–Poiseuille and Buckingham–Reiner equations, which are both derived from Newtonian and Bingham liquids. The shear that occurs inside the pipe can be obtained as follows:

$$\tau_w = \frac{\Delta p_{\text{tot}}}{L} \cdot \frac{R}{2} = \Delta p \cdot \frac{R}{2} \tag{10.1}$$

where τ_w is the shear stress, R is the distance from the center, and Δp_{tot} is the total pressure loss over distance L. According to Eq. (10.1), for CVC, there exists an area around the center of the pipe where the shear stress is less than the yield strength of the concrete mixture. This area is called the "plug", and its velocity is uniform (Feys et al., 2013). On the contrary, outside this plug zone, there exists a "lubrication layer (LL)" where the shear stress is higher than its yield strength, as shown in Figure 10.1. For some SCC, however, due to their low yield strength, it is possible to have a thin to none plug zone.

By integrating the velocity profile, discharge rate can be calculated, which is related to the pipe diameter, Newtonian viscosity, Bingham yield stress, and the Bingham plastic viscosity.

10.3 LUBRICATION LAYER

10.3.1 The formation of the lubrication layer

The formation of the LL results from the joint action of pumping pressure and pipe friction that creates a non-uniform shear field, and the shear force decreases linearly from the maximum value at the interface of pipe and material along the direction of pipe radius to zero at the center of the circle. The heterogeneity of concrete determines that the aggregate will move from the place with larger shear force to the place with smaller shear force, which

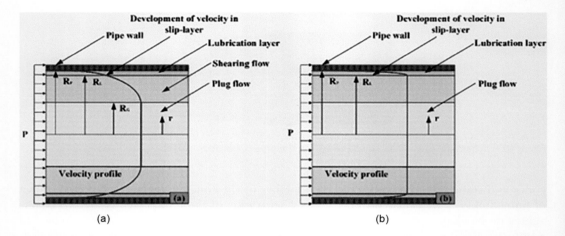

Figure 10.2 Flow rate profile of SCC (a) and CVC (b) in pumping pipeline (Yuan et al., 2018)

makes the slurry volume at the boundary higher than that of the main concrete, and then forms a LL. The formation of such a layer is well recognized by many researchers (Browne and Bamforth, 1977, Choi et al., 2013a, Jo et al., 2012, Morinaga, 1973, Phillips et al., 1992, Sakuta et al., 1979), whereas the determination of its properties and to which extent it influences the pumpability of the composite is still under debate. Sakuta et al. (1979) claimed that concrete pumpability is only related to the performance of LLs, rather than to the properties of the bulk concrete. However, the wide use of concrete with high fluidity, such as SCC, shadows question to this claim (Feys et al., 2014), due to the different flow rate profiles allowed, as shown in Figure 10.2.

10.3.2 The determination of the lubrication layer

Various instruments have been invented to quantify the properties of the LL, as it is paramount in determining the pumpability of concrete. There are mainly two types of instruments, i.e. the tribometer and the sliding pipe.

10.3.2.1 Tribometer

Different from the rheometer, a tribometer only measures the shearing conditions at the face of the rotating bob, according to the assumption that all shearing occurs in the lubrication layer, which holds for most concrete with moderate to low yield stress (Kaplan, 1999). Such an instrument was first invented by Kaplan by modifying the BTRheome, after which two other tribometers were introduced to improve the instrument's performance and reliability (Jolin et al., 2006, Ngo, 2009).

Despite these developments, the tribometer could not meet the needs of the industry till further improvement was made. With the development of concrete technology, an increasing number of high workability concrete (HWC) and SCC are employed for construction. However, the basic assumption of the previously described tribometers is not compatible with these highly flowable materials. This is because these tribometers assume a non-flowable bulk zone, while in reality, concrete with lower yield stress often has a sheared bulk zone, inside the LL but outside of the plug zone. A newly developed tribometer with a cone-bob addressed this problem, and the resulting viscosity agrees with the total flow resistance in the tribometer (Feys et al., 2014).

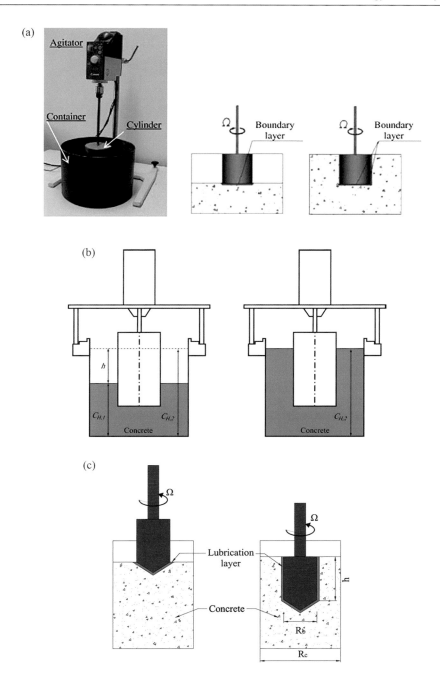

Figure 10.3 Diagram of various tribometers: (a) Ngo (Ngo et al., 2010), (b) Kwon (Kwon et al., 2013b), and (c) Feys (Feys et al., 2014).

10.3.2.2 Sliding pipe

Mechtcherine et al. (2014) invented the slide tube apparatus to evaluate the pumpability of concrete, as shown in Figure 10.4. The two parameters, a and b, measured by the slide tube tester correspond to the yield stress and viscosity constant of a lubricating layer in the tribometer.

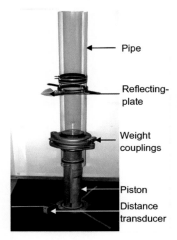

Figure 10.4 Sliding pipe designed by Mechtcherine et al. (2014).

The testing principle of the sliding pipe tester is to add concrete into the pipe, applying different weights on the upper part of the pipe to simulate the pumping pressure, to produce relative sliding between the pipe and concrete. The sensor is used to test the concrete flow rate under different weights. Then, the pumping prediction based on Kaplan is used to calculate the relationship between pumping pressure and pumping rate in the actual pumping process.

This method is similar to that of a tribometer, and it can stimulate the flow of concrete in a pump pipe more realistically. However, due to the limited length of the pipe, it is difficult to achieve the shear rate of concrete in the process of pumping. The sliding pipe apparatus only considers the case of ordinary concrete, that is, the shear zone in the concrete bulk is not included in its scientific base.

10.3.2.3 Other methods

Besides tribometers and sliding pipes, other apparatus and methods are employed to investigate the characteristics of LLs, including the ultrasonic velocity profiler (UVP) (Choi et al., 2013b), Particle Image Velocimetry (PIV), colored concrete method, and mortar method (Le et al., 2015). These methods have not only produced great results on the thickness, rheological properties, and composition of the LL, but also have their limitations in terms of cost or accurate quantification of their results.

10.4 PREDICTION OF PUMPING

The ease of pumping of the concrete is closely related to the rheological properties of both the bulk concrete and the lubricating layer. The prediction methodology of pumping pressure is based on the rheological properties of the bulk concrete and LL as well as the thickness of the LL (Choi et al., 2013a, Kaplan et al., 2005, Kwon et al., 2014). According to the rheological properties of concrete and its flow characteristics in pipelines, Kaplan proposed an analytical formula for calculating the pressure loss of concrete in horizontal pipelines (Kaplan et al., 2005). Kwon et al. (2014) derived an equation for the calculation of the flow rates by considering the thickness and rheological properties of the lubricating layer. Khatib (2013) proposed an analytical model for concrete that follows a Bingham flow behavior, where the rheological characteristics of the bulk concrete and the lubricating layer are taken into account. The model was developed by analyzing the velocity distribution of the fresh concrete located at various cross-sectional areas of the pipeline according to shear strain rates of the concrete located at different distances from the central axis of the pump pipe, which is used to calculate the flow rate during the pumping process. The analytical model was validated using full-scale pumping circuits and shows that the predicted pressure loss values are in good agreement with measured values.

10.4.1 Empirical model for pumping prediction

Starting from the fifties of the last century, empirical methods are discovered and used to provide a rough prediction of pressure losses during concrete pumping, and some pumping construction standards/codes still refer to these methods.

Based on the prediction diagram by Eckardstein, ACI recommended a method to predict the pumping pressure (ACI Committee 304, 1998). This method uses slump test to evaluate the pumpability of concrete, but it lacks a multi-lateral standard to evaluate the pumping quality and effect of pumping on the rheological properties of concrete. Therefore, several testing methods are used to facilitate the pumping evaluation procedure.

Furthermore, the technical code for construction of concrete pumping published by the Chinese Ministry of Housing and Urban-Rural Development recommends the Morinaga method for pressure loss prediction, as shown below:

$$\Delta P_H = \frac{2}{r}\left[K_1 + K_2\left(1 + \frac{t_2}{t_1}\right)V_2\right]a_2 \tag{10.2}$$

where

$$K_1 = 300 - S_1 \tag{10.3}$$

$$K_2 = 300 - S_2 \tag{10.4}$$

where ΔP_H is the pressure loss per meter within the pipe, Pa/m; r is the radius of the pipe, m; K_1 is the viscous coefficient, Pa; K_2 is the velocity coefficient, Pa-s/m; S_1 is the slump of the concrete, mm; t_2/t_1 is the ratio of switching time of concrete pump distribution valve to the time of piston pushing concrete which can be assumed to be 0.3 when the equipment is unknown; V_2 is the average velocity of the concrete mixture in the pipe, m/s; and a_2 is the ratio of radial pressure to axial pressure, which can be set to 0.9 for normal concrete (JGJ/T10–2011 Technical Specification for Construction of Concrete Pumping.

Last but not least, the pressure loss prediction method recommended by the construction guide for concrete pumping method of Japanese Architectural society is as follows:

$$P = P_v H + P_b(L + 3B + 2T + 2F) \tag{10.5}$$

$$P_v = 0.015 + 0.057\eta \tag{10.6}$$

$$P_b = 0.057 + 0.032\eta \tag{10.7}$$

where P_v is the pressure loss of vertical transport pipe, MPa; P_h is the pressure loss of horizontal transport pipe, MPa; H is pumping height, m; L is the length of the horizontal pipe, m; B is the bend length, m; T is the length of the conical tube, m; F is the hose length, m; and η is the plastic viscosity, Pa.s.

To sum up, the empirical formulas are put forward based on conventional vibrated concrete in the past decade. However, for highly flowable concrete such as self-compacting concrete, which is widely used in pumping engineering, the applicability of empirical methods is questionable because its rheological properties are different from those of conventional vibrated concrete.

10.4.2 Numerical model for pumping prediction

One of the most considered numerical models for pumping prediction was developed by Kaplan, linking the physical action of concrete transportation in cylindrical pipes to the rheological properties of the concrete.

To better predict the flow behavior under pumping, Kaplan raised two equations to calculate the pressure loss during pumping. Two scenarios are considered: the first is used for concrete with high yield strength when both plug and LL exists in the system, and the second scenario is for concrete-like self-consolidating concrete within which the LL is much thicker than CVC and a shear zone exists in the system. In the Kaplan model, the thickness of the LL is unknown, so a new parameter, viscous constant, is introduced to describe the ratio between shear stress and linear velocity of the fluid, in facilitating the description of the viscosity of the lubrication layer in the absence of a known value of the thickness (Kaplan 1999):

$$\Delta p = \frac{2}{R}\left(\frac{Q}{\pi R^2}\eta_{LL} + \tau_{0,LL}\right) \tag{10.8}$$

$$\Delta P = \frac{2}{R}\left(\frac{\dfrac{Q}{\pi R^2} - \dfrac{R}{4\mu_p}\tau_{0,LL} + \dfrac{R}{3\mu_p}\tau_{0,LL}}{1 + \dfrac{R}{4\mu_p}\eta_{LL}}\eta_{LL} + \tau_{0,LL}\right) \tag{10.9}$$

As the innovation of concrete design progresses, a growing number of high workability concrete has been utilized by the construction industry.

One of the mysteries of the Kaplan method avoided solving is the thickness of the LL. As there are very few affordable methods to quantify the exact thickness for each concrete design, the Kaplan method takes a detour to use the constant of viscosity for the LL instead of using the thickness and the viscosity of the LL. Distinct from it, Kwon et al. (2013a) provided another method describing the pressure loss of concrete transported in a pipe, assuming a universal LL thickness of 2 mm according to their previous measurements, and dividing the cross section of the mixture in the pipe into three zones, i.e. the plug, the LL, and the shear zone, which only exists when the shear stress is higher than the fluid's yield strength. The pumping rate prediction equation is as follows:

$$Q = 3600\frac{\pi}{24\mu_s\mu_P}\left[3\mu_P\Delta P\left(R_P^4 - R_L^4\right) - 8\tau_{S,0}\mu_P\left(R_P^3 - R_L^3\right) + 3\mu_S\Delta P\left(R_L^4 - R_G^4\right) - 8\tau_{P,0}\mu_S\left(R_L^3 - R_G^3\right)\right] \tag{10.10}$$

Moreover, as the recently often-used self-consolidating concrete is reported to have apparent thixotropic features, De Schryver and De Schutter (2020) offer an alternative dimensionless discharge formulation as shown below:

$$Q = \frac{\pi}{48}\left[\frac{\tau_0 R^3}{\mu}\right]\frac{1}{Pn^3}\Delta\begin{pmatrix} 6P_n^4 - 8P_n^3 + 2(1+\zeta^3)(1-3\zeta) + 6\varepsilon^3\sigma\Big|^R + 24\varepsilon^2\sigma\Big|^R + 3\varepsilon\sigma\Big|^R(4-\zeta^2) \\ -24\zeta^2\sigma\Big|^R - 8\sigma^3\Big|^R - 3\zeta^2(4+\zeta^2)\ln\left(\frac{\varepsilon+\sigma}{\zeta}\right)\Big|^R \end{pmatrix}$$

(10.11)

where τ_0 is the yield stress, R is the radius of the pipe, μ is the plastic viscosity, P_n is the pressure number, ζ is the dimensionless thixotropy addition, ε is the dimensionless corrected shear stress, and σ is the dimensionless thixotropically corrected shear stress. By introducing a thixotropy factor in determining the velocity changes along the cross-sectional direction of the flow, this extended Poiseuille flow formulation can simulate a concrete pumping flow that closely resembles the actual flow in the pipe, comparing methods that ignore the thixotropic feature of concrete. It is worth noting that no LL is assumed to form during pumping in this formulation because the aim of this is to discover the effect of thixotropy on the relation among the flow rates, the plug radius, wall shear rate, and pumping pressure loss. Certainly, this formulation could be extended to include more complicated pumping phenomena such as the formation of a thixotropic LL in the future.

10.4.3 Computer simulations for pressure loss predictions

Computer simulations are often used to calibrate and verify the effectiveness of the above numerical models. The Computational Fluid Dynamics (CFD) and the Discrete Element Method (DEM) are the most adopted ones. Besides, methods such as the Smoothed Particle Hydrodynamics (SPH) and the viscous granular material (VGM) models are also used.

Due to the importance of the LL in the pumping mechanism, several computer simulations are designed to predict the pumping performance. One of them is based on the Computational Fluid Dynamics (CFD) technique. Based on the hypothesis of shear-induced particle migration, the CFD approach produced flow rate and the velocity profile agreed with the results of large-scale pumping tests (Choi et al., 2013a). Another simulation using CFD produced a single-fluid numerical model for simulating Slipper tests, modeling both the plug zone and the LL based on Chateau–Ovarlez–Trung and Krieger–Dougherty models. The LL properties by this virtual Slipper simulation match those by laboratory experiments, especially for mixtures with lower water-to-cement ratios (Nerella and Mechtcherine, 2018).

Combing a Navier–Stokes solver, and a pressure gradient force model, a Discrete Element Method (DEM) is programmed to predict the flow problem in the concrete pumping process. The ability of DEM to model particle interactions and the kinematics of the discrete particles enables it to simulate not only fluid behavior in a straight pipe but also inside a bent or an elbow, where many analytical models or computer simulations cannot. This DEM analysis simulates the flow rate and the velocity profile well, and its result matches data from large-scale pumping tests (Tan, 2012).

10.5 EFFECT OF PUMPING ON THE FRESH PROPERTIES OF CONCRETE

10.5.1 Air content

The effect of pumping on the air content of fresh concrete lacks a unanimous trend due to the various mixture compositions, placement methods, and pumping conditions (Hover, 1989, Lessard et al., 1996, Yingling et al., 1992). Figure 10.5 shows the cases of horizontal pumping, in which a major portion of cases displayed a gain in the air content after pumping, but still, in a minor case, the air content of fresh concrete declined after pumping. For vertical pumping, most studies reported a loss in the air content (Lessard et al., 1996, Yingling et al., 1992). The complex mechanism during placement and the different use of the material may be the sources of the different results.

For horizontal pumping, the length of the pipe may be related to the degree of changes in the air content. A study by Jang et al. (2018) suggested a negative linear relationship between the horizontal pumping distance and the degree of changes in the air content (Figure 10.6, yellow triangle). The reason for this might be that while the vacuum effect is similar due to the end-hose configuration, the longer pumping distance requires a higher pressure, which is roughly equal to a larger pressure effect, so that concrete with a longer pumping distance experiences a much larger pressure effect, where the vacuum effect cannot count as much as it can for shorter pipes. Therefore, concrete in a longer pipe has a smaller or even negative gain in the air contents. The vacuum effect occurs when the fluid is propelled by pump pistons (Jolin et al., 2006) or by gravitational force in the case of a relatively slow vertical pumping (Elkey et al., 1994, Yingling et al., 1992). There is a negative pressure after the bulk concrete creates a vacuum effect, which increases the size of surviving air voids. These newly coalesced voids are more prone to bursting by consolidations and thus result in a loss of air throughout the entire mixture (Elkey et al., 1994, Vosahlik et al., 2018). The pressure effect is based on the theory that the internal pressure of air bubbles is inversely proportional

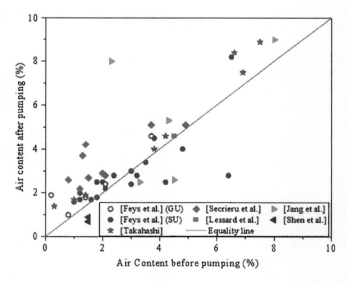

Figure 10.5 Air content for fresh concrete before and after pumping for horizontal circuits. (Data from Feys et al., 2016b, Jang et al., 2018, Lessard et al., 1996, Secrieru et al., 2018, Shen et al., 2021a, Takahashi, 2014.) (GU=Experiments done at Ghent University from Feys et al., 2016b; SU = Experiments done at Universite de Sherbrooke from Feys et al., 2016b) (Li et al., 2022).

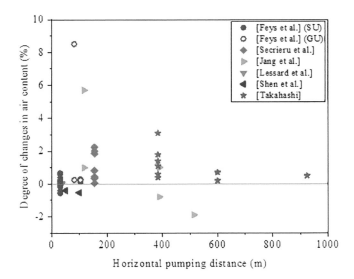

Figure 10.6 Degree of changes in air content concerning the horizontal pumping distance. (Data from Feys et al., 2016b, Jang et al., 2018, Lessard et al., 1996, Secrieru et al., 2018, Shen et al., 2021a, Takahashi, 2014), and the gray line is the indication line to illustrate the unchanging air content.) (Li et al., 2022).

to their sizes, with a sharp increase for bubbles smaller than 100 µm in diameter (Mielenz et al., 1958). According to the Henry's law, the concentration of the dissolved gas in solution at equilibrium is linearly proportional to the partial pressure of the gas, indicating that the smaller air bubbles are more prone to dissolution under pressure. After the concrete is expelled out of the pipe, the depressurization also affects both the air void distribution and the air content of mixtures. As soon as the mixture exits the pipe, there will be a sudden removal of pressure, creating a depressurization effect with the dissolved air escaping as a result. The number of bubbles will not increase much, but the size and interval between bubbles will be increased, which affects the overall air void system of the fresh concrete and has a detrimental effect on the freeze–thaw resistance after hardening, which will be discussed in later sections. On the other hand, the sudden loss of pressure increases the air content of fresh concrete. Whether such an increase will outweigh the decrease due to pressurization depends on the consolidation state of materials before measurements (Elkey et al., 1994).

However, due to the different material constituents and the end-hose configurations, there is no exact relationship between the pumping distance and changes in the air content in different studies.

Meanwhile, for vertical pumping, the placement method seems to be a paramount parameter determining the change in the air content by pumping. By comparing losses in the air content after the vertical pumping and the free falling in their laboratory, Yingling et al. (1992) concluded that the impact effect was a major factor in removing air during the vertical pumping (Figure 10.7). The impact effect hypothesis describes bubble escape caused by the physical impact between materials and external objects, including pipe elbows, formworks, and other items when concrete flows out from the pipe. It was reported that even a drop as shallow as 120 mm was enough to discover a visible drop in the air content (Elkey et al., 1994). However, there is not much knowledge about the size of a bubble that is more susceptible to such an impact effect, despite the buoyancy effects (which could be a possible explanation of this impact), which states that bigger bubbles are more likely to burst first.

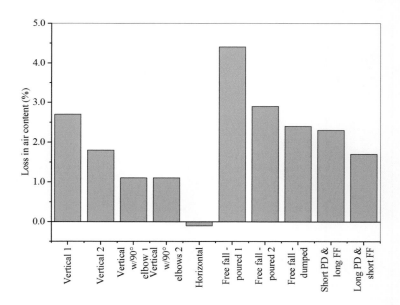

Figure 10.7 Change in the air content of the CVC after different discharge methods. PD=pipe dropping; FF=free falling. (Data from Yingling et al., 1992.) (Li et al., 2022).

Moreover, comparing vertical and horizontal pumping, Elkey et al. (1994) found that for conventional 5000 psi (34.47 MPa) concrete using fly ash (FA), an air-entrainment agent (AE), and a water-reducing agent (WRA), a 36 m vertical pumping slightly increased the air content by 0.25%, possibly due to a shift in the air bubble distribution toward smaller bubbles, which became easier to be detected by the pressure meter, or due to the consolidation of air after pumping but before the measurement. In contrast, Hover and Phares (1996) reported a 0.5%–2% reduction in the air content after an 18 m vertical pumping with an unconstrained free fall of 1.2 m and the tested ordinary concrete included a WRA and an AE. Additionally, Lessard et al. (1996) tested conventional concrete with a WRA and an AE and found that the horizontal pumping slightly increased the air by 0.1%, whereas vertical pumping (an approximately 20 m pipe plus a free fall of 0.15–0.3 m) reduced the air content by 0.9% and 0.5% when no ending reducer or a 0.9 m reducer was attached, respectively. An experiment by Vosahlik et al. (2018) revealed that both the horizontal and vertical pumping resulted in a 3%–4% drop in the air contents for both SCC and HPC.

In terms of the horizontal tests, two papers (Feys et al., 2009, Feys et al., 2016b) on SCC without AE revealed that horizontal pumping-induced changes in the air content were not constant. The only group with AE shows a slightly more pronounced decrease (0.7%) after horizontal pumping. However, one group with a different coarse aggregate reported a 1.2% rise in the air content, but the authors did not explain this phenomenon. It seems that higher initial air contents due to the addition of AE lead to a greater loss of the air contents after pumping, but the residual air is still much higher than concrete without AE. Researchers have performed similar tests on wet-mix shotcrete (Chen et al., 2019, Talukdar and Heere 2019), which was transported with a roughly horizontal short pipeline and sprayed onto the surface. In most cases, the air content decreased after pumping (Figure 10.8), and the decrease was the most significant for concrete with a high initial air content, and for those that included AE or WRA.

In addition to the transportation method and material constituents, the rheological properties affect the air bubble formation and stabilization (Du and Folliard, 2005). The movement of air bubbles can be inhibited by an increased viscosity due to viscous drag forces,

Rheology and Pumping 271

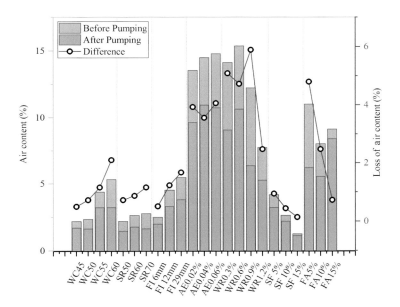

Figure 10.8 Effect of shotcrete on the air content with various mix designs (Chen et al., 2019). Lines connecting data points function only as visual guidance. (In this figure only: WC=water to bonding material ratio, SR=fine aggregate to total aggregate percent, FI=polypropylene fiber, AE = air-entraining agent, WR=water reducer, SF=silica fume, FA=fly ash) (Li et al., 2022).

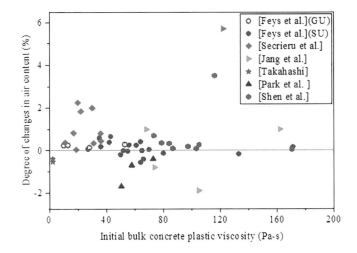

Figure 10.9 The initial viscosity appears versus changes in air content after pumping. (Data from Feys et al., 2016b, Jang et al., 2018, Park et al., 2010, Secrieru et al., 2018, Shen et al., 2021b, Takahashi, 2014.) (Li et al., 2022).

and the viscosity can limit the motion of air in two ways: the cushion effect, which absorbs shocks by outside disorder, and the barrier effect, which reduces the coalition of nearby air bubbles. The above relationship has been observed by Feys et al. (2016b), Jang et al. (2018), Park et al. (2010), Secrieru et al. (2018), and Takahashi (2014), as shown in Figure 10.9. In general, the increase in the initial concrete plastic viscosity is related to the degree of changes in the air content. The more viscous the fluid, the less likely that the air is released or gained during placement.

Additionally, an exploration of the effect of pressurization on fresh concrete using Portland cement was conducted, and it was determined that increasing pressure level reduces the total air content in the fresh concrete. However, for hardened concrete, the different initial air contents have different trends in the changes in the air content under varying pressure levels (Elkey et al., 1994) (Figure 10.10). When the exposed pressure is lower than 1 MPa (<150 psi), the air content increases regardless of the initial air amount. However, a noticeable drop in the air content occurs after 1 MPa in all groups, and at 2.1 MPa (300 psi), the contents again increase for groups with an initial total air of less than 5%.

Moreover, air content is influential to the yield stress, but the trend of change cannot be easily determined without the knowledge of the size of air bubbles inside the material (Feys et al., 2009, Secrieru et al., 2018). Generally, more small bubbles increase the yield stress for their ability to maintain spherical shapes during pumping, whereas larger ones are more prone to deformations and thus contribute to a decrease in overall material yield stress (Lessard et al., 1996). Based on the results of a 30 m-long pumping test (Feys et al., 2016b), as shown in Table 10.1, it seemed that although the addition of AE affects the initial rheological properties, it did not guarantee a pronounced change in yield stress.

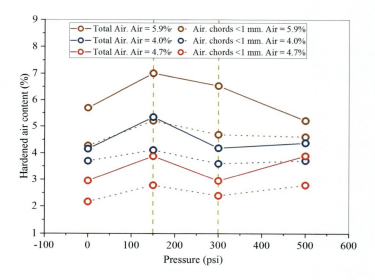

Figure 10.10 Various fresh air contents influence the hardened air content at different applied pressures (Elkey et al., 1994). (The lines connecting data points and the vertical lines function as visual guidance and do not have true physical meanings.) (Li et al., 2022).

Table 10.1 Changes in the rheological properties and air content (Feys et al., 2016b)

No.	Chemical admixture (exp. WRA)	Air content (%)			Concrete bulk yield stress (Pa)		
		Before	After	Change (%)	Before	After	Change (%)
SCC 10 (ref)		1.8	1.8	0.00%	24	41	70.8%
SCC 13 (AEA)	AE:0.67	6.5	8.2	26.2%	33	37	12.1%
SCC 14 (VMA[a])	VMA:0.05	3.8	4.5	18.4%	33	25	93.9%
SCC 15 (ref)		6.4	2.8	−56.3%	50	64	2.0%
SCC 18 (ref)		1.2	1.7	41.7%	36	48	133.3%

[a] VMA is the abbreviation of the viscosity-modifying agent.

10.5.2 Rheology

10.5.2.1 Yield stress and plastic viscosity

Many studies have been conducted to reveal the changing pattern of the yield stress and the plastic viscosity of concrete after pumping, as shown in Figure 10.11. The pump-induced rheological changes cannot be fully captured by rheometers during the property measurement process, as the shear rate in a pumping pipeline is much higher than that by laboratory devices, as suggested by Fataei et al. (2020). Therefore, full-scale pumping tests are needed to study this question. The pipeline length and the selection of material compositions of past studies were extensive, covering common situations faced in experiments and actual construction jobs. Since 2010, horizontal circuits ranging from 30 (Feys et al., 2016a) to 1000 m (Jang and Choi, 2019, Kwon et al., 2013a) have been used to test post-pumping rheological property changes. In addition, a wide range of cementitious components have been tested for the same purpose. The concrete tested had a w/cm ratio as low as 0.125 (Kwon et al., 2013a) and as high as 0.5 (Secrieru et al., 2018), and the supplementary cementitious materials used included silica fume, fly ashes, limestone filler, ground granulated blast furnace slag, and zirconia silica fume (Feys et al., 2016a, Kwon et al., 2013a, Park et al., 2020). Additionally, a large portion of the materials used contain a water reducer (Feys et al., 2016a, Shen et al., 2021b), sometimes as high as 3.54%, and some of the materials also contain air-entraining agents as well as a viscosity-modifying admixture (Feys et al., 2016a, Secrieru et al., 2018). These studies have drawn a clear pattern on the pump-induced changes of concrete rheological properties.

Pumped concrete was reported to be less viscous by most researchers (Feys et al., 2016a, Jang et al., 2018, Jang and Choi, 2019, Kwon et al., 2013a, Park et al., 2010, 2020, Secrieru et al., 2018, Shen et al., 2021b, Takahashi, 2014). Regardless of the material composition and pumping setup, the pumped concrete lost averaged 31.79% of its viscosity, and the degree of change in the viscosity increased as the initial viscosity increased. Due to the

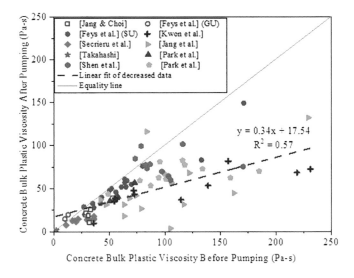

Figure 10.11 Comparison of the concrete bulk plastic viscosity before and after pumping. (Data from Shen et al. (2021b) are the initial tangential viscosity by the modified Bingham model instead of the plastic viscosity. Therefore, Shen et al. (2021b) is only included for references and is not included in the linearly fitted line.) (Data from Feys et al., 2016a, Jang et al., 2018, Jang and Choi, 2019, Kwon et al., 2013a, Park et al., 2010, 2020, Secrieru et al., 2018, Shen et al., 2021b, Takahashi, 2014.) (Li et al., 2022).

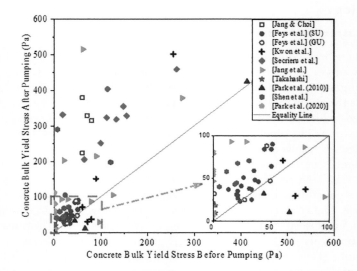

Figure 10.12 Comparison of the concrete bulk yield stress before and after pumping. (Data from Feys et al., 2016a, Jang et al., 2018, Jang and Choi, 2019, Kwon et al., 2013a, Park et al., 2010, 2020, Secrieru et al., 2018, Shen et al., 2021b, Takahashi 2014.) (Li et al., 2022).

difference in the material composition and the pumping setup, it is very difficult to conclude a mathematical function about viscosity changes that could fit every experimental result. Some studies, however, claimed that the degree of viscosity change was positively related to the length of the pipe (Jang and Choi 2019), or the discharge rate (Shen et al., 2021a), based on their results (Shen et al., 2021c).

Although exceptions exist, the yield stress generally increased after pumping (Figure 10.12). Regardless of the material compositions and the pumping setup, on average pumped concrete gained 264.73% of its yield stress, and the degree of change in the yield stress increased as the initial viscosity increased. The median degree of change in the yield stress was 96.15%, indicating a strong left-skewness of the data. No study had offered a changing trend for the yield stress due to pumping with the material composition, although some studies had related such a changing trend to the pumping setup (Jang and Choi, 2019).

The changes in the rheological properties are most often reported to be related to the discharge rate and the pumping distance. Feys et al. (2015) ran tests consisting of pumping the concrete at different flow rates in descending order. The study found that for concrete with low initial flowability, such as CVC, the descending discharge rate was related to the increased yield stress and the almost unchanging viscosity. However, for concrete with high initial flowability, such as HWC and SCC, the descending discharge rate was related to the decreasing yield stress, and the changing pattern of the viscosity was unclear. Such a difference might be related to the amount of the WRA because when the WRA is abundant, more cementitious particles can attach to the available agent, lowering the yield stress, even if the discharge rate is lowered. On the other hand, for the effect of the pumping distance on the post-pumping properties, Figure 10.13 shows the relationship between the pumping length and the plastic viscosity change of bulk concrete with data retrieved from eight full-scale horizontal loop experiments. Since not all of them used different pipe lengths within each experiment, only four groups of data could be compared under similar conditions. Among them, Jang and Choi (2019) showed a significant negative linear correlation. However, Park et al. (2020) and Jang et al. (2018) did not show significant correlations due to the different mixture proportions of concrete mixtures used in their tests, while Kwon et al. (2013) even had a positive correlation. If the mixture proportions are consistent, a better relationship between the pumping length

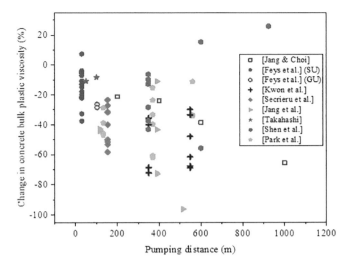

Figure 10.13 Pumping distance vs. the change in the concrete bulk plastic viscosity. (Data from Feys et al., 2016a, Jang et al., 2018, Jang and Choi, 2019, Kwon et al., 2013a, Park et al., 2020, Secrieru et al., 2018, Shen et al., 2021b, Takahashi, 2014.) (Li et al., 2022).

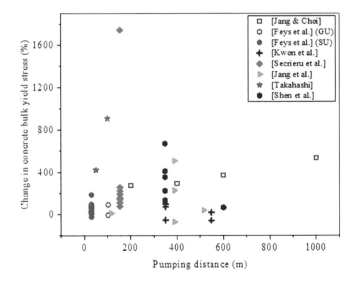

Figure 10.14 Pumping distance vs. the change in the concrete bulk yield stress. (Data from Feys et al., 2016a, Jang et al., 2018, Jang and Choi, 2019, Kwon et al., 2013a, Secrieru et al., 2018, Shen et al., 2021b, Takahashi 2014.) (Li et al., 2022).

and the plastic viscosity of mass concrete may be obtained. Unfortunately, there is only one paper that can fit this condition (Jang and Choi, 2019). Overall, the data of the eight experiments generally show a negative correlation between the pumping length and the loss of plastic viscosity of the bulk concrete, but we are not able to draw a convincing conclusion due to the different test conditions and the initial material properties.

Furthermore, Figure 10.14 is a diagram for the relationship between the pumping length and the yield stress change of the bulk concrete. Similar to this figure, only a few experiments tested samples with various pumping lengths. Among them, Jang and Choi (2019) showed an obvious positive linear correlation, and Kwon et al. (2013) showed some negative correlation. Jang et al. (2018) had no obvious correlation because the mixture proportions of each concrete used were different. Generally, data from all of these experiments show a positive correlation between the pump length and the plastic viscosity of large concrete, but it is difficult to conclude a mathematical function due to the difference between the test conditions and the initial material properties.

The experimental data from Jang and Choi (2019) show that the applied pressure can also have an important influence on the variation in the concrete bulk yield stress and plastic viscosity. A group of concretes with a w/cm ratio of 0.33 were tested with different levels of pressure in the horizontal pumping circuit, and the resulting rheological variations are presented in Figure 10.15.

The exact mechanism for the changes in the rheological properties due to pumping does not have a universal conclusion. Many researchers have provided partial assumptions and explanations, including the hydration of cement, the reverse of thixotropy, water absorption by aggregates, and dispersion of chemical admixtures (Feys et al., 2016a, Jang et al., 2018). However, there is a lack of a comprehensive analysis on all the possible mechanisms for this issue, so that this part provides a comparative study on the effect related to material constituents on the changes in rheological properties.

It is worth noting that due to the limited number of full-scale pumping tests and differences in the circuit setup, material design, and rheological apparatus, it is difficult to perform a straightforward across-the-board comparison. Different pumping setups, i.e., different pipe configurations, mixture compositions, and placement rates, can affect the pressure (Kaplan et al., 2005) and shear experienced by the material (Feys, 2019), which in turn affects the actual physical and chemical phenomena inducing the property changes.

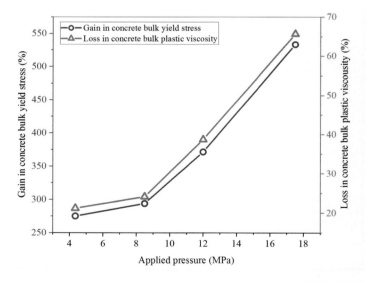

Figure 10.15 The variations in the concrete bulk yield stress and plastic viscosity are related to the applied pressure for a group of concretes with a w/c ratio of 0.33, 10% FA, and 45% blast furnace slag (BFS), as well as an addition of 0.7% WRA. The horizontal pumping distances varied from 200 to 1000 m. (Data from Jang and Choi 2019; the straight lines are visual guidance of the presented data points.) (Li et al., 2022).

Moreover, the observation results of such changes can vary due to different measurement schemes, although such differences can be reduced if the same measurement geometries are adopted and careful conversion schemes are applied, as suggested by Haist et al. (2020). Unfortunately, studies on the post-pumping characteristic changes often use rheometers of different types, and the conversion schemes suggested by Haist et al. (2020), which require certain geometrical boundary conditions to be fulfilled or involve a comparison of the measurement results to a reference substance, are not provided. Therefore, a direct numerical comparison of rheological measurements should be made very carefully. In addition, for concrete with a low yield stress, such as SCC, due to the use of the Bingham model, the initial yield stress of such concrete is close to 0, or even negative. However, when such a situation occurs, it is impossible to predict the pumping pressure loss based on the often-used prediction equations (Kaplan et al., 2005, Kwon et al., 2013a). Therefore, different authors used different initial yield stress approximations (mostly 0.1 or 1 Pa) for this case. Unfortunately, the yield stress of low-yield-stress concrete after pumping is generally much greater than 0 (mostly tens of Pa). In this way, the rate of change before and after pumping depends on the initial yield stress used, which has little practical engineering significance. Therefore, for concrete with high flowability, a more usable rheological model and, most importantly, a prediction method utilizing such a model need to be developed to address this issue. Moreover, for this review, if the initial yield stress of concrete is close to 0, it will not be included in discussions about the degree percentage of change in the yield stress for the above reasons.

As a vital part of concrete, cement paste affects the rheological properties through hydration and thixotropy during and after pumping. The water-to-cementitious material (w/cm) ratio and the binder composition are the most critical parameters related to the two effects. Unfortunately, distinctive setting times (from a few minutes to 2–3 hours) and pumping distances (from 30 meters to kilometers) of materials were used by different papers, making it difficult to directly compare the w/cm ratio or binder composition to show an obvious relationship to the rheology changes.

Hydration, as a process influencing both how densely particles are packed and how strong interparticle forces are, has a nontrivial influence on the rheological characteristics of concrete. During hydration, both the yield stress and the plastic viscosity will rise due to different mechanisms (Talero et al., 2017, Tattersall and Banfill, 1983). The hydration process encourages the growth of the minimum time-dependent viscosity due to the aggregation of particles (Otsubo et al., 1980). However, the increased number and strength of interparticle bonds fortify the microstructure flocculation that, at a macroscopic level, will be observed as increased yield stress (Struble and Lei, 1995). Uchikawa et al. (1985) noted that for cement paste, the development of the yield value within the first hours of hydration is characterized as the initial setting, which is controlled by the formation of ettringite and the hydration of calcium sulfate hemihydrate. Cement pastes with a w/c ratio of 0.3 show a 0.50–1.17 Pa/min gain in the yield stress in the first hour of setting, depending on the composition and features of clinkers. In addition, Struble and Lei (1995) found that the yield stress of cement paste ($w/c = 0.45$) only increased slowly until the hydration time reached 120 min, after which the growth in the yield stress rose rapidly in an exponential function. Such a critical time will be delayed as the w/c ratio increases. During the dormant period, the yield stress increases at approximately 0.42 Pa/min for cement paste. This conclusion coincides with Yang and Jennings, who reported that cement clusters remain unhydrated for at least 5 h if there is a lack of proper mixing and that the rheology of cement pastes in the first 2 h could be largely controlled by the mixing methods (Yang and Jennings, 1995). Moreover, for the changes in viscosity, Banfill et al. (1991) observed a moderate viscosity increase in the first hour of hydration on a class G oil well cement paste whose w/c ratio was 0.4.

On the other hand, although the duration of pumping is typically very short, the process is proven to accelerate cement hydration. The combination of accelerated hydration and better particle dispersion during pumping, which dismantles interparticle bonding and further fosters hydration, may provide part of the explanation for the increase in the yield stress after pumping. Shen et al. (2021a) found that after being pumped horizontally, the hydration process was accelerated, within which the dormant period and the setting period were shortened by 10.3%–27.3% and 15.6%–28.4%, respectively. A similar conclusion was reached for the mixing of cement pastes (Han and Ferron, 2016, Juilland et al., 2012, Saleh and Teodoriu, 2017).

Many studies have also found an obvious rise in temperature during pumping (Jang et al., 2018, Petit et al., 2008, Secrieru et al., 2018), which is believed to encourage hydration and thus increase the yield stress of fresh cementitious materials. Secrieru et al. (2018) found that after 154 m of pumping, ordinary concrete and HPC increased by 3°K and 8°K, respectively. Although it is difficult to distinguish the contribution of hydration and friction to the temperature rise, they concluded that the hotter material accelerated cement hydration and thus raised the yield stress. Future comparative studies should be performed to identify the causal relationship between the temperature rise and hydration acceleration due to pumping, as well as the quantitative contribution of temperature rise due to hydration and friction.

Additionally, numerical simulations have been used extensively on the influencing factors of the pumping pressure prediction and the accurate prediction of the concrete flow behavior during pumping (Fataei et al., 2021, De Schryver et al., 2021, De Schryver and De Schutter, 2020, Tavangar et al., 2022, Xu et al., 2022). The results of some studies can also demonstrate the effect of pumping on the concrete rheological properties using a computer simulation method (Fataei et al., 2021, Tavangar et al., 2022). Fataei et al. (2021) found that for shear-thickening concrete, such as SCC, the higher discharge rate causes an increase in the viscosity of the lubricating layer, increasing the particle diffusion toward the plug zone. In contrast, for CVC, the apparent viscosity in the lubricating layer is almost constant at the high shear rates reached in the lubricating layer. In addition, Tavangar et al. (2022) proved that a higher flow rate during pumping causes a thicker LL, indicating that the more intensive shearing induces more particle migrations, drives fewer coarse particles in the LL, and leaves a denser plug zone. In addition, Xu et al. (2022) used the moving particle semi-implicit method and found that the pumping process could cause as much as 1% variation in the volume fraction of coarse aggregate. As a result, the rheological properties of the concrete are changed. The particle migration induced is thus related to the change in the material properties during pumping.

10.5.2.2 Thixotropy

Thixotropy is a time-response of the microstructure related to the local spatial rearrangement of the material (Barnes, 1997). All concrete can show thixotropy, and this evolution can be reversed by shearing, such as by strong mixing or pumping (Roussel et al., 2012). Roussel et al. (2012) suggested that at least two types of thixotropy exist, namely, short-term thixotropy (approximately a few seconds), related to colloidal flocculation, and long-term thixotropy (approximately 20 min), due to early hydration. The calcium silicate hydrate (C–S–H) gel is the main source of the latter type of thixotropy, and thus it is also related to the acceleration of hydration due to, for example, the addition of binders with strong nucleating properties or an increase in temperature (Roussel et al., 2012). In addition, the interparticle colloidal force is another source of two types of thixotropy (Abebe and Lohaus, 2017).

The thixotropic nature of concrete (Barnes, 1997, Mewis and Wagner, 2009, Rahman et al., 2014) may be a reason for shear-induced changes in rheology (Lowke, 2018). The concrete may be affected by thixotropy both before and during pumping, which may lead to a change in its workability, especially for SCC (Feys and Asghari, 2019, Han and Ferron, 2016, De Schryver and De Schutter, 2020). Moreover, a horizontal pumping test for a CVC of C30/37 using FA and a naphthalene-based plasticizer also proved that the thixotropic effect occurred shortly after the pumping process was paused for 20 min (De Schutter 2017). A retarder was used to eliminate the influence of hydration during the experiment.

According to Lapasin et al. (1979), Li et al. (2018), and Lowke (2018), the thixotropic behavior of SCC is related to the *w/cm* ratio, the maximum packing density of the mixture, and the superplasticizer dosage, and the former two factors are contributed by the cement paste. The reverse of the thixotropy effect was also observed during pure shearing (Feys and Asghari, 2019) or pressure (De Schutter, 2017). Thixotropy causes reversible physical or chemical bonds between particles and thus influences rheological behaviors (Barnes, 1997). Both physical bonds and chemical bonds can tie cementitious particles together. Regardless of the distinct causalities of these bonds, they may both be broken down by shearing forces (Chateau, 2012). Some connections can be disconnected when a higher shear is introduced. This explains why even if a premixture is often applied before pumping, the pumping process can still cause dispersion and redispersion of particles, thereby resulting in concrete having different rheological properties. The dispersion and deflocculation of cement particles subsequently lead to increased surface areas, and hence an increase in the early hydration rate and extent. The thixotropic effect also explains why different *w/cm* ratios can affect the effectiveness of particle dispersion by shearing (Feys et al., 2016a). A lower *w/cm* ratio implies a higher relative particle fraction, which increases the interparticle forces, requiring even more shearing to break the connections.

A hypothesis verified by Kim et al. (2017) states that pressure on the concrete mixture can encourage particle dispersion to reverse the thixotropy effect, which changes the rheological properties of fresh concrete. Ouchi and Sakue (2008) first raised the hypothesis that particle dispersion by pressure acting on the concrete would create a larger total surface area of cement particles, which could release water and cause better adsorption of the WRA. By specifically studying the cement paste under high pressure, since cement pastes can be representative of the LL in the pumped concrete, Kim et al. (2017) engineered a testing process to isolate the two contributing factors during the actual pumping process: the pressure and the shearing action induced by the velocity gradient. Kim et al. (2017) discovered that the pressure effect on the rheological properties was obvious on cement pastes with a *w/cm* value of less than 0.40, but small on those with higher *w/cm* values. This coincides with Feys et al. (2016a), which might be explained by the same mechanism where a lower *w/cm* requires higher shearing energy to disperse the interparticle forces.

In addition, experimental data from Jang and Choi (2019) showed that the applied pressure also had a great influence on the variations in the yield stress and plastic viscosity of the bulk concrete. A group of concretes with a w/c ratio of 0.33, a 10% FA, and a 45% blast furnace slag (BFS) as well as a 0.7% superplasticizer addition were tested with different pressure levels in the horizontal pumping circuit, with the resulting rheological variations presented in Figure 10.16. However, the experimental results from the changes in the rheology of cement paste under pressure by Kim et al. (2017) argued that when the w/cm ratio is below 0.40 and the applied pressure is below 20 MPa, which was the case in the study of Jang and Choi (2019), both the yield stress and the plastic viscosity decreased. After pressurization, the cement pastes were subjected to shear, after which both properties also decreased. The effect of hydration was minimal in this experiment. According to Jang and Choi (2019) and Kim et al. (2017), even if it was related to the applied pressure, the gain of

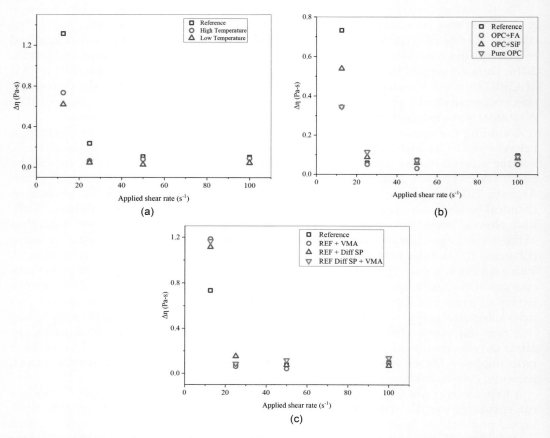

Figure 10.16 Change in the apparent viscosity with the applied shear rate (Feys and Asghari, 2019).

bulk yield stress may consequentially be more relevant to encouraged hydration that has an effect preceding the adverse effect by the reverse of thixotropy.

It was noted that for a well-dispersed concrete under a constant mixing speed, the temperature, chemical compositions of binder, and chemical admixtures were the reasons for the change in viscosity of a mixture under a varying applied shear rate (Feys and Asghari, 2019, Li et al., 2018). At a lower shear rate, the temperature and binder compositions still played important roles in holding the shear rate constant. A higher temperature could induce more intensive changes in the apparent viscosity before and after shearing compared to those at a high shear rate, and silica fume (SF) encouraged more increases than OPC alone and OPC plus FA, as shown in Figure 10.17. However, as the shear rate reached $21\,s^{-1}$, binder type posed a negligible effect on the viscosity change, and at a shear rate equal to $50\,s^{-1}$, the effect of the temperature damped out as well (Feys and Asghari, 2019).

In summary, the reverse of the thixotropy effect during pumping can be a source of the pump-induced rheology changes. Both the pressure and the shearing effect applied by the pumping process can contribute to the breaking of some bonds between cementitious particles. As a result, the deflocculation of cement clusters should be a joint effect of both high pressure and shearing force, both of which contribute to the loss of viscosity, but the gain in the yield stress may be more relevant to the accelerated cement hydration than to the reverse of thixotropy. In addition, the degree of loss in viscosity is related not only to the intensity of shearing but also to the temperature and the material compositions.

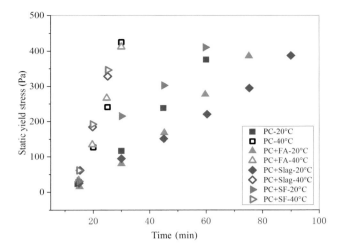

Figure 10.17 Evolution of the static yield stress of plain and blended cement pates at various temperatures (Huang et al., 2019, Li et al., 2022).

10.5.2.3 Supplementary cementitious materials (SCMs)

The incorporation of SCMs can affect the rheological properties through their physical characteristics and early-age reaction kinetics. The physical characteristics of solid particles can be used to predict the static yield stress through the Yodel model (Flatt and Bowen, 2006) and the apparent viscosity through the Krieger–Dougherty model (Krieger and Dougherty, 1959). Cement-based materials, as well as cement–SCM systems, show a link between the flowability characteristics of dry particles and the rheology of their suspensions. Meanwhile, the reaction kinetics of the binder system can be accelerated through the filler effect introduced by SCMs. Because some SCMs do not produce hydrates at the early stage of hydration, the actual *w/cm* ratio appears to be higher than that of the equivalent pure Portland cement system, resulting in a higher actual *w/cm* ratio for more hydrates to be formed. In addition, the extra space for the cement hydrates allows for extra surfaces for nucleation (Lothenbach et al., 2011). As a result, the acceleration of hydration can be achieved with a fine quartz filler (Fernandez Lopez, 2009), SF (Lothenbach et al., 2011), and slag (Utton et al., 2008, Wu et al., 1983).

Although a limited number of studies have investigated the effect of SCMs on the post-pumping characteristics of concrete, studies have been conducted regarding the effect of the mixing energy on the Portland cement (PC)–SCM material. First, the addition of SCM can modify the shear-thinning effect of cement pastes by the shearing action and the effect of SCM, depending on the nature of mineral properties, with metakaolin amplifying the shear-thinning effect, SF reducing it, and quartz or FA presenting little influence on it (Cyr et al., 2000). Jiao et al. (2018, 2019) found that continuous shearing can reduce thixotropy and thus retain some workability. Moreover, some SCMs, such as FA, harm the stability of concrete after a replacement mass of 20% of the total binder is added. Ground granulated blast furnace slag (GGBS) was also reported to encourage bleeding when replacing cement (Sun and Young, 2014). Nevertheless, other SCMs, such as metakaolin, can improve the stability of fresh concrete due to its fineness. Meanwhile, a ternary system with FA and metakaolin could achieve a higher fluidity without compensating segregation stability, like that found by Mehdipour et al. (2013).

Furthermore, a comparative study (Khalid et al., 2016) on the effect of different SCMs on the reaction of rheological properties to the mixing time showed that systems with fly ashes and marble powders were prone to a lower V-funnel flow time when subjected to a longer mixing time, while the SF system had an increased V-funnel flow time under prolonged mixing. The reason is that SF has much smaller particle sizes, so the mixing force is not intense enough to break the interparticle bonds, yet fly ashes and marble powders have weaker interparticle bonds that are breakable under prolonged mixing.

An increase in temperature has been reported in the pumping experiments (Jang et al., 2018, Petit et al., 2008, Secrieru, 2018), and cement systems with different SCMs react differently to the change in temperature due to the hydration kinetics by various SCM characteristics. Generally, as studied by Huang et al. (2019), an increase in temperature from 20°C to 40°C incited a structural build-up, and thus increased the static yield stress for cementitious materials (Figure 10.17). Compared to cement pastes with only plain Portland cement (PC), the addition of SF had the most drastic improvement in interparticle bonding, probably due to its small specific surface area. The addition of FA and slag could also improve the formation of the bonds. Additionally, comparing the data between 20°C and 40°C, it seemed that controlling the sampling time, the differences in the static yield stress, and the structural build-up time between material constituents were moderately diminished as the temperature increased.

Sorting out data from the released horizontal pumping tests, the contribution of different cementitious materials to the change in plastic viscosity per meter of mass concrete is slightly different (Figure 10.18). Compared with Portland cement (PC) powder, the addition of SF has the greatest effect on reducing the plastic viscosity, followed by limestone powder. The addition of FA has little difference. However, the addition of ground granulated blast furnace slag with silica fumes, zirconia silica fumes, FA with zirconia silica fumes, and FA with steel slag can inhibit the reduction of plastic viscosity.

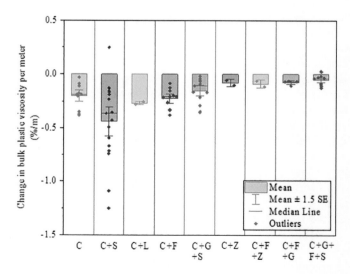

Figure 10.18 Variable changes in the concrete bulk plastic viscosity per meter using different binder compositions. (In this figure only: C=ordinary Portland cement; S=silica fume; L=limestone filler; F=fly ash; Z=zirconia silica fume; G=ground granulated blast furnace slag.) (Data from Feys et al., 2016a, Jang and Choi, 2019, Kwon et al., 2013a, Secrieru et al., 2018, Takahashi, 2014.) (Li et al., 2022).

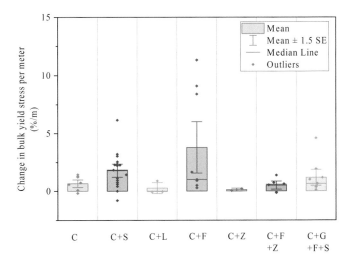

Figure 10.19 Variable change in the concrete bulk yield stress per meter by different binder compositions. (In this figure only: C = ordinary Portland cement; S = silica fume; L = limestone filler; F = fly ash; Z = zirconia silica fume; G = ground granulated blast furnace slag.) (Data from Feys et al., 2016a, Jang and Choi, 2019, Kwon et al., 2013a, Park et al., 2020, Secrieru et al., 2018, Shen et al., 2021b, Takahashi, 2014.) (Li et al., 2022).

Similarly, it can be determined that the yield stress of SCM systems is significantly reduced compared with PC alone, except for the combination of PC + SF, PC + FA, and PC + FA + steel slag (Figure 10.19). Especially in the case of PC + FA, the discreteness of the yield stress of mass concrete is very high, which may be due to the large difference between the FA content and *w/cm* ratio.

In summary, the addition of SCMs has an impact on post-pumping rheology changes. Based on laboratory studies, the effect of shearing and elevated temperature during pumping can both cause an increase in the viscosity and the static yield stress for mixtures with smaller particle sizes, such as silica fumes, but the temperature effect causes a smaller increase in the static yield stress for mixtures with larger particle sizes, and the shearing effect even causes a decrease in the viscosity for mixtures with larger particle sizes. As a complex combination of shearing and temperature increases, the results from full-size pumping show that the addition of different SCMs has different effects on the post-pumping rheology changes, and the reason for such differences can be traced back to the different physical and chemical characteristics of SCMs.

10.5.2.4 Water absorption by aggregates

Although a high pressure may drive a certain amount of water into the aggregate, Choi et al. (2016) proved that even applying a 250-bar pressure for 15 min could not pose noticeable increases in the absorption ratio of the crushed stone or coarse aggregate, along with or in HPC, even though the lightweight coarse aggregate showed an increase in the absorption ratio. Therefore, an elevated pressure during pumping may not cause significant increases in the aggregate absorption ratio, which leads to a change in the rheological properties, as the aggregate used is fully saturated before mixing.

10.5.2.5 Water-reducing agent

The addition of chemical admixtures, such as WRA, further complicates the post-pumping concrete property issue, since the residual dosage of the admixture before pumping determines the degree of particle dispersion that the pump-induced shearing action can effectively provide the system. If more than a sufficient amount of plasticizer is added before pumping, the deflocculated cement particles released during pumping can provide more adsorption sites for chemical admixtures and thus increase the flowability.

However, the demand for plasticizers increases after pumping because of the excessive shearing and rise in temperature, threatening the flowability of concrete if insufficient admixture is added. Intense shear not only releases more cement particles. It can also peel WRA molecules off from particles, leaving it in an "inactive" state, which further increases the scarcity of active WRA (Takahashi, 2014). What escalates the WRA demands is that the rate of admixture adsorption increases with temperature (Plank et al., 2010), and a temperature rise is commonly observed in pumped concrete. In addition, different WRAs may react differently toward temperature changes, since PNS or PMS shows a correlation between the rheological properties and time regardless of temperature. However, the rheological properties of PCP-incorporated cement pastes depend on the temperature (Petit et al., 2010).

Studies have been conducted regarding the relationship between WRA adsorption and the flowability of fresh cementitious materials under different mixing speeds, intensities, and durations, but further studies should be conducted to investigate the role of pumping in such a relationship. Juilland et al. (2012) found that higher shear forces dispersed cement agglomerates and removed early hydrates to reveal more adsorption sites. Additionally, the anhydrous dissolution actions are also changed by the shearing effect. Moreover, the extended mixing time alleviates the over-fluidification effect by some WRAs. On the other hand, the existence of WRA can amplify the results of extended and prolonged mixing, which promotes an increase in viscosity. Additionally, the longer mixing time turns the WRA-cement system from an equilibrium of aggregation and disaggregation into a flocculation state, and the retardation effect of WRA on cement hydration can be hindered by extended mixing (Han and Douglas, 2015).

Although the potential effects can project to the pumping of concrete, by analyzing three horizontal pumping experiments involving the usage of WRA (Feys et al., 2016a, Jang and Choi, 2019, Kwon et al., 2013a), no obvious relationship could be established for the dosage of WRA and the change in concrete bulk rheological properties, with or without the influence of pumping distance.

10.6 SUMMARY

The concrete pumping technique is now one of the most common practices in the field of construction. The scientific evaluation of concrete pumpability is the foundation of concrete pumping technology. Definition and determination of pumpability have been developed for decades, and the evaluation of the lubrication layer is a vital part of assessing the pumpability of a certain concrete. Moreover, to determine the pumpability of concrete, many types of prediction methods have been proposed. Empirical models are often used for their ease of calculation, but their compatibility with modern concrete, which has high flowability, is questioned by scientists. Numerical models can accurately predict the pumping behavior of concrete in the straight pipelines, but the flow behavior of concrete in the elbow and hose can only be predicted based on empirical judgment. Additionally, computer simulations are also developed to predict the pumping performance, but further researches need to be performed to accurately simulate the complex concrete pumping process.

The pumping process has an inevitable effect on the air content, rheological properties, and mechanical properties of concrete. The change in the air content is more likely related to the transportation methods rather than the concrete mixture. In addition, the initial concrete viscosity may have implications on the air content change, but such a correlation needs to be further studied. Usually, the viscosity decreases, and the yield stress increases due to a high pressure and shearing rate during pumping. The accelerated hydration and the reverse of the thixotropy contribute to the rheological changes and are the main mechanisms influencing the rheology change.

REFERENCES

Abebe Y. and Lohaus L. (2017). "Rheological characterization of the structural breakdown process to analyze the stability of flowable mortars under vibration." *Construction and Building Materials*, 131, 517–525

ACI Committee 304. (1998). ACI 304.2R-96 placing concrete by pumping methods.

ARTESIA. (n.d.). "Oil well cement." Retrieved from https://theartesiacompanies.com/oil-well-cement/.

Banfill, P. F. G., et al. (1991). "Simultaneous rheological and kinetic measurements on cement pastes." *Cement and Concrete Research*, 21(6), 1148–1154.

Barnes, H. A. (1997). "Thixotropy - a review." *Journal of Non-Newtonian Fluid Mechanics*, 70, 1–33.

Browne, R. D., and Bamforth, P. B. (1977). "Tests to establish concrete pumpability." *Journal of the American Concrete Institute* 74(5), 193–203.

Chateau, X. 2012. "Particle packing and the rheology of concrete," in N. Roussel (ed.), *Understanding the rheology of concrete*. Woodhead Publishing, pp. 117–143. http://www.sciencedirect.com/science/article/pii/B9780857090287500066.

Chen, L., et al. (2019). "Effect of pumping and spraying processes on the rheological properties and air content of wet-mix shotcrete with various admixtures." *Construction and Building Materials*, 225, 311–323.

Choi, M., et al. (2013a). "Lubrication layer properties during concrete pumping." *Cement and Concrete Research*, 45(1), 69–78.

Choi, M., et al. (2013b). "Numerical prediction on pipe flow of pumped concrete based on shear-induced particle migration." *Cement and Concrete Research*, 52, 216.

Choi, M. S., et al. (2014). "Prediction of concrete pumping using various rheological models." *International Journal of Concrete Structures and Materials*, 8(4), 269–278.

Choi, Y., et al. (2016). "Absorption properties of coarse aggregate according to pressurization for development of high fluidity concrete under high pressure pumping." *Journal of the Korea Institute for Structural Maintenance and Inspection*, 20(3), 122–129.

Coussot, P., and C. Ancey. (1999). "Rheophysical classification of concentrated suspensions and granular pastes." *Physical Review E - Statistical Physics, Plasmas, Fluids, and Related Interdisciplinary Topics*, 59(4), 4445–4457.

Cyr, M., et al. (2000). "Study of the shear thickening effect of superplasticizers on the rheological behaviour of cement pastes containing or not mineral additives." *Cement and Concrete Research*, 30(9), 1477–1483.

De Schryver, R., et al. (2021). "Numerical reliability study based on rheological input for bingham paste pumping using a finite volume approach in openfoam." *Materials*, 14(17), 5011.

De Schryver, R., and De Schutter, G. (2020). "Insights in thixotropic concrete pumping by a Poiseuille flow extension." *Applied Rheology*, 30(1), 77–101.

De Schutter, G. (2017). "Thixotropic effects during large-scale concrete pump tests on site." *Advances in Construction Materials and Systems*, 188, 26–32.

Du, L., and Folliard, K. J. (2005). "Mechanisms of air entrainment in concrete." *Cement and Concrete Research*, 35(8), 1463–1471.

Elkey, W., et al. (1994). Concrete Pumping Effects on Entrained Air Voids. Washington State Department of Transportation, Washington State Transportation Commission, Transit, Research, and Intermodal Planning (TRIP) Division, http://www.wsdot.wa.gov/research/reports/fullreports/313.1.pdf

Fataei, S., et al. (2020). "Experimental insights into concrete flow-regimes subject to Shear-Induced Particle Migration (SIPM) during pumping." *Materials*, 13(5), 1233.

Fataei, S., et al. (2021). "A first-order physical model for the prediction of shear-induced particle migration and lubricating layer formation during concrete pumping." *Cement and Concrete Research*, 147, 106530.

Fernandez Lopez, R. (2009). "Calcined clayey soils as a potential replacement for cement in developing countries", PhD Thesis, Ecole Polytechnique Fédérale de Lausanne, France.

Feys, D., et al. (2009a). "Interactions between rheological properties and pumping of self-compacting concrete", PhD Thesis, University of Gent, Belgium.

Feys, D., et al. (2009b). "Rheology and pumping of self-compacting concrete." American Concrete Institute, ACI Special Publication (261 SP), 83–99.

Feys, D., et al. (2013). "Extension of the Reiner-Riwlin equation to determine modified bingham parameters measured in coaxial cylinders rheometers." *Materials and Structures/Materiaux et Constructions* 46(1–2), 289–311.

Feys, D., et al. (2014). "Development of a tribometer to characterize lubrication layer properties of self-consolidating concrete." *Cement and Concrete Composites*, 54, 40–52.

Feys, D., et al. (2015). "Prediction of pumping pressure by means of new tribometer for highly-workable concrete." *Cement and Concrete Composites*, 57, 102–115.

Feys, D., et al. (2016a). "Changes in rheology of self-consolidating concrete induced by pumping." *Materials and Structures/Materiaux et Constructions* 49(11), 4657–4677.

Feys, D., et al. (2016b). "How do concrete rheology, tribology, flow rate and pipe radius influence pumping pressure?" *Cement and Concrete Composites*, 66, 38–46.

Feys, D. (2019). "How much is bulk concrete sheared during pumping?" *Construction and Building Materials*, 223, 341–351.

Feys, D., and Asghari, A. (2019). "Influence of maximum applied shear rate on the measured rheological properties of flowable cement pastes." *Cement and Concrete Research*, 117, 69–81.

Flatt, R. J., and Bowen, P. (2006). "Yodel: A yield stress model for suspensions." *Journal of the American Ceramic Society*, 89(4), 1244–1256.

Haist, M., et al. (2020). "Interlaboratory study on rheological properties of cement pastes and reference substances: Comparability of measurements performed with different rheometers and measurement geometries." *Materials and Structures/Materiaux et Constructions* 53(4), 1–26.

Han, D., and Douglas, R. (2015). "Effect of mixing method on microstructure and rheology of cement paste." *Construction and Building Materials*, 93, 278–288.

Han, D., and Ferron, R. D. (2016). "Influence of high mixing intensity on rheology, hydration, and microstructure of fresh state cement paste." *Cement and Concrete Research*, 84, 95–106.

Hover, K. (1989). "Some recent problems with air-entrained concrete." *Cement Concrete Aggregates*, 11(1), 67–72.

Hover, K. C., and Phares, R. J. (1996). "Impact of concrete placing method on air content, air-void system parameters, and Freeze-Thaw durability." *Transportation Research Record*, 1532, 1–8.

Huang, H., et al. (2019). "Temperature dependence of structural build-up and its relation with hydration kinetics of cement paste." *Construction and Building Materials*, 201, 553–562.

Jang, K. P., et al. (2018). "Experimental observation on variation of rheological properties during concrete pumping." *International Journal of Concrete Structures and Materials*, 12, 1–5.

Jang, K. P., and Choi, M. S. (2019). "How affect the pipe length of pumping circuit on concrete pumping." *Construction and Building Materials*, 208, 758–766.

JGJ/T10-2011 Technical Specification for Construction of Concrete Pumping (in Chinese), 2011.

Jiao, D., et al. (2018). "Influences of shear-mixing rate and fly ash on rheological behavior of cement pastes under continuous mixing." *Construction and Building Materials*, 188, 170–177.

Jiao, D., et al. (2019). "Time-dependent rheological behavior of cementitious paste under continuous shear mixing." *Construction and Building Materials*, 226, 591–600.

Jo, S. D., Park, C. K., Jeong, J. H., Lee, S. H., and Kwon, S. H. (2012). "A computational approach to estimating a lubricating layer in concrete pumping." *Computers Materials and Continua*, 27(3), 189–210.

Jolin, M., et al. (2006). "Pumping concrete: A fundamental and practical approach". In *Proceedings of the 10th International Conference on Shotcrete for Underground Support*, Edited by Dudley R.M., Harvey W.P., British Columbia, Canada, 270–284.

Juilland P., et al (2012). "Effect of mixing on the early hydration of alite and OPC systems." *Cement and Concrete Research*, 42(9), 1175–1188.

Kaplan, D. (1999). "Pompage Des Bétons". PhD Thesis, ENPC, Champs-sur-Marne, France.

Khalid, A. R., et al. (2016). "Effect of mixing time on flowability and slump retention of self-compacting paste system incorporating various secondary raw materials." *Arabian Journal for Science and Engineering*, 41, 1283–1290.

Khatib, R. (2013). "Analysis and prediction of pumping characteristics of high-strength self-consolidating concrete", PhD Thesis, Université de Sherbrooke, Canada.

Kim, J. H., et al. (2017). "Rheology of cement paste under high pressure." *Cement and Concrete Composites*, 77, 60–67.

Krieger, I. M., and Dougherty, T. J. (1959). "A mechanism for non-newtonian flow in suspensions of rigid spheres." *Transactions of the Society of Rheology*, 3(1), 137–152.

Kwon, H., et al. (2014). "Prediction of concrete pumping: Part II — Analytical prediction and experimental verification." *ACI Materials Journal*, 110, 657.

Kwon, S. H., et al. (2013a). "Prediction of concrete pumping: Part II-Analytical prediction and experimental verification." *ACI Materials Journal*, 110(6), 657–667.

Kwon, S. H., et al. (2013b). "Prediction of concrete pumping: Part I-Development of new tribometer for analysis of lubricating layer." *ACI Materials Journal*, 110(6), 647–655.

Lapasin, R., et al. (1979). "Thixotropic behaviour of cement pastes." *Cement and Concrete Research*, 9, 309–318.

Le, H.D., et al. (2015). "Effect of lubrication layer on velocity profile of concrete in a pumping pipe." *Materials & Structures*, 48(12), 3991–4003.

Lessard, M., et al. (1996). "Effect of pumping on air characteristics of conventional concrete." *Transportation Research Record*, 1532, 9–14.

Li, F., et al. (2022). "An overview on the effect of pumping on concrete properties." *Cement and Concrete Composites*, 129, 104501.

Li, Z., et al. (2018). "Numerical method for thixotropic behavior of fresh concrete." *Construction and Building Materials*, 187, 931–941.

Lothenbach, B., et al. (2011). "Supplementary cementitious materials." *Cement and Concrete Research*, 41(12), 1244–1256.

Lowke, D. (2018). "Thixotropy of SCC—A model describing the effect of particle packing and superplasticizer adsorption on thixotropic structural build-up of the mortar phase based on interparticle interactions." *Cement and Concrete Research*, 104, 94–104.

Mechtcherine, V., et al. 2014. "Testing pumpability of concrete using sliding pipe rheometer." *Construction and Building Materials*, 53, 312–323.

Mehdipour, I., et al. (2013). "Effect of mineral admixtures on fluidity and stability of self-consolidating mortar subjected to prolonged mixing time." *Construction and Building Materials*, 40, 1029–1037.

Mewis, J., and Wagner, N. J. (2009). "Thixotropy." *Advances in Colloid and Interface Science*, 147–148(C), 214–227.

Mielenz, R. C., et al. 1958. "Origin, evolution, and effects of the air void system in concrete part 4- the air void system in job concrete." *ACI Journal Proceedings*, 30(55), 507–517.

Morinaga, S. (1973). "Pumpability of concrete and pumping pressure in pipeline." *RIELM Seminar Held in Leeds*.

Nerella, V. N., and Mechtcherine, V. (2018). "Virtual sliding pipe rheometer for estimating pumpability of concrete." *Construction and Building Materials*, 170, 366–377.

Ngo, T. T. (2009). Influence of Concrete Composition on Pumping Parameters and Validation of a Prediction Model for the Viscous Constant. Doctoral dissertation, University Cergy-Pontoise, Cergy-Pontoise, France.

Ngo, T. T., et al. (2010). "Use of tribometer to estimate interface friction and concrete boundary layer composition during the fluid concrete pumping." *Construction and Building Materials*, 24(7), 1253–1261.

Otsubo, Y., et al. (1980). "Time-dependent flow of cement pastes." *Cement and Concrete Research*, 10(5), 631–638.

Ouchi, M., and Sakue, J. (2008). "Self-compactability of fresh concrete in terms of dispersion and coagulation of particles of cement subject to pumping." *Proceedings of the 3rd North-American Conference on the Design and Use of Self-Consolidating Concrete*, Chicago.

Park, C. K., et al. (2010). "Development of Low Viscosity High Strength Concrete and Pumping Simulation Technology". *Research report for Institute of Construction Technology, Samsung C&T Corporation*, Korea.

Park, C. K., et al. (2020). "Analysis on pressure losses in pipe bends based on real-scale concrete pumping tests." *ACI Materials Journal*, 117(3), 205–216.

Petit, J. Y., et al. (2008). "Methodology to couple time-temperature effects on rheology of mortar." *ACI Materials Journal*, 105(4), 342–349.

Petit, J. Y., et al. (2010). "Effect of temperature on the rheology of flowable mortars." *Cement and Concrete Composites*, 32, 43–53.

Phillips, R. J., Armstrong, R. C., Brown, R. A., Graham, A. L., and Abbot, J. R. (1992). "A constitutive equation for concentrated suspensions that accounts for shear-induced particle migration." *Physics of Fluids*, 4, 30–40.

Plank, J., et al. (2010). "Experimental determination of the thermodynamic parameters affecting the adsorption behaviour and dispersion effectiveness of PCE superplasticizers." *Cement and Concrete Research*, 40, 699–709.

Rahman, M. K., et al. (2014). "Thixotropic behavior of self compacting concrete with different mineral admixtures." *Construction and Building Materials*, 50, 710–717.

Roussel, N., et al. (2012). "The origins of thixotropy of fresh cement pastes." *Cement and Concrete Research*, 42(1), 148–157.

Sakuta, M., et al. (1979). "Pumpability and rheological properties of fresh concrete." In *Proceedings of the Conference on Quality Control of Concrete Structures*, Stockholm, Sweden. pp. 125–132.

Saleh, F. K., and Teodoriu, C. (2017). "The mechanism of mixing and mixing energy for oil and gas wells cement slurries: A literature review and benchmarking of the findings." *Journal of Natural Gas Science and Engineering*, 38, 388–401.

Secrieru, E., et al. (2018a). "Changes in concrete properties during pumping and formation of lubricating material under pressure." *Cement and Concrete Research*, 108, 129–139.

Secrieru, E. (2018b). Pumping Behaviour of Modern Concretes–Characterisation and Prediction. PhD. Thesis, Technische Universität Dresden, Germany.

Shah, S. P., and Lomboy, G. R. (2014). "Future research needs in self-consolidating concrete." *Journal of Sustainable Cement-Based Materials*, 4(3), 154–163.

Shen, W., et al. (2021a). "Influence of pumping on the resistivity evolution of high-strength concrete and its relation to the rheology." *Construction and Building Materials*, 302, 124095.

Shen, W., et al. (2021b). "Change in fresh properties of high-strength concrete due to pumping." *Construction and Building Materials*, 300, 124069.

Shen, W., et al. (2021c). "How do discharge rate and pipeline length influence the rheological properties of self-consolidating concrete after pumping?" *Cement and Concrete Composites*, 124, 104231.

Struble, L., and Lei, W.-G. (1995). "Rheological changes associated with setting of cement paste." *Advanced Cement Based Materials*, 2(6), 224–230.

Sun, Z., and Young, C. (2014). "Bleeding of SCC pastes with fly ash and GGBFS replacement." *Journal of Sustainable Cement-Based Materials*, 3(3–4), 220–229.

Takahashi, K. (2014). "Effects of mixing and pumping energy on technological and microstructural properties of cement-based mortars", PhD Thesis, Technische Universität Bergakademie Freiberg, Germany.

Talero, R., et al. (2017). "Role of the filler on Portland cement hydration at very early ages: Rheological behaviour of their fresh cement pastes." *Construction and Building Materials*, 151, 939–949.

Talukdar, S., and Heere, R. (2019). "The effects of pumping on the air content and void structure of air-entrained, wet mix fibre reinforced shotcrete." *Case Studies in Construction Materials*, 11, e00288.

Tan, Y., et al. (2012). "Numerical simulation of concrete pumping process and investigation of wear mechanism of the piping wall." *Tribology International*, 46(1), 137–144.

Tattersall, H., and G. Banfill. (1983). "*The rheology of fresh concrete.*" Pitman Books Limited, London, England.

Tavangar, T., et al. (2022). "Novel tri-viscous model to simulate pumping of flowable concrete through characterization of lubrication layer and plug zones." *Cement and Concrete Composites*, 126, 104370.

Uchikawa, H., et al. (1985). "Influence of character of clinker on the early hydration process and rheological property of cement paste." *Cement and Concrete Research*, 15(c), 561–572.

Utton, C. A., et al. (2008). "Effect of temperatures up to 90°C on the early hydration of Portland-blast furnace slag cements." *Journal of the American Ceramic Society*, 91(3), 948–954.

Vosahlik, J., et al. (2018). "Concrete pumping and its effect on the air void system." *Materials and Structures/Materiaux et Constructions*, 51(4), 1–5.

Wu, X., et al. (1983). "Early stage hydration of slag-cement." *Cement and Concrete Research*, 13(2), 277–286.

Xu, Z., et al. (2022). "Numerical approach to pipe flow of fresh concrete based on MPS method." *Cement and Concrete Research*, 152, 106679.

Yang, M., and Jennings, H. M. (1995). "Influences of mixing methods on the microstructure and rheological behavior of cement paste." *Advanced Cement Based Materials*, 2(2), 70–78.

Yingling, J., et al. (1992). "Loss of air content in pumped concrete." *Concrete International*, 14, 57–61.

Yuan, Q., et al. (2018). "An overview on the prediction and rheological characterization of pumping concrete." *Materials Reports*, 32(17), 2976–2985.

Chapter 11

Rheology and 3D printing

11.1 INTRODUCTION TO 3D-PRINTING CONCRETE

11.1.1 Development of 3D printing technology

3D-printing technology, also known as additive manufacturing, is a new technology gradually rising in the late 1980s. It is a promising process to construct objects with the help of computer-aided design and automated operations. Since the invention of the first 3D printer in 1983, this technology has developed rapidly (Sakin and Kiroglu, 2017). 3D printing technology is widely used in various industries and has made many achievements in the pharmaceutical, medical, automotive, aerospace, and electronic industry (Petrovic et al., 2011, Horn and Harrysson, 2012, Bogue, 2013, Vaezi et al., 2013). The idea of 3D printing also brings new challenges and opportunities to the field of civil engineering. 3D printing concrete has attracted great attention in the past few years due to its potential to change traditional construction to produce building structures without formwork (Cesaretti et al., 2014, Khoshnevis et al., 2001, Khoshnevis, 2006, Lim et al., 2011). Also, 3D printing has the characteristics of flexible design, fast construction speed, and low labor and energy consumption (Paolini et al., 2019, De Schutter et al., 2018, Williams and Butler-Jones, 2019, Asprone et al., 2018, Hongyao et al., 2019). It is conceivable that this technology has great potential for large-scale application in the field of civil engineering. In 1997, Pegna (1997) made the first attempt at 3D printing cement-based material technology. The research and engineering cases of many experts and scholars have proved the feasibility and great development potential of 3D printing technology in civil engineering.

This innovative construction process mainly includes powder bed printing (also referenced as D-shape), contour crafting (CC), and concrete printing (Wu et al., 2016, De Schutter et al., 2018, Buswell et al., 2018, Asprone et al., 2018, Hager et al., 2016). Although the emphasis of the above three processes is different, the core principle is to construct 3D solid by layered printing.

D-shape technology, which was invented by Enrico Dini in 2007, is based on powder materials (Cesaretti et al., 2014). The printing process of the system of a D-shape object is shown in Figure 11.1. First, a layer of binder is laid, and then the aggregate is sprayed on it to bond with the binder. After printing the objects, the scattered aggregate can be recycled. Since the unbonded power supports the upper part being manufactured, this method can print complex structure models (Lowke et al., 2018), and it does not need to consider the setting time of materials. It was reported that compared with the traditional construction method, the D-shape process only needs 1/4 of construction time when printing components of the same size, which has obvious time-saving characteristics. The core of the D-shape is the printer's print head, which is made of a beam and equipped with 300 micro nozzles. There is a certain interval between the micro nozzles, and the adhesive is selectively sprayed

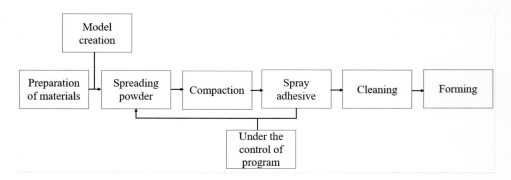

Figure 11.1 The process of D-shape.

to bond the dispersed fine sand when printing, while the unbound sand can be recycled for the next use. The flow rate of the binder at the nozzle is related to the nozzle size and the liquid pressure, see formula (11.1). For D-shape technology, European Space Agency studies the use of artificial lunar dust to print and construct lunar bases (Cesaretti et al. 2014):

$$Q = k_v \sqrt{\frac{\Delta P}{10}} \tag{11.1}$$

where Q is the flow rate of the binder (m³/h), Δp is the liquid pressure (Pa), and k_v is the flow constant (m³/h.Pa).

CC technology was first invented by Khoshnevis (2006), which has been successfully applied in practical engineering construction. In the process, the X-axis or Z-axis movement of the printing head is pulled and controlled by the crane, and the two side rails drive the printing head to move along the y-axis. The printing nozzle extrudes the concrete material continuously and evenly according to the pre-designed procedure so that it is piled up layer by layer. The feature of CC technology is that it adds a spatula to the extrusion device, as shown in Figure 11.2. The angle and direction of the trowel can be automatically adjusted to form different geometry and make the printed surface flat (Zareiyan and Khoshnevis, 2017a, Khoshnevis et al., 2006). Besides, the printable object is used as the external wall template, and then the reinforcement is arranged inside to form the element, which greatly saves the cost of the template. The use of contour crafting can print a 1-square-foot wall

Figure 11.2 Spatula for contour crafting (Zareiyan et al., 2017).

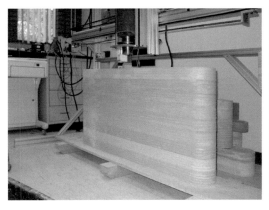 

Figure 11.3 Walls printed through contour crafting (Zareiyan et al., 2017).

for less than 20 s, and print a 200 m² house in a day (Lim et al., 2012). The walls printed through contour crafting are shown in Figure 11.3. The process can be summarized as four processes, namely, model design and treatment, raw material transportation, nozzle movement, and reinforcement. Model design is realized by computer software, and model processing is called slicing. At present, 3D printing model design is relatively simple, and modeling is mainly based on CAD (computer-aided design), CG (computer graphics), image-based modeling, scanning-based modeling, and so on. In the slicing process, it is necessary to input the designed printing parameters, such as the printing rate, and then the slicing software will calculate the optimal printing path according to the set parameters. The raw material transportation and nozzle movement speed should be matched to ensure that the fresh pulp can be smoothly transported to the extrusion device. At present, the outline process printers are mainly mechanical arm types and cable types. Compared with the traditional frame printer, the robot arm contour crafting has a larger printing space and faster printing speed by setting a guide rail at the bottom. Branch developed the world's largest c-fab mechanical arm 3D printer in 2015 and successfully printed the wall for the first time. Cable-type contour crafting is to connect mobile platforms or terminal equipment through a series of cables. It has the advantages of easy disassembly, easy reassembly, easy transportation, and low price. According to whether the cable can completely control the terminal equipment, the cable control operator is divided into full control type and non-full control type, and the full control type is suitable for the situation of high speed and high accuracy (Ma and Wang, 2020). In 2015, the World Advanced Saving Project (WASP) used a 3D printer with a height of 12 m and a width of 6 m to build an adobe structure with eco-friendly materials such as clay, straw, and soil on an open-air construction site. The printer was called BigDelta, and the elevator was connected to the scaffolding of the printer to transport the solid materials.

It is worth mentioning that contour crafting was used for assisting the rapid buildup of an initial operational capability lunar base. It is intended to use CC technology to draw up a detailed plan for a high-fidelity simulation at NASA's Desert Research and Technology Studies (D-RATS) facility, to construct certain crucial infrastructure elements in order to evaluate the merits, limitations, and feasibility of adapting and using the CC technology for the extraterrestrial application. Elements suggested to be built and tested include roads, landing pads and aprons, shade walls, dust barriers, thermal protection shields, dust-free platforms, and other built-up structures utilizing the in-situ resource utilization (ISRU) strategy.

Concrete printing was proposed by researchers of Loughborough University (Le et al., 2012b, Lim et al., 2011). The process of concrete printing is similar to contour crafting, which belongs to the extrusion type. The printing nozzle is installed on the beam and can move freely along the X, Y, and Z axes. The fresh slurry is transported to the extrusion head through the pipeline and moves according to the set route and then stacked layer by layer. Compared with the contour process, the 3D printing deposition resolution of concrete is smaller, which can better control the surface quality of components. The printer and large-scale span arch structure printed through concrete printing are shown in Figure 11.4. Lim et al. (2011) developed a concrete 3D printing system with a 5.4 m×4.4 m×5.4 m steel frame, and the maximum moving speed of the printing head was 5 m/min. In 2014, WinSun, a Chinese company, used concrete and glass fiber to print ten houses with 200 m^2 floor areas in one day with a 150.0 m×10.0 m×6.6 m printer. Also, the company has built the world's first 3D printing villa with 1100 m^2 floor areas and the world's highest 3D printing building, namely a 5-story building. Rudenko (2015) used cement and sand to print structural components separately, which were assembled into the building. In 2016, Gosselin et al. (2016) invented a six-axis manipulator 3D printing device with an extrusion head. Hebei University of Technology has built a concrete 3D printing bridge with a single span of 18.04 m and a total length of 28.1 m in 2019, which is also the largest fabricated 3D printing bridge in the world.

Each of the three 3D printing technologies has its unique advantages, as shown in Table 11.1. However, 3D printing technology is not compatible with both printing speed and resolution. D-shape process has high printing resolution because of its small diameter and slow printing speed. Contour crafting and concrete printing are equipped with separate large printing heads, so the printing speed is fast but the printing resolution is low. In the aspect of printing size, the manipulator of CC technology can move flexibly and print out large-scale components. However, due to the limitation of the printing frame, D-shape and concrete printing cannot print oversized objects.

Figure 11.4 The printer and large-scale span arch structure printed through concrete printing (Wang et al., 2021).

Table 11.1 Comparison of 3D printing processes for cementitious materials (Ma and Wang 2020)

Parameter	D Shape	Contour crafting	Concrete printing
Process	Selective adhesion	Extrusion	Extrusion
Printing materials	Sand	Adhesive materials	Concrete
support system	Unbound powder	Beam	Gypsum
Resolution	High	Low	Low
Thickness of layer	4–6 mm	13 mm	5–25 mm
Printing speed	Slow	Fast	Fast
Printing size	Limited by print frame	Huge	Limited by print frame
Number of nozzles	Hundreds	1	1
Diameter of nozzles	0.15 mm	15 mm	9~20 mm

At present, 3D printing technology is developing rapidly. Although 3D printing can speed up the construction process, help build complex building shapes and reduce labor and energy consumption, there are still some problems with this new technology. First, due to the limitation of printer size, the frame of printing equipment limits the printing range of components. When the aggregate in the mixture is too large, it will be difficult to transport and difficult to extrude. The existing research focuses on the 3D printing of cement paste and mortar. Second, compared with the traditional casting process, the strength of 3D printing components will be reduced, which will put forward higher performance requirements for the materials of this process. Third, 3D printing requires the concrete to satisfy the printability, which is completely different from the workability of normal concrete.

11.1.2 Requirement for 3D printing concrete

Although 3D printing technology has great potential in the construction system, it is still a challenging task for the performance of concrete in the fresh state and the mechanical properties in the hardened state. Unlike conventional construction, 3D printing concrete is used to build objects without the formwork. The performance requirements of 3D printing materials are quite different from that of traditional concrete. In most cases, 3D printing materials must go through four production steps, as shown in Table 11.2.

Firstly, materials need to be pumped to the printhead. This requires the pumpability of 3D printing materials, and the pumpability has been discussed in Chapter 10. The main challenge of 3D printing concrete is to have extrudability and buildability simultaneously. The mixture needs good fluidity to ensure that the continuous concrete is pumped and extruded through the nozzle, and the concrete requires good support capacity to withstand

Table 11.2 Production steps, and underlying physics relevant within the extrusion-based additive manufacturing with cement-based materials

Production step	Action and underlying physics of extrusion-based 3DPC
Transportation	Pumping or gravitational flow with or without energy input, 3DPC needs to flow in pipe.
Printhead process	Extrusion using primary motivation (pumping), ram extrusion, or screw extrusion. Complicated shearing action in this process.
Deformation of 3DPC during deposition	Gravitational flow, viscoelastic-plastic deformations, and complicated shearing action in this process. 3DPC needs to remain smooth and stable.
Behavior of build material after deposition	Deformations due to self-weight and kinetic energy of deposition; early-age shrinkage, early-age creep, thermal dilation, layered bonding.

the load generated by subsequent concrete layers (Buswell et al., 2007, Perrot et al., 2016). Fluidity and buildability are two key factors affecting the working performance of 3D printing cement-based materials, which are closely related to their rheological properties. The rheological properties of materials, including yield stress, plastic viscosity, and thixotropy, are very important for concrete printing since the print quality and the hardened properties are directly related to its fresh properties. The yield stress corresponds to the shear stress required for the fresh concrete to start flowing, while the plastic viscosity measures the resistance after flowing. Therefore, the rheological theory is applied to evaluate the working performance of 3D printing cement-based materials, that is, the rheological property of cement-based materials is the key to achieve smooth printing.

Marchon et al. (2018) proposed a four-step hardening evolution of printed concrete based on the physiochemical concept: (1) high-shearing zone, (2) deposition, (3) green strength growth, and (4) rapid strength development. Step I includes all the shearing history that the material undergoes from the initiation of the mixing to the deposition time (e.g., mixing, pumping, and extrusion). Therefore, the evolution of structuration can be broken down or rebuilt depending on the shear exposed to the material along the processing path. Step II represents the yield stress after the extrusion, where the material is not subjected to a shear flow anymore. However, there exists a mishmash of various forces including colloidal interactions and the formation of bonds between early reaction products. The interaction force is determined by a potential energy well for every particle, which tends to reach equilibrium by dislocation to where the energy is minimum between its neighbors. Step III is called green strength and indicates the duration that particles have already flocculated and reached equilibrium on thixotropic forces; the material undergoes a linear structural buildup. Subsequently, the percolated network of particles is rigidified, and a nonlinear growth in the hardening evolution is observed. The structural buildup becomes almost irreversible.

Besides, some scholars have studied the relationship between printing properties and rheological parameters of 3D printing geopolymer materials. Controlling the rheological properties of fresh concrete is of paramount importance in concrete extrusion, and it defines the success or failure of the printing process. Le et al. (2012b) proposed that the yield stress of 3D printing paste in the range of 178.5–359.8 Pa can be extruded. Alghamdi et al. (2019) proposed that yield stress of 700 Pa is the upper limit of printing performance of geopolymer paste, while Panda and Tan (2018) found that the yield stress of 0.6–1.0 kPa is the favorable range of mortar extrusion. The difference in conclusions in the literature might be due to the difference between rheological equipment and extrusion device. More sophisticatedly, Ranjbar et al. (2021) proposed a straightforward analysis procedure which enables an accurate prediction of concrete printability through dynamic-mode rheological measurements. It was found that two following sequential shearing steps could model the hardening evolution of geopolymer mortar during and after deposition: (1) modeling the shearing history of the material by pre-shearing with an oscillatory strain more than the smaller critical strain, γ_c, and (2) applying small oscillations with a strain below the γ_c to model the concrete after deposition.

11.2 PRINTABILITY OF 3D PRINTING CONCRETE

11.2.1 The definition of printability

The printability of 3D printing concrete is the first key issue to be solved. The definition of printability is the capacity for a material to be successfully processed via the 3D printing process. The material should be "extrudable" during the period when the material is completely mixed and printed, which corresponds to a wide range of flowability values above a certain critical value. Also, the fluidity of the material should not lead to the separation of components. Once deposited by the printer head, the material needs to be extruded and immediately formed, corresponding to another range of flowability values below that certain critical value. The material first needs to maintain its shape after extrusion and can support some weight without excessive deformation when subsequent layers are deposited on it. The ideal material should be flowable before deposition, be both extrudable and buildable at the time of deposition, and rapidly harden after deposition. Lim et al. (2011) first proposed that the four key properties of 3D printing cement-based materials are pumpability, printability, buildability, and open time. Among these, the open time is defined as the period where the above properties are consistent within acceptable tolerances. They pointed out that extrudability, buildability, and interlayer bonding are the key factors to determine printability. Besides, extrudability and buildability are the two most critical and competing properties for 3D printing concrete (Le et al., 2012b). This is because high workability promotes extrudability, while buildability requires low workability of materials. To achieve printability, these two parameters must be properly balanced. Soltan and Li (2018) believed that an idealized behavior of a "printable" cementitious material can be summarized in terms of flowability over time, as shown in Figure 11.5. Zhang et al. (2019) considered that rheological properties (viscosity, yield stress, thixotropy) and open time are two critical wet properties to control the printable property of concrete material, and the printability includes pumpability, extrudability, and buildability. Kazemian et al. (2017) put forward three criteria to judge the surface quality of 3D printing concrete, including the flawless surface, clear square boundary of the printing layer, and uniformity. They found that the initial setting time was not suitable for characterizing the printability window. Besides, the characterization method is too subjective to use the basic physical parameters to characterize printability. Generally, most researches on printability focus on extrudability and buildability, which are closely related to the rheological properties of 3D printing materials.

11.2.2 The test for printability

At present, most of the research on printability mainly focuses on fluidity, extrudability, and buildability. However, due to the lack of relevant standards in the field of 3D printing of cement-based materials, researchers mostly refer to or directly use the standard experimental methods in the field of cement-based materials or even other fields to characterize the 3D printability of cement-based materials.

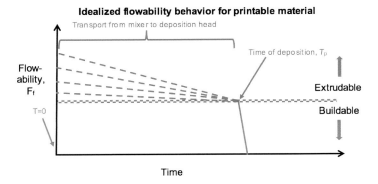

Figure 11.5 The relationship between flowability, extrudability, and buildability of printable materials (Soltan and Li, 2018).

Fluidity is one of the important parameters to evaluate the printability of concrete materials. Good fluidity can ensure the paste to pass through the transmission device and extrude smoothly. At present, the evaluation methods of liquidity tests mainly include flow table test, slump test, T_{50} slump test, V-shaped funnel test, and L-shaped box test (Paul et al., 2018, Soltan and Li, 2018, Zhang et al., 2018, Ma et al., 2018), as shown in Figure 11.6.

The extrudability of concrete refers to the ability of the material to extrude continuously and uniformly from the nozzle, which is the key index to determine its printability. At present, there is no special equipment to test and evaluate the extrudability of concrete. Le et al. (2012b) judged the extrudability of fresh concrete by observing that there is no blockage and interruption in the extrusion process; that is, it can reach 4500 mm continuous extrusion length. Soltan and Li (2018) used a modified glue gun to test the extrudability of fresh concrete, as shown in Figure 11.7, but this method has the defects of unstable extrusion force and insufficient extrusion force. The time from the start of mixing to the rise of shear stress of 0.3 kPa is used to characterize the open time (Le et al., 2012b).

The buildability of cement-based materials refers to the ability of the printed materials to maintain shape stability under the action of self-weight, which is another important index to evaluate the printability of cement-based materials. Many researchers express

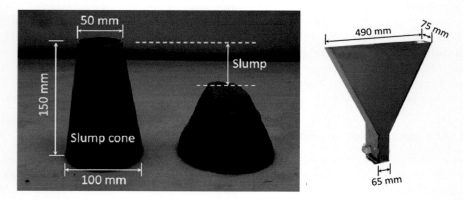

Figure 11.6 The tests for flowability (Ma et al., 2018).

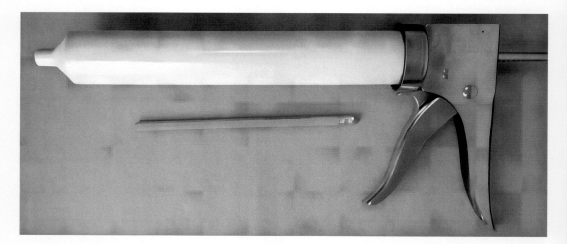

Figure 11.7 The modified gun for extrudability (Soltan and Li, 2018).

buildability by testing the longitudinal deformation value of printed materials (Zhang et al., 2018c, Ma et al., 2018a). Zhang et al. (2018c) characterized the buildability by testing the wet embryo strength of the materials. Layer settlement test and cylinder stability were used to evaluate the buildability, and the penetration resistance of different hydration times was tested by concrete penetrometer (ASTM C403) (Kazemian et al., 2017). The number of layers while the material maintains its shape is used to characterize the buildability (Le et al., 2012b). Tay et al. (2019) measured the buildability of samples by slump test to determine printable areas. They suggested that the mixture with a slump between 4–8 mm and spread between 150–190 mm has good buildability. However, 192.5–269 mm initial spread diameter was considered the optimum range of flowability to meet the buildability (Zhang et al., 2019). In the buildability test of 3D printing concrete, Wolfs et al. (2018) tested the compressive strength and Young's modulus of elasticity of mixtures in a fresh state and established the relationship between destructive compressive strength and ultrasonic nondestructive testing. Also, the evaluation methods of buildability tests include the plastome test and the direct shear test. In the plastome test, the relationship between the extrusion force F and the mortar height h in the ideal plastic model can be established. A direct shear test is usually used to measure the mechanical parameters of soil. In recent years, some researchers use the direct shear test to measure the rheological properties of concrete mortar. Besides, setting behavior is of great significance to control the printing process, which is an important parameter related to the stiffness development of concretes. Vicat needle test, wave transmission method, and wave reflection method are often used to monitor the setting property of concrete (Ma and Wang, 2018).

11.2.3 Criteria to evaluate the loading bearing capacity

3D printing is a fast construction way, and the materials are stacked on top of fresh materials. If the bottom materials are not strong enough to bear the weight of upper layers, failures may happen. At present, two mechanisms have been recognized as causes of collapse in extrusion-based, layered 3D concrete printing during manufacturing, i.e. material failure (Figure 11.8a), and loss of stability (Figure 11.8b).

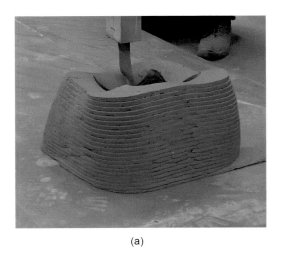

(a) (b)

Figure 11.8 Collapse in extrusion-based additive manufacturing: (a) material failure and (b) stability failure (Suiker et al., 2020).

Either mechanism is triggered by a combination of material, support, load, and geometry conditions. The difference is that the former one is generally constant, but the latter does change over time due to the gradual buildup of the object during manufacturing. The resistance to failure during printing is dependent not only on the material's characteristics but also on the object design, e.g., size, geometry, and process parameters such as print speed and extrusion speed.

To define the initiation of material failure, i.e., fracture, yielding, etc., many criteria have been proposed, but it is hard to say which is the best one for extrusion-based 3D concrete printing. This is primarily due to the transition that printable cementitious materials undergo from a non-Newtonian fluid state to a solid state during the manufacturing process. This process involves many complicated chemical and physical interactions, and mechanical issues.

Perrot et al. (2016) presented a buildability approach based on the rheological yield stress of the print mortar as the material failure criterion:

$$\alpha_{geom} \tau_0(t) \geq \rho g h(t) \tag{11.2}$$

where α_{geom} is a geometrical factor, $\tau_0(t)$ is the time-dependent yield stress of the mortar, ρ is the density (kg/m³), g is the gravitational acceleration (9.8 N/kg), and $h(t)$ is the time-dependent object height (m).

Wangler et al. (2016) applied a von Mises plasticity criterion, by introducing a factor $\sqrt{3}$ into the buildability equation, which yields:

$$\tau_0(t) \geq \rho g h(t)/\sqrt{3} \tag{11.3}$$

In more sophisticated cases, both material and stability failure are taken into account (Wolfs et al., 2019b). This method is based on FEM (Finite Element Method) expanded with time-dependent properties, including the development of experimental procedures, and the adoption of the Mohr–Coulomb failure criterion to determine the full failure envelope. The dual failure criterion of plastic yielding and elastic buckling requires the experimental determination of five time-dependent material properties: apparent Young's modulus $E(t)$, Poisson's ratio $\nu(t)$, cohesion $C(t)$, angle of internal friction $\varphi(t)$, and dilatancy angle $\psi(t)$. Suiker (2018) introduced a parametric mechanistic model that focuses on the competition between elastic buckling and plastic collapse and can be used to predict structural failure of straight, free-standing walls during 3D printing.

Roussel (2018) proposed another method combining the rheology of material with stability, and the criteria give:

$$E_0(t) \geq 3\rho g h(t)^3 / 2\delta^2$$

$$h_t = 2\delta \sqrt{\frac{1+\nu}{3\sqrt{3}\gamma_c}} \tag{11.4}$$

where E is Young's modulus, ν is Poisson's ratio, γ_c is the critical shear strain, and δ is the width of one linear meter of a wall.

Finally, an analytical model was proposed by Jeong et al. (2019), who took the age of the material along the print path length into account. The shear strength (solid mechanics) was adopted as a failure criterion without extensive elaboration.

11.2.4 Printable cement-based materials

The main challenge for 3D printing concrete materials to build objects ranging in scale from little elements to large components is to meet the printability. A printable concrete needs good flowability that ensures to pump and squeeze through the printer, and it needs a good supporting capacity to resist the load caused by subsequent concrete layers without large deformation or collapse. In this regard, a thixotropic material was developed and applied to 3D printing applications. Generally, fluidity and buildability are two important factors for cement-based materials to realize printing. Therefore, controlling the rheological properties of cement-based materials is the key to solve the printing process.

A few cementitious materials have been explored in the 3D printing building components. Portland cement is the most widely used cementing material in 3D printing concrete. Gibbons et al. (2010) explored the use of rapid hardening Portland cement (RHPC) to build structures through the D-shape process. Maier et al. (2011) proved that calcium aluminate cement (CAC) has high feasibility in 3D printing structures and has good fresh and hardening properties. The physical properties and printability of concrete for 3D printing largely depend on the composition and properties of its components (Perrot et al, 2016). At present, fine sand with a maximum particle size of less than 2 mm is mostly used as aggregate (Nerella et al., 2016, Le et al., 2012a). Rushing et al. (2017) studied the application of concrete with coarse aggregate sizes of 9.5 mm in 3D printing. Zhang et al. (2019) proposed that the rheological properties were inevitably affected by open time in the different sand to cement ratios, especially thixotropy. They found the thixotropic agent increased the thixotropy of concrete and can also improve initial flowability. A high-performance mortar consisting of 54% sand, 36% cementitious compounds, and 10% water for concrete printing is developed (Lim et al., 2012). Gosselin et al. (2016) proposed a new type of high-performance concrete paste composed of 30–40 wt.% Portland cement, 40–50 wt.% crystalline silica, 10 wt.% silica fume, and 10 wt.% limestone filler, which is suitable for a large-scale 3D printing system. Shakor et al. (2017) developed a new modified cement powder material as a 3D printing material. Ma et al. (2018c) found that the tailing sand to the river sand ratio of 2:3 is the most suitable parameter for printing.

Some researchers try to add fiber to cement-based materials. Le et al. (2012b) proposed a high-performance fiber-reinforced 3D printing cement-based material to meet the printability, and the 28 d compressive strength of the printing material can reach 110 MPa. He found that the hardened performance of 3D printing layered structure was inevitably affected by anisotropy, and the reduction in compressive strength depends on the orientation of the loading relative to the layers (Le et al., 2012a). Hambach and Volkmer (2017) attempted to prepare 3D-printed composites doped with short fibers, such as carbon fiber, glass fiber, and basalt fiber. The printing material is composed of 61.5% P.O52.5R Portland cement, 21% silica fume, 15% water, and 2.5% water reducer, and the water-cement ratio is 0.3. To improve the dispersibility of fiber and the adhesion between fiber and cement matrix, the fiber was treated at a high temperature before the test. The way of fiber orientation is to make the short fiber parallel to the moving path of the nozzle during the extrusion process with the help of the printing nozzle whose diameter is far less than the fiber length. The bending strength and compressive strength of the material can reach 30 and 80 MPa, respectively. Panda et al. (2017b) prepared a glass fiber-reinforced geopolymer for extrusion 3D printing, which used 23% F-grade fly ash, 5% slag, 3% silica fume, 47% fine river sand, 15% liquid potassium silicate, 2% hydroxypropyl methylcellulose, and 5% tap water. The results show that the mechanical properties of the printed material can be improved by adding fiber, but the effect of reinforcement is not significant, which may be due to the weak bonding force between fiber and cement matrix. The phenomenon of fiber loosening and fiber pullout detected by the test results also proves this point. Hou et al. (2018) found that the blended fibers show unique orientation characteristics along the printing path during extrusion. Soltan and Li (2018) proposed a kind of printable engineered cementitious composites (ECC), which include dispersed short polymer fibers to generate robust tensile strain-hardening. These ECC materials are designed to eliminate the need for steel reinforcement in structures and provide more freedom and efficiency for 3D printing engineering applications.

The composition of geopolymer is more variable than that of Portland cement, including various precursors and activators. Thus, the factors affecting its printability are very complex and its properties can be tailored for different purposes. The composition of raw materials has a great influence on the rheology of 3D printing geopolymer materials. Panda et al. (2019a) found that the yield stress, thixotropy, and strength of alkali-activated fly ash increase with the increase of slag content. Besides, they proposed that the minimum thixotropy value obtained from the area held between the up and down curve of torque and rotation speed for the 3D printing geopolymer mix should be 10,000. Below the minimum value of thixotropy, the material is not suitable for extrusion-based printing applications (Panda et al., 2017a). Zhang et al. (2018b) studied the effect of steel slag content on the rheological properties of 3D printing steel-slag-based geopolymer paste and found that with the increase of steel slag content, the shear stress and yield stress of the fresh paste increased, and the plastic viscosity decreased. Panda and Tan (2018) studied the influence of sand to cement ratio and water to solid ratio on the printability of 3D printing alkali-activated mortar. They found that when the sand to cement ratio was high, the hose could not be extruded due to the increase of yield stress caused by the increase of particle friction. Besides, the yield stress and plastic viscosity of 3D printing alkali-activated materials decrease sharply with the increase of water to solid ratio. Researches (Panda and Tan, 2018, Panda et al., 2019b) show that the influence of clay on the fresh properties of geopolymer largely depends on the water to cement ratio and the amount of activator, and a higher amount of activator will reduce the thixotropic effect of clay.

In addition to raw materials, alkaline activator also plays an important role in the rheology of 3D printing geopolymer. It is pointed out by Palacios et al. (2008) and Alghamdi et al. (2019) that the rheological properties of geopolymer depend on the properties of the activator used, such as the chemical composition, type, and concentration of the activator. Some researchers (Puertas et al., 2014, Panda and Tan, 2018) pointed out that the rheology and fluidity of sodium silicate-activated slag are related to the SiO_2/Na_2O (S/N) molar ratio and Na_2O concentration. Panda and Tan (2018) found that the yield stress and apparent viscosity of mortar increase with the increase of the modulus of the activator. Zhang et al. (2018a) found that the Si/Na ratio of alkali activator has an important influence on the extrudability and buildability of the material. The viscosity, yield stress, and their growth rate of the fresh paste are reduced by the alkali activator with a high Si/Na ratio, which affects the extrusion performance and buildability of the material. Palacios et al. (2008) found that the increase in Na_2O concentration increased the yield stress of alkali-activated slag. However, Alghamdi et al. (2019) proposed that the increase of activator concentration will lead to the decrease of shear yield stress and the enhancement of cohesiveness of fresh paste. Kashani et al. (2014) found that the slag activated by KOH and NaOH had higher yield stress, while the yield stress of slag activated by sodium silicate remained unchanged. Poulesquen et al. (2011) showed that the reaction speed of metakaolin activated by NaOH was faster than that of metakaolin activated by KOH.

Generally, literature only focuses on the mixture proportion of 3D printing concrete with smaller particle sizes. However, 3D printing concrete with coarse aggregate puts forward higher requirements for the printer extrusion head, and the related research is very limited. At present, trial and error or single factor variable methods were often used to optimize the mix proportion of 3D printing concrete to meet the requirement of printability. Also, the rheological parameters proposed for printability depend on the test setup, mix design, and shear rate applied by the rheometer. The existing research results are far from enough to provide a reference for the preparation of 3D printing materials. There is no consensus on the selection of materials, design procedures, the interaction between components, and the influence of their combination on the printability of concrete mixtures.

11.3 INTERLAYER BONDING AND RHEOLOGY

11.3.1 The characterization of interlayer bonding

Layer-by-layer printing is the main feature of 3D printing technology, and the integrity of two continuous layers is worse than that of integral casting construction. Therefore, the interlayer bonding performance of printed components is one of the major concerns for 3D printing cement-based materials. Basically, the interlayer bonding performance can be characterized by its mechanical properties, durability, and interlayer microstructure, which are different from those of bulk materials.

Mechanical properties are the most important properties for interlayer bonding, which generally include the tensile strength test (Nematollahi et al., 2018, Sakka et al., 2019), splitting tensile test (Zareiyan and Khoshnevis, 2017a, Zareiyan and Khoshnevis, 2017b), slant shear test (Diab et al., 2017, Megid and Khayat, 2017), and cross method (Ma et al., 2020). Each method has its own characteristics. The tensile strength test operation is relatively simple, and the specimen is directly drawn, which is suitable for the specimen with low interlayer bond strength. When the bond strength is high, there will be problems in the fixation of specimens. Because the specimen is fixed on the testing machine by adhesive, the interlayer interface may not be damaged and the specimen and the testing machine may fall off, so that the interlayer bond strength cannot be accurately tested. The tensile strength can directly reflect the interlayer bond strength, which is one of the most widely used methods. The splitting tensile test, which is also called brasil splitting test, is to apply the force perpendicular to the interface to make the interface produce the failure form from both ends to the middle, and the interlayer bonding strength can be calculated from the applied loading. It needs to be accurately aligned in the test; otherwise, it will cause a large error. Slant shear test is a common method to test the bond strength of new and old concretes. The cross method can test the shear strength and tensile strength of the interlayer at the same time, which is an effective method to test the mechanical properties of the interlayer. The loading diagrams of the above tests are shown in Figure 11.9.

Since the interlayer is weaker than the bulk material, the durability of the interlayer related to permeability is worse than that of integral casting components. As a result, the penetration-related durability of printed interlayer will also be a concern. Penetration-related durability mainly includes chloride ion penetration (Qian and Xu, 2018), water penetration (Iris et al., 2018, Megid and Khayat, 2019), and carbon dioxide penetration. In real engineering, the penetration rate of chloride ions, water, and carbon dioxide is extremely slow. Therefore, various techniques are employed to accelerate the laboratory-based test, and fast results can be obtained in a relatively short time period. In the rapid chloride ion penetration test, the voltage is applied to accelerate the penetration, and the water penetration test can also accelerate the penetration by exerting high water pressure. The carbonation test is to put the specimen in the environment with a high concentration of carbon dioxide. The above tests can be used to characterize the penetration resistance of interlayer by comparing the penetration depth of substance in interlayer and bulk zones. It is worth noting that penetration resistance to different substances may be different because of different penetration media and penetration mechanisms.

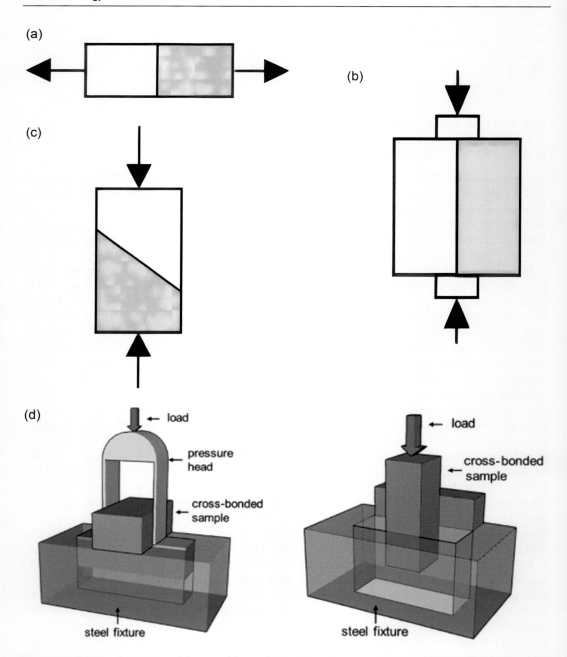

Figure 11.9 Schematic diagram of the test: (a) tensile test, (b) splitting test, (c) slant shear test, and (d) cross-bonded test.

It is well known that the mechanical properties and durability of components are affected by their microstructure. From the micro point of view, it can better explain the reasons for the weak interlayer bonding. The microstructure of the interlayer can be examined by optical microscope, scanning electron microscope, computed tomography scanning, and pore structure analysis techniques, mainly from the following aspects: morphology, phase, cracks, pores, and so on.

11.3.2 The effect of rheological properties on interlayer bonding

3D printing concrete is constructed by pumping extrusion, which requires that the material has the fluidity required for pumping extrusion, and at the same time, it can obtain strength quickly to maintain its shape and bear the upper superimposed load. All these requirements are related to the rheology of 3D printing materials, which mainly studies the relationship between stress, shear rate, and time (Jiao et al., 2017). The rheological properties of cement paste change with time, while 3D printing concrete is constructed by layer-by-layer printing, and there is a time interval between adjacent printing layers. Therefore, the rheological properties of materials will affect the bonding effect between adjacent layers, and then directly affect the interlayer bonding strength. The requirements of 3D printing concrete for rheological properties include yield stress, viscosity, and structural buildup rate (Yuan et al., 2019). The yield stress of slurry can be divided into dynamic yield stress and static yield stress. The minimum shear stress to maintain the flow of slurry is considered as dynamic yield stress. The critical stress to destroy the structure of slurry and make the slurry change from a static state to a flow state is static yield stress (Chen et al., 2020). The material with good fluidity has low yield stress, which is conducive to extrusion, and high interlayer bonding strength, but poor shape retention ability. It is found that the yield stress of materials with high structural buildup increases rapidly and the shape retention ability is strong. However, if the structural buildup rate is too fast and the yield stress exceeds the critical value, there is no mixing between the two adjacent layers of concrete at all, and a "cold joint" is formed at the interlayer, which affects the interlayer bonding (Roussel, 2018, Sakka et al., 2019).

Generally, the structural buildup rate of concrete can be controlled by using an accelerator and a retarder at the same time. From the theological point of view, printing paste should have an appropriate viscosity. On the one hand, low viscosity facilitates the process of pumping and extrusion; however, the viscosity should be high enough to prevent bleeding and segregation (Rahul et al., 2019, Nazar et al., 2020). On the other hand, the increase in viscosity will lead to an increase in extrusion pressure, and high viscosity causes better shape retention ability (Chaves Figueiredo et al., 2019). To obtain suitable rheological properties, it is necessary to adjust the material proportion to achieve a balance in yield stress, viscosity, and structural buildup/thixotropy. Cellulose polymer, superplasticizer, accelerator, and retarding agent are often used to adjust the properties of 3D printing materials. The hydroxyl and ether bonds on the cellulose polymer chain are easy to combine with water through hydrogen bonds, gradually reducing free water, thus enhancing the yield stress and plastic viscosity of cement mortar. Superplasticizer and lithium carbonate can be adsorbed on the surface of cement particles, slowing down the hydration rate of cement, increasing the setting time of cement paste, and reducing the yield stress and plastic viscosity (Chen et al., 2018). Adding fiber to the slurry can enhance the stiffness of the slurry and reduce the interlayer deformation, but the porosity will be more than that of the slurry without fiber, resulting in a decrease in the interlayer bonding strength (Nematollahi et al., 2018).

Thixotropy of materials is an important factor affecting the bonding properties between layers. The thixotropy refers to the property that the viscosity of material gradually decreases under the action of shear force and gradually recovers when the shear force is removed (Yuan et al., 2017). The structure of cement-based materials is mainly formed by the reversible structure change caused by thixotropy and the irreversible structure change caused by cement hydration (Ferron et al., 2007). Among them, the reversible structure change caused by thixotropy was dominant in the early stage, and the interaction between cement particles resulted in flocculation. When a cementitious suspension is sheared, the structure is destroyed. When the shear effect disappears, the formation of weak physical bonds between particles makes the structure reconstructed, which is reversible. This rebuilding may come from Brownian motion, which may lead to the slow rearrangement of particle configuration or the evolution of colloidal interaction between particles (Roussel and Cussigh, 2008, Flatt, 2004, Khayat et al., 2002). The thixotropy of the material can be enhanced by adding nano-particles and/or accelerators to the material to amplify flocculation (Roussel, 2018, Marchon et al., 2018). Thixotropy is essential to obtain sufficient bearing capacity. Of course, it is noted that the higher the thixotropy, the better the performance. The materials with high thixotropy have a higher structural buildup rate, which is not conducive to the development of interlaminar bond strength (Jiao et al., 2021). Moreover, 3D printing concrete should have a certain fluidity. Therefore, for a given material, there should be an optimum thixotropic behavior. By adjusting the proportion of materials, we can get the performance we need. Sakka et al. (2019) found that the medium thixotropic rate (0.48–0.64 Pa/s) of materials has the highest interlayer bond strength, and the addition of styrene–butadiene rubber (SBR) emulsion can also effectively enhance the interlayer bond strength. Nano clay can induce the connection between cement particles, increase the size of flocs in the cement slurry, and enhance the yield stress of the slurry. It is one of the commonly used admixtures to increase the thixotropy of 3D printing concrete (Ferron et al., 2013, Qian and Schutter, 2018, Tregger et al., 2010). Besides, the thixotropy of concrete can be increased by increasing the total amount and fineness of powders and reducing the water to cement ratio. For coarse aggregate, thixotropy is also related to the volume fraction of coarse aggregate. Coarse aggregate can enhance the internal friction and particle interlocking, thus enhancing thixotropy (Banfill, 2003, Iris et al., 2018).

High thixotropic concrete is characterized by rapid structural buildup, promoting the connection between cement particles, making the upper surface of concrete dry before the upper printing material is covered, and reducing the bonding effect between layers. Meanwhile, the static yield stress of concrete with high thixotropy increases rapidly, and the connection between layers gradually weakens with the increase of the stiffness of a single printed layer. Therefore, the interlayer bond strength of a high-thixotropy mixture is lower than that of a low-thixotropy mixture. The structural construction rate should be lower than a certain threshold to allow the subsequent print layer to mix with the previous print layer. A too high structural buildup rate leads to the lack of mechanical consolidation between the two layers of concrete, which leads to the obvious boundary at the layer (Iris et al., 2018, Megid and Khayat, 2017, Megid and Khayat, 2019). For example, there is a critical time for casting two adjacent layers of self-compacting concrete. Once the layer interval exceeds this critical time, the mechanical strength between interlayers will be significantly reduced by up to 40%, and this critical time is also related to the thixotropy of concrete (Roussel and Cussigh, 2008). The thixotropy of concrete is related to the composition of materials. According to the construction conditions, choosing the appropriate thixotropy can obtain the best interlayer bonding (Roussel and Cussigh, 2008).

It has been mentioned that the increase of the rest time will change the rheological properties of the concrete, and then affect the interlayer bonding properties. The interlayer time interval is the key process parameter of 3D printing technology. A reasonable printing time interval is the key technical way to ensure good interlayer bonding properties. With the extension of the time interval, the hydration reaction continues, increasing the stiffness of the monolayer paste and decreasing the bonding effect of the interlayer paste, and the bonding performance gradually decreases. The paste with high thixotropy and viscosity has a large fluidity loss before the next printing layer is covered, and the printing layers cannot be well contacted and mixed so that the interlayer bonding strength is lower. Generally, the shorter the time interval, the better the interlayer bonding. With the increase in time interval, the interlayer bond strength decreases gradually (Le et al., 2012a, Wolfs et al., 2019, Tay et al., 2018). Wolfs et al. (2019) found that the bending strength and splitting tensile strength of the specimens with a time interval of 24 h were 16% and 21% lower than those with a time interval of 15 s. The results of Tay et al. (2018) show that the interlayer tensile strength decreases most obviously in the time interval of 1–5 min, and then decreases slowly. At different time intervals, the interlayer bond strength presents different trends. The results of Qian and Xu (2018) show that the bond strength of the interlayer is lower than 80% of the casting specimen when the interval time exceeds the initial setting time. At the same time interval, the interlayer bond strength of the high-thixotropy materials decreased significantly. A similar conclusion can be seen in the research of Yuan et al. (2022) that the printing time interval which is less than 10 min is more crucial for the formation of interlayer bonding. In addition, Yuan et al. (2022) found that the interlayer shear bond strength can be expressed as a function of the printing time interval and approximately has a negative exponent relation with the printing time interval.

Also, the rheological properties affect the surface water content of the concrete, and surface moisture is also one of the factors affecting the interlayer bonding properties. Some researchers have studied the relationship between surface humidity and interlayer bonding properties, but their conclusions are not consistent due to the different materials and test methods used. Gillette (1963) found that the existence of free water on the surface of concrete can reduce the interlayer bond strength, while Austin et al. (1995) believed that the drying of interlayer surfaces is more conducive to the increase of interlayer bond strength. At the same time, some studies show that there is no obvious correlation between interlayer bonding strength and surface humidity (Pigeon and Saucier, 1992). The above is about the research of traditional cast-in-the-mold concrete, but 3D printing concrete is different. In the printing process, 3D printing concrete is in a fresh state. Due to the evaporation of water, the hydration of cement at the interlayer is not complete, which leads to high porosity (Sanjayan et al., 2018, Keita et al., 2019). The greater the water loss, the worse the interlayer adhesion. Sanjayan et al. (2018) found that the specimens with high surface humidity have higher interlayer bond strength. It is not the time interval between layers, but the water content on the surface of the printed layer that affects the interlayer bonding strength. Due to the bleeding phenomenon of cement paste, the interlayer bond strength of specimens with long-time intervals may be higher than that of specimens with a short time interval. At the same time, the evaporation of water at the interface is also affected by environmental conditions. Therefore, controlling the water loss at the interface can effectively reduce the loss of interlayer bonding strength. According to Roussel's research, the interlayer strength of the specimen was still higher than that of the reference even at several hours' intervals when the material was protected from drying (Roussel, 2018). Due to the lack of protecting formwork, 3D printing concrete loses water faster. Adding superabsorbent polymer to printing concrete to provide internal maintenance can effectively reduce the loss of water in concrete (Liu et al., 2018). Wu et al. (2019) found that the appropriate printing time is before the starting time of the cement hydration induction period. Beyond this time, the number, continuity, and length of interlaminar cold joints increase, and the interlayer bond strength decreases.

In studies by Roussel (2018) and Wangler et al. (2016), a formula (11.5) of the maximum printing time interval is proposed:

$$T_{max} = \frac{\sqrt{\frac{(\rho g h_0)^2}{12} + \left(\frac{2\mu_p v}{h_0}\right)^2}}{A_{thix}} \tag{11.5}$$

where T_{max} is the maximum time interval, μ_p is the plastic viscosity, v is the printing speed, and A_{thix} is the rate of structural buildup. The maximum time interval is affected by the properties of the material and printing parameters. The higher the structure construction rate of the material, the shorter the operational time. When the printing time interval exceeds the maximum time interval, the interlayer bonding will decrease significantly, which will affect the integrity and stability of the printing component.

11.3.3 The effect of rheological properties on interface durability

The durability of printed elements, especially carbonation resistance and chloride ion resistance, depends on the transportation of ionic species. The more pores connected at the interlayer, the easier it is to be penetrated by liquid or other substances. High yield stress and plastic viscosity will produce a high bond strength, and low bond strength is related to the low durability of the interface due to the high porosities. Therefore, the change of rheological properties has an impact not only on the mechanical properties of interlayer bonding but also on the interface durability.

The dynamic yield stress means the minimum force required to maintain the flow of the cement paste. It reflects the flow behavior of the 3DPC when the cement paste is printed out of the nozzle. 3DPC, with a low dynamic yield stress, will automatically adjust its shape under self-weight to match the rough surface of the bottom layer, making the upper layer and the bottom layer stick together (Geng et al., 2020). However, if the dynamic yield stress of 3DPC is too large, it will lead to undesirable contact between the upper and bottom layers (Buswell et al., 2018). In this case, there will be a part of air trapped between the upper and bottom layers, which will reduce the actual contact area between the upper and bottom layers, thus affecting the transportation of ionic species of the interface. Based on the experimental results (Yao et al., 2022, Yuan et al., 2022), the interface has poor resistance to chloride ion erosion and carbonization. With the increase of dynamic yield stress, the penetration depth of chloride was deeper, which indicated that the durability at the interface became worse. The effect mechanism was due to the existence of harmful pores at the interface, which led to the smaller actual contact area at the interface of 3DPC, and hence reduced the durability of the cement interface.

Structural buildup rate determined by the hydration of cement and physical cross-linking flocculation further influences the surface moisture of the interlayer, and affects the interlayer durability. Higher interlayer hydration results in better interlayer bonding of two rough surfaces that are connected and bonded together by hydrates (Geng et al., 2020). Furthermore, before the layers contact, the bottom-printed layer with too high structural buildup rate has a diverse effect on the interaction between the upper and bottom-printed layers of cement, thus affecting the interface properties. In this case, the higher structural buildup rate leads to the weaker mechanical consolidation between the two layers of 3DPC (Iris et al., 2018), resulting in a clear boundary at the layers and poor interface durability.

The change in interlayer durability is consistent with its mechanical properties that have been widely reported in the literature. The higher interlayer bond strength results in better durability. Both the durability and the mechanical property of interlayer depend on the quality of interlayer bonding, while the quality of the bond generated between superposed layers is mostly influenced by the rheological and thixotropic properties of the material used (Panda et al., 2021, Akbar et al., 2019, Sanjayan et al., 2018). Based on a case study, Zhang et al. (2021) investigated the durability of large-scale, 3D-printed, cement-based materials and found that 3DPC was more resistant to sulfate attack and carbonation than mold-cast cement-based materials, but had lower resistance to frost damage and chloride ion penetration. They explained that the printed filaments were pressed tightly by the upper large-scale printed components, which left fewer voids for CO_2 to pass through. Yuan et al. (2022) found that when the carbonation depth and chloride penetration depth reached a high value, the shear bond strength of the interface must be low which showed the opposite trend to them, and confirmed that both the carbonation depth and chloride penetration depth were highly correlated with the shear bond strength by gray relational analysis.

The change of material rheological properties and the adjustment of printing parameters all contribute to the generation of interface pores and cracks. The existence of interface pores and cracks creates a preferential path for water and other corrosive infiltration and hinders durability (Roussel and Cussigh, 2008, Assaad and Issa, 2016). Megid and Khayat (2019) used the residual water seepage resistance to characterize the anti-permeability of the interlayer. The higher the residual water seepage resistance, the better the anti-permeability. The results show that the interlayer impermeability of high-thixotropy materials is much lower than that of low-thixotropy materials. For example, when the time interval is 20 min, the residual water permeability resistance of low-thixotropy materials is 86%, while that of high-thixotropy materials is only 29%. The material with high thixotropy has a higher yield stress growth rate, enhances its shape retention ability, and can effectively reduce the formwork pressure for the construction project requiring formwork. However, the loss of interlayer bonding performance caused by high thixotropy cannot be ignored. In addition to thixotropy, the influence factors of interlayer bond strength discussed above also have effects on the durability of the interlayer. For example, a longer printing time interval will cause higher porosity at the interface of 3DPC, which has a negative influence on the interlayer durability (Yuan et al., 2022).

11.3.4 The effect of rheological properties on interface microstructure

The reason why the mechanical properties and durability of the 3D printing component interlayer are weak can be reflected by the interface microstructure. The rheological properties of slurry change with time, which will affect the interlayer contact effect and microstructure.

Rheology affects the porosity at the interface of layers. Some studies have found that 3DPC produces a large number of pores due to air at the interface, thus damaging the interlayer bonding properties (Xiao et al., 2021, Weng et al., 2021, He et al., 2020). The degree of hydration and porosity at the interface is closely related to the dynamic yield stress and structural buildup rate. Higher yield stress and thixotropy increase the stiffness of concrete, making it difficult for successive printing layers to adjust according to the rough surface of the bottom layer and thus enhance the density of the pores at the interface (Geng et al., 2020). In the research of Yao et al. (2022), the number of pores at the interface increased obviously and some longer cracks appeared when the dynamic yield stress of 3DPC increased. These phenomena were in accordance with the results of the slant shear strength test: whether it was 3, 7 or 28 d. The cement paste with the highest dynamic yield stress had the lowest slant shear strength, and with the increase of the dynamic yield stress, the slant shear strength decreased in each period. In the experiment on slant shear strength, the failure sites of specimens were all at the interface between layers, which indicated that the decline of slant shear strength was caused by the decrease in interface microstructure performance.

Rheology affects the formation of cracks at the interface. The hydration products with interlaminar microstructure have compact structures and better mechanical properties and durability. If the rheological properties of the printing paste are not well controlled, the fluidity of the paste will become worse and the yield stress will increase before the upper layer of printing material is covered, and the two layers cannot be well contacted and mixed, and the "cold joint" may appear in the printing component. Through the early microstructure detection, it can be observed that there are obvious weak connections between the interlayer interfaces. As shown in Figure 11.10, the cracks between two adjacent printing layers are

Figure 11.10 Crack of interface structure (Xu et al., 2021).

long and wide, which leads to a decrease of the interlayer bonding area. Also, stress concentration is easy to appear at the edge of pores, which reduces the interlayer bond strength. The stress concentration at the edge of the pore increases proportionally with the increase of the crack size. Even if the pore is below the interface area, the bond strength is low. With the increase in pore size and number, the interlayer bonding property decreases (Zareiyan and Khoshnevis, 2017b). The higher the construction rate, the higher the possibility of a cold joint to appear and the worse the interlayer bonding.

The interlayer bonding performance is lower than that of the matrix because the microstructure is not as dense as the matrix. Some researchers believe that the bonding interlayer between new and old concretes can be divided into three layers: penetrating layer, strongly affected layer, and weakly affected layer (Xie et al., 2002). The penetrating layer is located in the old concrete, in which a large amount of C–S–H and a small amount of ettringite or calcium hydroxide can be observed, which has no adverse effect on the interlayer strength. The strongly affected layer is located between new and old concretes, which is the weakest layer on the interlayer, affecting the interlayer bonding. The weakly affected layer is located in the new concrete, which has the similar crystal shape and quantity with the new concrete and has the same characteristics. The strength is higher than that of the strong influence layer, and its thickness depends on the properties of the new concrete. It can be seen from the above model that the contact position of two layers of concrete has the most significant effect on the interlayer bonding.

At the same time, there is a phenomenon of self-healing between concrete layers. With the development of cement hydration, the hydration products grow, and the early cracks in the structure are filled continuously, which can enhance the interlayer bonding to a certain extent. There are two kinds of hydration products for filling structural cracks: one can bridge microstructure and make up for structural defects, mainly the C–S–H phase, as shown in Figure 11.11a, and the other is only able to fill cracks and cannot provide effective connection, as shown in Figure 11.11b. The latter is a combination of calcite, ettringite, and/or Portland stone. Because it cannot bridge the interlayer structure well, the mechanical properties of the interlayer are not as good as the former.

As is known that, dynamic yield stress and plastic viscosity influence concrete pumping and extrusion stages, whereas static yield stress, thixotropy, and structuration rate define shape retention and buildability after the extrusion (Roussel, 2018). However, it is noteworthy that the rheological properties also affect the interface microstructure, which determines the mechanical properties and durability of the interface. If rheological properties are improper, interlayer structures will have many defects such as large holes and cracks. Due to the lack of chemical connection and mechanical interlock between interlayers, the mechanical properties and durability of interfaces are also poor.

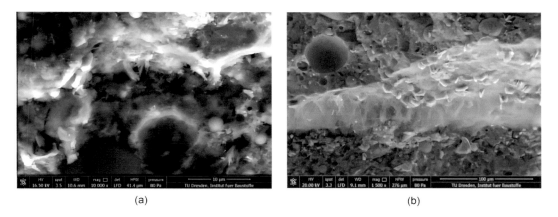

(a)   (b)

Figure 11.11 Hydration products of microstructure (Nerella et al., 2019).

11.4 SUMMARY

Digitalization and automation construction bring great potential in terms of high productivity and in compensating for shortages of skilled labor. In addition, the formwork used for traditional reinforced concrete (RC) construction accounts for one-third to half of the construction cost and one-third to half of the construction time. Large-scale 3D concrete printing (3DCP) is the most promising new concrete technology for implementing digital data from the planning phase and ultimately to actual automated production in factories and on construction sites without formwork.

Extrusion-based technology is the most promising technology for 3D printing cement-based materials. Materials are extruded from a nozzle and automatically deposited layer by layer according to a digitally designed printing route without using any formwork.

Therefore, 3D printing technology puts forward completely new requirements on 3D printing materials. Rheological properties become the most important aspects for both fresh properties and hardened properties of 3D printing materials. 3DCP is still in its very early stage and far from the large-scale application. Research on the rheology of cement-based materials is desperately needed to advance the 3DCP technology.

REFERENCES

Akbar, A., Javid, S., et al. (2019). "A new photogrammetry method to study the relationship between thixotropy and bond strength of multi-layers casting of self- consolidating concrete." *Construction and Building Materials*, 204, 530–540.

Alghamdi, H., Nair, S.A.O., et al (2019). "Insights into material design, extrusion rheology, and properties of 3D-printable alkali-activated fly ash-based binders." *Materials & Design*, 167, 107634.

Asprone, D., Auricchio, F., et al. (2018). "3D printing of reinforced concrete elements: Technology and design approach." *Construction and Building Materials*, 165, 218–231.

Assaad, J., and Issa, C. (2016). "Preliminary study on interfacial bond strength due to successive casting lifts of self-consolidating concrete—effect of thixotropy." *Construction and Building Materials*, 126, 351–360.

Austin, S., Robins, P., et al. (1995). "Tensile bond testing of concrete repairs." *Materials and Structures*, 28, 249–259.

Banfill, P. F. (2003). "The rheology of fresh cement and concrete-a review." In *Proceedings of the 11th international cement chemistry congress*, Edited by Justnes, H., Durban, South Africa, 50–62.

Bogue, R. (2013). "3D printing: The dawn of a new era in manufacturing." *Assembly Automation*, 33, 307–311.

Buswell, R. A., de Silva, L., et al. (2018). "3D printing using concrete extrusion: A roadmap for research." *Cement and Concrete Research*, 112, 37–49.

Buswell, R. A., Soar, R.C., et al. (2007). "Freeform construction: Mega-scale rapid manufacturing for construction." *Automation in Construction*, 16, 224–231.

Cesaretti, G., Dini, E., et al. (2014). "Building components for an outpost on the lunar soil by means of a novel 3d printing technology." *Acta Astronautica*, 93, 430–450.

Chaves Figueiredo, S, Romero Rodrıguez, C. R, et al. (2019). "An approach to develop printable strain hardening cementitious composites." *Materials & Design*, 169, 107651.

Chen, M., Li, L., et al. (2018). "Rheological and mechanical properties of admixtures modified 3D printing sulphoaluminate cementitious materials." *Construction and Building Materials*, 189, 601–611.

Chen, M., Liu, B., et al. (2020). "Rheological parameters, thixotropy and creep of 3D-printed calcium ulfoaluminate cement composites modified by bentonite." *Composites Part B Engineering*, 186, 107821.

De Schutter, G. D., Lesage, K., et al. (2018). "Vision of 3D printing with concrete—Technical, economic and environmental potentials." *Cement and Concrete Research*, 112, 25–36.

Diab, A., Abd Elmoaty, A., et al. (2017). "Slant shear bond strength between self compacting concrete and old concrete." *Construction and Building Materials*, 130, 73–82.

Ferron, R. D., Shah, S., et al. (2013). "Aggregation and breakage kinetics of fresh cement paste." *Cement and Concrete Research*, 50, 1–10.

Ferron, R. P., Gregori, A., et al. (2007). "Rheological method to evaluate structural buildup in self-consolidating concrete cement pastes." *ACI Materials Journal*, 104, 242–250.

Flatt, R. J. (2004). "Dispersion forces in cement suspensions." *Cement and Concrete Research*, 34, 399–408.

Geng, Z. F., She, W., et al. (2020). "Layer-interface properties in 3D printed concrete###Dual hierarchical structure and micromechanical characterization." *Cement and Concrete Research*, 138, 106220.

Gibbons, G. J., Williams, R., et al. (2010). "3D printing of cement composites." *Advances in Applied Ceramics*, 109, 287–290.

Gillette, R. W. (1963). "Performance of bonded concrete overlay." *Journal Proceedings*, 60, 39–49.

Gosselin, C., Duballet, R., et al. (2016). "Large-scale 3D printing of ultra-high performance concrete – A new processing route for architects and builders." *Materials & Design*, 100, 102–109.

Hager, I., Golonka, A., et al. (2016). "3D printing of buildings and building components as the future of sustainable construction." *Procedia Engineering*, 151, 292–299.

Hambach, M., Volkmer, D. (2017). "Properties of 3D-printed fiber-reinforced Portland cement paste." *Cement and Concrete Composites*, 79, 62–70.

He, L. W., Chow, W. T., et al. (2020). "Effects of interlayer notch and shear stress on interlayer strength of 3D printed cement paste." *Additive Manufacturing*, 36, 101390.

Hongyao, S., Lingnan, P., et al. (2019). "Research on large-scale additive manufacturing based on multi-robot collaboration technology." *Additive Manufacturing*, 30, 100906.

Horn, T. J., and Harrysson, O. L. A. (2012). "Overview of current additive manufacturing technologies and selected applications." *Science Progress*, 95, 255.

Hou, Z., Tian, X., et al. (2018). "3D printed continuous fibre reinforced composite corrugated structure." *Composite Structures*, 184, 1005–1010.

Iris, G., Bel, G., et al. (2018). "Thixotropy and interlayer bond strength of self-compacting recycled concrete." *Construction and Building Materials*, 161, 479–488.

Jeong, H., Han, S. J., et al. (2019). "Rheological property criteria for buildable 3D printing concrete." *Materials*, 12, 657.

Jiao, D, De Schryver, R., et al. (2021). "Thixotropic structural build-up of cement-based materials: A state-of-the-art review." *Cement and Concrete Composites*, 122, 104152.

Jiao, D., Shi, C., et al. (2017). "Effect of constituents on rheological properties of fresh concrete-a review." *Cement and Concrete Composites*, 83, 146–159.

Kashani, A., Provis, J. L., et al. (2014). "The interrelationship between surface chemistry and rheology in alkali activated slag paste." *Construction and Building Materials*, 65, 583–591.

Kazemian, A., Yuan, X., et al. (2017). "Cementitious materials for construction-scale 3D printing: Laboratory testing of fresh printing mixture." *Construction and Building Materials*, 145, 639–647.

Keita, E., Bessaies-Bey, H., et al. (2019). "Weak bond strength between successive layers in extrusion-based additive manufacturing: Measurement and physical origin." *Cement and Concrete Research*, 123, 105787.

Khayat, K. H., Saric-Coric, M., et al. (2002). "Influence of thixotropy on stability characteristics of cement grout and concrete." *ACI Materials Journal*, 99, 234–241.

Khoshnevis, B. (2014). "Automated construction by contour crafting-related robotics and information technologies." *Automation in Construction*, 13, 5–19.

Khoshnevis, B., Bukkapatnam, S. T., et al. (2001). "Experimental investigation of contour crafting using ceramics materials." *Rapid Prototyping Journal*, 7, 32–42.

Khoshnevis, B., Hwang, D., et al. (2006). "Mega-scale fabrication by contour crafting." *International Journal of Industrial and Systems Engineering*, 1, 301.

Le, T. T., Austin, S. A., et al. (2012a). "Hardened properties of high-performance printing concrete." *Cement and Concrete Research*, 42, 558–566.

Le, T. T., Austin, S. A., et al. (2012b). "Mix design and fresh properties for high-performance printing concrete." *Materials and Structures*, 45, 1221–1232.

Lim, S., Buswell, R., et al. (2011). "Development of a viable concrete printing process." *28th International Symposium on Automation and Robotics in Construction, (ISARC2011)*, Seoul, South Korea, 665–670.

Lim, S., Buswell, R., et al. (2012). "Developments in construction-scale additive manufacturing processes." *Automation in Construction*, 21, 262–268.

Liu, G., Lu, W., et al. (2018). "Interlayer shear strength of roller compacted concrete (RCC) with various interlayer treatments." *Construction and Building Materials*, 166, 647–656.

Lowke, D., Dini, E., et al. (2018). "Particle-bed 3D printing in concrete construction – Possibilities and challenges." *Cement and Concrete Research*, 112, 50–65.

Ma, G., Li, Z., and Wang, L. (2018a). "Printable properties of cementitious material containing copper tailings for extrusion based 3D printing." *Construction and Building Materials*, 162, 613–627.

Ma, G, Salman, N. M., Wang, L, et al. (2020). "A novel additive mortar leveraging internal curing for enhancing interlayer bonding of cementitious composite for 3D printing." *Construction and Building Materials*, 244, 118305.

Ma, G., Wang, L., et al. (2018b). "State-of-the-art of 3D printing technology of cementitious material—An emerging technique for construction." *Science China (Technological Sciences)*, 61, 475–495.

Ma, G., and Wang, L. (2018c). "A critical review of preparation design and workability measurement of concrete material for largescale 3D printing." *Frontiers of Structural and Civil Engineering*, 12, 382–400.

Ma, G., and Wang, L. (2020). *3D printing key technologies for cementitious materials*. China's Building Material Industry Press, Beijing, China.

Maier, A. K., Dezmirean, L., et al. (2011). "Three-dimensional printing of flash-setting calcium aluminate cement." *Journal of Materials Science*, 46, 2947–2954.

Marchon, D., Kwawashima, S., et al. (2018). "Hydration and rheology control of concrete for digital fabrication: Potential admixtures and cement chemistry." *Cement and Concrete Research*, 112, 96–110.

Megid, W. A., and Khayat, K. H. (2017). "Bond strength in multilayer casting of self-consolidating concrete." *ACI Materials Journal*, 114, 467–476.

Megid, W. A., and Khayat, K. H. (2019). "Effect of structural buildup at rest of self-consolidating concrete on mechanical and transport properties of multilayer casting." *Construction and Building Materials*, 196, 626–636.

Nazar, S., Yang, J., Thomas, B. S., et al., (2020). "Rheological properties of cementitious composites with and without nano-materials: A comprehensive review." *Journal of Cleaner Production*, 272, 122701.

Nematollahi, B., Xia, M., et al. (2018). "Effect of type of fiber on inter-layer bond and flexural strengths of extrusion-based 3D printed geopolymer." *Materials Science Forum*, 939, 155–162.

Nerella, V. N., Hempel, S., et al. (2019). "Effects of layer-interface properties on mechanical performance of concrete elements produced by extrusion-based 3D-printing." *Construction and Building Materials*, 205, 586–601.

Nerella, V. N., Krause, M., et al. (2016). "Studying printability of fresh concrete for formwork free concrete on-site 3D printing technology (CONPrint3D)." *25th Conference on Rheology of Building Materials*, Regensburg, Germany.

Palacios, M., Banfill, P. F. G., et al. (2008). "Rheology and setting behavior of alkaliactivated slag pastes and mortars: Effect if organic admixture." *ACI Materials Journal*, 105, 140–148.

Panda, B., Paul, S. C., et al. (2017a). "Additive Manufacturing of geopolymer for sustainable built environment." *Journal of Cleaner Production*, 167, 281–288.

Panda, B., Paul, S. C., et al. (2017b). "Anisotropic mechanical performance of 3D printed fiber reinforced sustainable construction material." *Materials Letters*, 209, 146–149.

Panda, B., Singh, G. B., et al. (2019a). "Synthesis and characterization of one-part geopolymers for extrusion based 3D concrete printing." *Journal of Cleaner Production*, 220, 610–619.

Panda, B., Sonat, C., et al. (2021). "Use of magnesium-silicatehydrate (M-S-H) cement mixes in 3D printing applications." *Cement and Concrete Composites*, 117, 103901.

Panda, B., and Tan, M. J. (2018). "Experimental study on mix proportion and fresh properties of fly ash based geopolymer for 3D concrete printing." *Ceramics International*, 44, 10258–10265.

Panda, B., Unluer, C., et al. (2019b). "Extrusion and rheology characterization of geopolymer nanocomposites used in 3D printing." *Composites Part B Engineering*, 176, 107290.

Paolini, A., Kollmannsberger, S., et al. (2019). "Additive manufacturing in construction: A review on processes, applications, and digital planning methods." *Additive Manufacturing*, 30, 100894.

Paul, S. C., Yi, W. D. T., et al. (2018). "Fresh and hardened properties of 3D printable cementitious materials for building and construction." *Archives of Civil and Mechanical Engineering*, 18, 311–319.

Pegna, J. (1997). "Exploratory investigation of solid freeform construction." *Automation in Construction*, 5, 427–437.

Perrot, A., Rangeard, D., et al. (2016). "Structural built-up of cement-based materials used for 3D-printing extrusion techniques." *Materials and Structures*, 49, 1213–1220.

Petrovic, V., Gonzalez, J. V. H., et al. (2011). "Additive layered manufacturing: Sectors of industrial application shown through case studies." *International Journal of Production Research*, 49, 1061–1079.

Pigeon, M., and Saucier, F. (1992). "Durability of repaired concrete structures." in Malotra (ed.), *Advances in concrete technology*, Energy, Mines and Resources, Canada, pp. 741–773.

Poulesquen, A., Frizon, F., et al. (2011). "Rheological behavior of alkali-activated metakaolin during geopolymerization." *Journal of Non-Crystalline Solids*, 357, 3565–3571.

Puertas, F., Varga, C., et al. (2014). "Rheology of alkali-activated slag pastes. Effect of the nature and concentration of the activating solution." *Cement and Concrete Composites*, 53, 279–288.

Qian, P., and Xu, Q. (2018). "Experimental investigation on properties of interface between concrete layers." *Construction and Building Materials*, 174, 120–129.

Qian, Y., and Schutter, G. D. (2018). "Enhancing thixotropy of fresh cement pastes with nanoclay in presence of polycarboxylate ether superplasticizer (PCE)." *Cement and Concrete Research*, 111, 15–22.

Rahul, A. V., Santhanam, M., et al. (2019). "3D printable concrete: Mixture design and test methods." *Cement and Concrete Composites*, 97, 12–23.

Ranjbar, N., Mehrali, M., et al. (2021). "Rheological characterization of 3D printable geopolymers." *Cement and Concrete Research*, 147, 106498.

Roussel, N. (2018). "Rheological requirements for printable concretes." *Cement and Concrete Research*, 112, 76–85.

Roussel, N., and Cussigh, F. (2008). "Distinct-layer casting of SCC: The mechanical consequences of thixotropy." *Cement and Concrete Research*, 38, 624–632.

Rudenko, A. (2015). "3D printed concrete castle is complete." 3D Concrete House Printer. http://www.totalkustom.com/.

Rushing, T. S., Alchaar, G., et al. (2017). "Investigation of concrete mixtures for additive construction." *Rapid Prototyping Journal*, 23, 74–80.

Sakka, F. E., Assaad, J. J., et al. (2019). "Thixotropy and interfacial bond strengths of polymermodified printed mortars." *Materials and Structures*, 52, 79.

Sanjayan, J., Nematollahi, B., et al. (2018). "Effect of surface moisture on inter-layer strength of 3D printed concrete." *Construction and Building Materials*, 172, 468–475.

Shakor, P., Sanjayan, J., et al. (2017). "Modified 3D printed powder to cement-based material and mechanical properties of cement scaffold used in 3D printing." *Construction and Building Materials*, 138, 398–409.

Soltan, D. G., and Li, V. C. (2018). "A self-reinforced cementitious composite for building-scale 3D printing." *Cement and Concrete Composites*, 90, 1–13.

Suiker, A. (2018). "Mechanical performance of wall structures in 3D printing processes: Theory, design tools and experiments." *International Journal of Mechanical Sciences*, 137, 145–170.

Suiker, A., Wolfs, R., et al. (2020). "Elastic buckling and plastic collapse during 3D concrete printing." *Cement and Concrete Research*, 135, 106016.

Tay, Y., Qian, Y., et al. (2019). "Printability region for 3D concrete printing using slump and slump flow test." *Composites Part B Engineering*, 174, 106968.

Tay, Y., Ting, G., et al. (2018). "Time gap effect on bond strength of 3D-printed concrete." *Virtual and Physical Prototyping*, 14, 1500420.

Tregger, N. A., Pakula, M. E., et al. (2010). "Influence of clays on the rheology of cement pastes." *Cement and Concrete Research*, 40, 384–391.

Vaezi, M., Seitz, H., et al. (2013). "A review on 3D micro-additive manufacturing technologies." *International Journal of Advanced Manufacturing Technology*, 67, 1721–1754.

Wang, L., et al. (2021). "Cementitious composites blending with high belite sulfoaluminate and medium-heat Portland cements for largescale 3D printing." *Additive Manufacturing*, 46, 102189.

Wangler, T., Lloret, E., et al. (2016). "Digital concrete: Opportunities and challenges." *RILEM Technical Letter*, 1, 67–75.

Weng, Y. W., Li, M. Y., et al. (2021). "Synchronized concrete and bonding agent deposition system for interlayer bond strength enhancement in 3D concrete printing." *Automation in Construction*, 123, 103546.

Williams, H., and Butler-Jones, E. (2019). "Additive manufacturing standards for space resource utilization." *Additive Manufacturing*, 28, 676–681.

Wolfs, R. J. M., Bos, F. P., et al. (2018). "Correlation between destructive compression tests and non-destructive ultrasonic measurements on early age 3D printed concrete." *Construction and Building Materials*, 181, 447–454.

Wolfs, R. J. M., Bos, F. P., et al. (2019a). "Hardened properties of 3D printed concrete: The influence of process parameters on interlayer adhesion." *Cement and Concrete Research*, 119, 132–140.

Wolfs, R., Bos, F., et al. (2019b). "Triaxial compression testing on early age concrete for numerical analysis of 3D concrete printing." *Cement and Concrete Composites*, 104, 103344.

Wu, H., Jiang, Y., et al. (2019). "Research on interlayer bonding properties of 3D printing cement-based materials." *New Building Materials*, 12, 5–8.

Wu, P., Wang, J., et al. (2016). "A critical review of the use of 3-D printing in the construction industry." *Automation in Construction*, 68, 21–31.

Xiao, J. Z., Liu, H. R., et al. (2021). "Finite element analysis on the anisotropic behavior of 3D printed concrete under compression and flexure." *Additive Manufacturing*, 39, 101712.

Xie, H., Li, G., et al. (2002). "Microstructure model of the interfacial zone between fresh and old concrete." *Journal of Wuhan University of Technology*, 17, 64–68.

Xu, Y. Q., Yuan, Q., et al. (2021). "Correlation of interlayer properties and rheological behaviors of 3DPC with various printing time intervals." *Additive Manufacturing*, 47, 102327.

Yao, H., Yuan, Q., et al. (2022). "The relationship between the rheological behavior and interlayer bonding properties of 3D printing cementitious materials with the addition of attapulgite." *Construction and Building Materials*, 316, 125809.

Yuan, Q., Li, Z., et al. (2019). "A feasible method for measuring the buildability of fresh 3D printing mortar." *Construction and Building Materials*, 227, 116600.

Yuan, Q., Zhou, D., et al. (2017). "On the measurement of evolution of structural build-up of cement paste with time by static yield stress test vs. small amplitude oscillatory shear test." *Cement and Concrete Research*, 99, 183–189.

Yuan, Q., Xie, Z., Yao, H., Huang, T., Li, Z., and Zheng, X. (2022). Effect of polyacrylamide on the workability and interlayer interface properties of 3D printed cementitious materials. *Journal of Materials Research and Technology*, 19, 3394–3405

Zareiyan, B., and Khoshnevis, B. (2017a). "Effects of interlocking on interlayer adhesion and strength of structures in 3D printing of concrete." *Automation in Construction*, 83, 212–221.

Zareiyan, B., and Khoshnevis, B. (2017b). "Interlayer adhesion and strength of structures in contour crafting - Effects of aggregate size, extrusion rate, and layer thickness." Automation in Construction, 81, 112–121.

Zhang, D.W., Wang, D.M., et al. (2018a). "The study of the structure rebuilding and yield stress of 3D printing geopolymer pastes." *Construction and Building Materials*, 184, 575–580.

Zhang, D. W., Wang, D. M., et al. (2018b). "Effect of steel slag content on rheological properties of 3D printing geopolymer materials." *Journal of Basic Science and Engineering*, 26, 596–604.

Zhang, Y., Zhang, Y., et al. (2018c). "Fresh properties of a novel 3D printing concrete ink." *Construction and Building Materials*, 174, 263–271.

Zhang, Y., Zhang, Y., et al. (2019). "Rheological and harden properties of the high-thixotropy 3D printing concrete." *Construction and Building Materials*, 201, 278–285.

Zhang, Y., Zhang, Y. S., et al. (2021). "Hardened properties and durability of large-scale 3D printed cement-based materials." *Materials and Structures*, 54, 45.

Index

3D printing concrete 169, 291, 295–297, 299, 301–302, 305–307
Abrams cone 75–76, 79, 81–82
addition sequence 67–69
additive manufacturing 291, 295, 299
ADRHEO 138, 147
aggregate spacing 159–161, 172–173
aggregate volume fraction 51–55, 59, 61, 69–70, 163
air bubble distribution 270
air content 40, 42, 200, 260, 268–272, 285
air-entraining agent 40, 43, 185, 193, 273
alkali hydroxides 217
alkali-activated fly ash 217–219, 221, 224, 230–231, 234
alkali-activated materials (AAMs) 215, 302
alkali-activated slag 228, 233, 302
alkaline activators 217
aluminosilicate 215, 217–218, 223
Andreasen & Andersen Model 61
anti-permeability 309
anti-thixotropy 13
apparent viscosity 8, 30, 199, 221, 236–246, 278–281, 302
aspect ratio 62–65, 70, 104, 169, 240–242, 244, 246–248, 251

barrier effect 271
Bingham model 5–6, 10, 95, 101, 124, 129–132, 139, 155, 215–216, 233–234, 249, 273, 293
blast furnace slag 31, 33–34, 37, 192, 195–196, 228, 273, 276, 279, 281–283
bleeding resistance 165–166
BML Viscometer 124–125, 133
Boltzmann constant 26–27
Brownian 25–27, 43, 156, 306
BTRHEOM 18–19, 79, 124–125, 137–139, 142–143, 147, 151–152, 262
Buckingham-Reiner equation 261
buildability 95, 198, 295–302, 310
bulk concrete 262, 264–265, 268, 274–275, 279

calcium aluminate cement 216, 301
calibration coefficient 108–109
Cannon-Fenske viscometer 104–105
capillary viscometer 104–108, 117
carbon fiber 243–245, 301
carbonation resistance 308
C-A-S-H gel 218, 221, 225
Casson model 11
Cauchy stress principle 128
cellulosic fiber 245
CEMAGREF-IMG rheometer 140–141
cement paste backfilling 234, 250
Chateau-Ovarlez-Trung model 55–56, 69
chemical admixture 25, 27, 38–40, 59, 183, 185, 189, 190, 201, 225–232, 249–251, 260–261, 276, 280, 284
chemical composition 30, 52, 217, 235, 280, 302
chloride ion resistance 308
coarse aggregate 15, 17, 20, 53, 56, 59, 61–62, 66, 70, 82, 124, 132, 136–137, 140, 161, 163–164, 167–170, 174, 176, 186, 194–195, 198–201, 204, 206–208, 260–261, 270, 278, 283, 301–302, 306
coaxial cylinder rheometer 18, 96, 100, 125–126, 128–129, 133, 139–140
cold joint 305, 307, 309–310
collapsed slump 77–78
colloidal interaction 20, 25–27, 29, 37–38, 43, 198, 296, 306
Computational Fluid Dynamics 267
concentrated suspension 28, 51–52, 54, 167, 248
concrete rheology method 164, 169, 173
construction and demolition waste 195
ConTec 18–19, 69, 75, 124–127, 132–133, 136, 145, 150, 152
continuous model 60
contour crafting 291–295
conventional vibrated concrete 51, 157, 183–185, 208, 266

convexity 62–64, 70
corrosion inhibitors 198
Couette 125–126, 128, 130, 133
Coulomb's friction law 260
critical strain 110–111, 113, 115, 296
critical volume fraction 28, 55, 241, 244, 249
CRTS III slab ballastless track 189–190
C-S-H 19, 42, 222, 236, 278, 310
cushion effect 271

dead zone 97–98
deflocculation 40, 279–280
deformation 1–7, 20, 39, 76, 90, 111–112, 116, 123, 155, 209, 272, 295, 297, 299, 301, 305
deposition 207, 294–297
depressurization 269
differential viscosity 8
diffuse double layers 26
dilatant 10
dilute regime 248
direct contact 18, 28, 54–55, 205
discharge rate 260–261, 274, 278
Discrete Element Method 267
dissolution 43, 215, 217–219, 221, 225, 231–232, 250, 269, 284
draining vessel 124
drum 42, 68, 96, 143
D-shape 155, 291–292, 294, 301
dune sand 62, 198
dynamic stability 204, 206–208
dynamic viscosity 3, 8, 105–106, 109, 238

efficiency of fibers 240–241, 243
Einstein equation 52
electrostatic 25–26, 38, 185, 194, 219, 223, 225, 228, 237
empirical model 54, 67, 265, 284
end effect 127–128, 131, 133
end hose 268–269
equivalent spherical particle 160, 170
excess mortar film thickness 163
excess paste theory 56, 60, 69, 166
excess paste thickness 52, 57–58, 156, 166, 178
extrudability 67, 95, 295, 297–298, 302
extrusion-based 3D printing 12, 221

falling sphere viscometer 107–109, 117
Farris model 53
fiber content 241, 244–247, 251
fiber distribution 243
fiber length 241, 243–246, 301
fiber orientation 242–243, 248, 301
fiber reinforced cement-based materials 240, 242, 251
fiber types 244
fine aggregate 15, 53, 61–62, 161, 164, 167–168, 170, 174, 176, 186, 200, 271
fineness modulus 62, 176, 186, 195

flocculation 28, 32, 34, 36–37, 41–43, 144, 148, 193, 199, 201, 220, 222, 225, 249, 251, 277–278, 284, 305–306, 308
flow curve 7–10, 100, 109–110, 126, 143–146, 149
flow rate 89, 104, 106, 262, 264–265, 267, 274, 278, 292
flow regime 20, 51
flow viscosity ratio 161
Flow-through Test 207
formwork characteristics 200–201
formwork pressure 198, 200–204, 208, 309
Fourier transform 111–112
frequency sweep 113
FRESHWIN 145–146
Fuller-Thompson model 60
full-scale pumping circuit 265

geopolymer 246, 296, 301–302
glass fiber 68, 244–246, 294, 301
gradation 34, 59–61, 70, 169
granular skeleton 54, 205, 208, 241–242, 248
ground clay bricks 197

Hagen-Poiseuille 261
Hamaker constant 25–28
Henry's Law 269
Herschel-Bulkley model 11, 55, 130–131, 139, 155, 205
high-performance concrete 16, 30, 38, 51, 123, 156, 167, 178, 190, 232, 260, 301
Hooke's law 2–4, 111
horizontal pumping 268–270, 276, 279, 282, 284
hybrid fiber system 246
hydration products 19, 41, 95, 115, 198, 222, 235, 237–239, 309–311
hydrodynamic force 25, 27–28, 43, 156, 200
hydrodynamic pressure 136–137
hysteresis loop 12, 150, 220

IBB rheometer 18–19, 124–125, 140
ICAR rheometer 18–19, 69, 75, 124–126, 140, 143–144, 146, 149–150
initial sulfate concentration 236, 239–240
inner cylinder 96–97, 125–129, 132–133, 136, 140, 145–146, 148
interlayer bonding 297, 303–310
International Center for Aggregate Research (ICAR) 126
interparticle force 25–26, 28, 205, 223, 225, 235, 237, 250, 277, 279
interstitial solution 20, 27, 29, 43
intrinsic viscosity 28, 52–53, 56, 163
irregular and elongated 66

J-ring 85, 89–91

Kaplan 207, 261–262, 264, 266, 276, 277

Kelvin–Voigt model 4–5
kinematic viscosity 8, 105–106
Krieger-Dougherty model 28, 53, 69, 267, 281

laminar 7, 96, 99, 104, 128, 207, 306–307, 309
large amplitude oscillatory shear 111
Laskar 31–32, 34–35, 37, 133–135, 143, 216, 227, 230–231
lateral pressure 199–204
layer settlement test 299
L-box 75, 85–88, 91, 155
LCPC box 85, 87–88
limestone powder 31, 36–37, 43, 86, 163, 169–170, 173, 184, 198–199, 201, 233, 282
linear viscoelastic region (LVER) 113
Lissajous Curves 111–112, 114–115
loading bearing capacity 299
loss modulus 111–114, 221, 223
lubrication layer 142, 185, 261–262, 266

material composition 273–274
material failure criterion 300
maximum packing fraction 28, 52–54, 57, 205
Maxwell model 4–6
measuring geometry 51, 69
mechanical properties 39, 155, 168–169, 187, 192, 204, 225, 238, 241, 243, 246, 285, 295, 301, 303–304, 308–310
metakaolin 184, 192, 196, 198–199, 221, 225–226, 232, 281, 302
mineral admixture 25, 31, 36–37, 43, 96, 197, 201
mixer type 67–68
mixing process 67, 143, 237
modified Bingham model 130–131, 155, 215–216, 273
molecular weight 39, 115, 193, 227–228, 231–232, 250
Mooney equation 53, 106

nano-silica 197
NaOH concentration 217–219, 230
naphthalene sulphonate-based superplasticizer 38
narrow gap 96, 104, 125, 132
National Institute of Standards and Technology (NIST) 69, 123
natural aggregate 195, 198–200
Newtonian flow 7–9
Nielsen model 64
non-contact forces 25
non-flowable bulk zone 262
non-Newtonian flow 7, 9–11, 250
nozzle 12, 291–295, 298, 301, 308, 311
numerical model 266–267, 284

outer cylinder 18, 96–98, 100, 104, 109, 125–127, 129, 132, 135, 140, 146

parallel plate rheometer 99–100, 102–104, 117, 124, 133, 137–138, 142
Particle Image Velocimetry 98, 264
particle migration 131–132, 206, 267, 278
particle morphology 62, 70
particle shape 25, 43, 53, 64–66, 70, 159, 186, 241
particle size distribution 28, 32–33, 36–37, 43, 53, 60–61, 159, 167, 186–187, 208, 236
paste rheology criteria 156, 158, 160–161, 169, 178
paste-to-sand volume fraction 56
penetration resistance 299, 303
percolation volume fraction 28
pipe length 260, 273–274
pipeline 109, 234–235, 238–239, 250, 262, 264–265, 270, 273, 284, 294
plug flow 98, 104, 131–132, 139–140
plug zone 128, 132, 261–262, 267, 278
Poiseuille flow formulation 267
polycarboxylate-based superplasticizer 35, 38, 169, 197, 225
polymer structure 39, 193, 225
polymerization 193, 215, 219, 221–222, 249
polypropylene fiber 243–244, 246, 271
porcelain polishing residue 197
porosity 16, 31, 167, 199, 241, 305, 307, 309
power law model 12, 28
pozzolan 31, 197, 232
precursor 215–219, 221, 223–228, 232, 249–250, 302
prediction 20, 51, 79–80, 201, 203, 242, 248, 264–267, 277–278, 284, 296
pressure and gravity capillary viscometer 104
pressure loss 260–261, 264–267, 277
pressurization 43, 269, 272, 279
print head (Printhead) 291, 295
printability 51, 296–298, 301–303
protocol 43, 109–110
pseudo-plastic 10–11
pumpability 39–41, 51, 156, 198, 201, 251, 259–263, 265, 284, 295, 297
pumped concrete 259–260, 273–274, 279, 284

Rabinowitsch-Mooney equation 106
random dense packing fraction 55
rate-controlled 125, 140
raw materials of SCC 184, 189
recycled aggregate 186, 190, 195–196, 199
Reiner-Riwlin equation 97, 130–131
retarder 231, 279, 305
rheograph 156–158, 178
rheological equations for non-Newtonian fluids 12
rheology divergence packing fraction 56, 65
rice husk ash (RHA) 190
river sand 66, 173, 176, 198, 301
Robinson 52

Roscoe 52
rotational speed 17, 42, 68–69, 99, 127–128, 131, 133, 136, 138–140, 144, 146–150, 152, 197–198
rough surface 51, 69, 98, 101–102, 196, 308–309
roundness 63
Runge-Kutta method 89

Searle-type 125, 130, 133
segregation factor 146
self-compacting high-performance concrete (SCHPC) 190
semi-dilute regime 248
shape retention ability 305, 309
shape stability 298
shear history 42, 51, 69, 147
shear rate 3, 6–11, 18, 27, 28, 40–43, 68, 95–110, 123, 125, 128, 132–140, 149, 155–156, 185, 199–200, 206, 215, 232, 236–237, 240, 243, 245, 264, 267, 273–280, 303–305
shear slump 77–78
shear strain 4, 7, 110, 243, 265, 300
shear stress 7–12, 18, 40, 42, 51, 53, 75, 79, 95–98, 101–106, 109–111, 123, 125, 128–129, 131–140, 148–149, 155–157, 197–200, 220, 231, 235–239, 261, 266–267, 297–298, 302, 305
shear thickening 10–13, 28, 40, 68–69, 95, 102, 109, 196–200, 223, 237–238, 246, 278
shear thinning 10–12, 28, 39–40, 68, 95, 109, 115–117, 198–200, 236–237, 281
shear zone 128–129, 132, 264, 266
shear-induced sedimentation 140
shrinkage reducing agent 231
side chain 39, 193–194, 227–228, 230, 250
Sieve Segregation Test 206–208
silica fume (SF) 20, 31, 34–37, 39, 43, 59, 184, 191, 226, 232, 271, 273, 280, 282–283, 301
simplex centroid design 156, 167–169, 176, 178
simulation 79, 96, 98, 103, 136, 143, 267, 278, 284, 293
slant shear test 303–304
sliding pipe 262–264
slippage 96, 101–102, 104, 128, 136–137, 140
slipper 267
slump test 17, 75–77, 80–81, 91, 123, 150, 169, 198, 231, 265, 298–299
small-amplitude oscillatory shear (SAOS) 111–113, 115–117
Smoothed Particle Hydrodynamics 267
sodium silicates 219–220, 222
solid volume fraction 13, 28–29, 55, 59–60, 65, 156, 295, 224, 233
stabilization 234, 270
static stability 156, 204, 206

static yield stress 12, 42, 51, 110, 115, 126, 148, 196, 198, 202–203, 216, 281–283, 305–306, 310
statistical method 186–187
steady state 69, 127, 156
steel fiber 161, 164, 169–171, 173, 192, 242–244, 246–248, 251
steel slag 216, 232, 282–283, 302
steric hindrance 25, 38–39, 185, 194, 228, 231, 237
Stokes equation 109
storage modulus 12, 42, 111–117, 218–219, 221, 223, 225
strain sweep 112–113, 221
strain-controlled rheometer 112–113
stress growth test 99, 126, 143–144, 148
stress-controlled 125, 140
structural build-up 29, 156, 223, 236, 282
structuration rate 36, 198–199, 310
supplementary cementitious materials 43, 163, 191, 260, 273, 281
surface moisture 307–308
surface quality 294, 297
surface roughening 98, 102, 117

tailings 234–236, 238–240, 250–251
Tattersall two-point rheometer (MK II) 124–125
temperature 41, 43, 68, 82, 95, 100–101, 104, 107–109, 117, 140, 194–195, 201, 203, 239–240, 260, 278, 280–284, 301
ternary binder 37, 197
thixotropy 11–13, 28–29, 39–43, 95, 104, 109–110, 106, 138, 142, 146–150, 198–199, 201, 203, 216, 220, 223, 232, 235, 238–239, 251, 267, 276–281, 296–302, 305–311
time sweep 113
time-dependent 5, 11–13, 41, 43, 106, 113, 235, 238, 251, 277, 300
torque-rotational speed diagram 147
transformation equation 125–126, 128, 130–131, 133, 139, 150
transition volume fraction 28, 55
tribometer 262–264
true slump 77–78

UIUC concrete rheometer 142
ultra-high-performance AAM concrete 243, 246
ultrasonic velocity profiler 264
unsheared zone 129

van der Waals forces 25–26, 28, 110, 217
vectorized-rheograph approach 156, 158, 178
velocity profile 134–135, 261, 264, 267
vertical pumping 268–270
V-funnel 75, 82–85, 88–89, 91, 155, 158, 169, 282

viscoelasticity 4, 19, 104, 111–112, 115, 218–219, 223
viscosity modifying agent (VMA) 34, 39–40, 183, 185, 208, 232, 288
viscous drag force 26–27, 270
viscous granular material 267
Viskomat XL 140, 152
Visual Stability Index (VSI) 206–207

wall slip 96, 98, 101–102, 117, 150
water film thickness 30, 37, 166–167

water reducing agent 16, 41, 115, 230–231, 238, 242, 270, 284
Weissenberg-Rabinowitsch equation 102
welan gum 39–40
workability box 156–158

Yodel 28, 281
Young's modulus 299–300

zero slump 77–78, 202
zeta potential 33, 193, 217–219, 228, 240